Carbon Capture and Storage

Carbon Capture and Storage

Stephen A. Rackley

AMSTERDAM • BOSTON • HEIDELBERG • LONDON • NEW YORK • OXFORD
PARIS • SAN DIEGO • SAN FRANCISCO • SINGAPORE • SYDNEY • TOKYO
Butterworth-Heinemann is an imprint of Elsevier

Butterworth-Heinemann is an imprint of Elsevier
30 Corporate Drive, Suite 400, Burlington, MA 01803, USA
Linacre House, Jordan Hill, Oxford OX2 8DP, UK

Library of Congress Cataloging-in-Publication Data
Application submitted

British Library Cataloguing-in-Publication Data
A catalogue record for this book is available from the British Library.

ISBN: 978-1-85617-636-1

For information on all Butterworth-Heinemann publications visit our Web site at www.elsevierdirect.com

Printed in the United States of America
09 10 11 12 13 10 9 8 7 6 5 4 3 2 1

for
Adam and Jenny
and their
children's children's
children

Contents

Preface

The seed from which this book has grown was planted by the launch of the Virgin Earth Challenge by Sir Richard Branson and former U.S. Vice President Al Gore on Feb. 9, 2007. The aim of the Challenge is to encourage the development of commercially viable new technology, processes, and methods that can remove significant volumes of anthropogenic greenhouse gases from the atmosphere and contribute materially to the stability of the earth's climate.

With emissions from fossil fuel combustion running at 6.0–6.5 Gt-C per year (Gt-C = 10^9 metric tonnes of carbon), a material contribution to climate stability implies the potential for deployment on a scale of 1 Gt-C per year, roughly a thousand times larger than any currently operating project. While these volumes seem prodigious, anthropogenic emissions pale into insignificance beside the natural fluxes such as terrestrial photosynthesis, at ~120 Gt-C year, and oceanic uptake and release at ~90 Gt-C per year.

A diverse range of carbon capture and storage (CCS) technologies are currently at various stages of research, development, and demonstration. While a few of these technologies have reached the deployment stage, many still require significant further development work to improve technical capabilities and reduce costs. Although front-runners are already emerging, it is likely that the long-term potential of CCS will be achieved through the application of a broad portfolio of different technologies. These could range from the current favorite—solvent-based capture from coal-fired power plants with geological storage—to the decarbonation of fuels ahead of combustion, the manipulation of ecological factors such as microbial populations or ocean fertility to increase carbon inventories in soils and in the oceans, and many others.

The aim of this book is to contribute in small part to the progress of this endeavor by providing a comprehensive, technical, but nonspecialist overview of technologies at various stages of maturity that, it is hoped, will provide technical background for decision makers and encourage a coming generation of students and young engineers to tackle the 21st century's most important technological challenge.

The book is presented in five parts, dealing in turn with fundamentals, capture, storage and monitoring, transportation, and information resources.

The three chapters of Part I establish some fundamentals. Chapter 1 describes the global carbon cycle and outlines the perturbing impact of anthropogenic carbon dioxide emissions on carbon fluxes and sinks. In Chapter 2, a brief initial overview of CCS technologies is given, taking each of the main industrial sources of carbon emissions as the starting point. Since capture from power generation plants will be a major focus of early CCS implementation, Chapter 3 provides a fairly comprehensive introduction to power generation technologies. The emphasis here is on the current state of the art and on systems under development that are likely to be deployed during the period in which CCS technologies mature.

With these foundations established, Part II provides a more detailed description of carbon capture technologies. The first two chapters are written from an industry perspective, for the power industry (Chapter 4) and other industries (Chapter 5), and the next five chapters from a technology perspective, covering absorption, adsorption, membrane, cryogenic, and mineral carbonation technologies.

Part III then addresses the storage of captured CO_2 and related monitoring requirements, covering geological storage (Chapter 11), ocean storage (Chapter 12), and storage in terrestrial ecosystems (Chapter 13). The final chapter in Part III describes opportunities to increase industrial usage of CO_2 in ways that can significantly contribute to global CCS objectives, such as low-carbon cement and biofuel production.

The transportation of CO_2 between capture and storage sites, either by pipeline infrastructure or by marine transport, is covered in Part IV.

The book concludes in Part V with a compendium of information resources, including units and conversion factors, a list of key abbreviations, and a glossary of some of the key technical terms encountered.

While the focus of this book is on the technical aspects of CCS, many other factors will play a part in determining the extent to which CCS technologies are eventually deployed—chief among them being costs. Apart from some general indications of currently estimated or target costs of some CCS options, this book avoids any analysis of the cost of implementation of the various technologies discussed. The capital and operating costs and the economics of individual CCS projects will be highly case-dependent, with exchange rate volatility further complicating any general analysis. Future reductions in the costs and energy requirements of CCS technologies can also be expected, pending the outcome of further R&D efforts and the learning from early demonstration projects. The extent and timing of these improvements and their impact on overall capture costs are highly uncertain, so that current costs are a poor guide to either actual or relative future CCS implementation prospects or costs.

Various chapters of the book have benefited from review by a number of scientists and other professionals who are engaged in the broad range of technologies described here. My special thanks are due to Dr. John Benemann (Benemann Associates), Dr. Somayeh Goodarzi (University of Calgary), Rob and Karin Lavoie (Calpetra Research & Consulting), Dr. Klaus Lorenz (Ohio State University), Dr. Antonie Oosterkamp (Research Foundation Polytec), Dr. Edward Peltzer (Monterey Bay Aquarium Research Institute), Prof. James Ritter (University of South Carolina), Prof. Anja Schuster (Universität Stuttgart), Dr. Takahisa Yokoyaka (Central Research Institute of Electric Power Industry [CRIEPI]), and Prof. Ron Zevenhoven (Åbo Akademi University).

Their critical input, generously provided, is reflected in these pages; the responsibility for the remaining shortcomings, errors, and omissions remains with the author. Any comments, suggestions, or other feedback from readers will be most welcome; please send them to ccst2010@gmail.com.

It has been a pleasure to work with the team at Elsevier in bringing this book to fruition, and my thanks are due to Ken McCombs and Irene Hosey, who shepherded and supported it from concept to completion, and to the production team—notably Donald Whitehead of MPS Content Services and Anne McGee—ably led by Maria Alonso.

A final word of thanks is due to Serge, Brin, and Jimmy, without whose vision this project would have been a far greater challenge.

In the two decades since the 1990 publication by the UN Intergovernmental Panel on Climate Change of its First Assessment Report, in the face of an increasing body of evidence and understanding, the Panel's careful language of uncertainty has been progressively strengthened to the point where the Fourth Assessment Report was able to state with very high confidence that "the net effect of human activities since 1750 has been one of warming. Most of the observed increase in global average temperatures since the mid-20th century is very likely due to the observed increase in anthropogenic GHG concentrations" (IPCC AR4, 2007).

Looking beyond AR5, due in 2014, with new evidence mounting daily that the climatic impact of our activities is at the upper end of the range of predictions, the task before us is to ensure that, at the end of our finite window of opportunity for change, we do not conclude ...

"This earth is ruined! We gotta get a new one." (Fey, T. (2007). Greenzo, *30Rock*, **2** (5)).

Stephen A. Rackley
August 2009

Acknowledgements

ECO_2® is a registered trademark of Powerspan Corp.
Econamine™ is a registered trademark of Fluor Corp.
Generon® is a registered trademark of the Dow Chemical Company.
Inconel® is a registered trademark of Special Metals Corp.
Prism® is a registered trademark of the Monsanto Company.
Selexol® is a registered trademark of Union Carbide Corp.
Separex™ is a registered trademark of UOP-Honeywell Inc.
Skymine® is a registered trademark of Skyonic Inc.
Teflon® is a registered trademark of E. I. du Pont de Nemours and Company.

Every effort has been made to acknowledge registered names and trademarks. Any omissions should be advised to the publisher and will be corrected in a future edition.

Part I

Introduction and overview

1 Introduction

The fossil fuel resources of our planet—estimated at between 4000 and 6000 gigatonnes of carbon (Gt-C)—are the product of biological and geologic processes that have occurred over hundreds of million of years and continue today. The carbon sequestered in these resources over geological time was originally a constituent of the atmosphere of a younger earth—an atmosphere that contained ~1500 parts per million (ppm) CO_2 at the beginning of the Carboniferous age, 360 million years ago, when the evolution of earth's first primitive forests began the slow process of biogeological sequestration.

Since the dawn of the industrial age, circa 1750, and particularly since the invention of the internal combustion engine, ~5% of these resource volumes have been combusted and an estimated 280 Gt-C released back into the atmosphere in the form of CO_2. In the same period a further ~150 Gt-C has been released to the atmosphere from soil carbon pools as a result of changes in land use. The atmospheric, terrestrial, and oceanic carbon cycles have dispersed the greater part of these anthropogenic emissions, locking the CO_2 away by dissolution in the oceans and in long-lived carbon pools in soils. During the period since 1750, the CO_2 concentration in the atmosphere has increased from 280 ppm to 368 ppm in 2000, and ~388 ppm in 2010, the highest level in the past 650,000 years and one that is not likely to have been exceeded in the past 20 million years, where "likely" reflects the Intergovernmental Panel on Climate Change (IPCC) judgment of a 66–90% chance.

This increase in atmospheric CO_2 concentration ($[CO_2]$) influences the balance of incoming and outgoing energy in the earth-atmosphere system, CO_2 being the most significant anthropogenic greenhouse gas (GHG). In its Fourth Assessment Report (AR4), published in 2007, the IPCC concluded that global average surface temperatures had increased by 0.74 ± 0.18°C over the 20th century (Figure 1.1), and that "most of the observed increase in global average temperatures since the mid-20th century is very likely (*>90% probability*) due to the observed increase in anthropogenic GHG concentrations."

Although anthropogenic CO_2 emissions are relatively small compared to the natural carbon fluxes—for example, photosynthetic and soil respiration fluxes, at ~60 Gt-C per year, are 10 times greater than current emissions from fossil fuel combustion—these anthropogenic releases have occurred on a time scale of hundreds rather than hundreds of millions of years. Anthropogenic change has also reduced the effectiveness of certain climate feedback mechanisms; for

Doi:10.1016/B978-1-85617-636-1.00001-8

example, changes in land-use and land-management practices have reduced the ability of soils to build soil carbon inventory in response to higher atmospheric CO_2, while ocean acidification has reduced the capacity of the oceans to take up additional CO_2 from the atmosphere.

The energy consumption of modern economies continues to grow, with some scenarios predicting a doubling of global energy demand between 2010 and 2050. Fossil fuels currently satisfy 85% of global energy demand and fuel a similar proportion of global electricity generation, and their predominance in the global energy mix will continue well into the 21st century, perhaps much longer. In the absence of mitigation, the resulting emissions will lead to further increase in atmospheric [CO_2], causing further warming and inducing many changes in global climate. Even if [CO_2] is stabilized before 2100, the warming and other climate effects are expected to continue for centuries, due to the long time scales associated with climate processes. Climate predictions for a variety of stabilization scenarios suggest warming over a multicentury time scale in the range of 2°C to 9°C, with more recent results favoring the upper half of this range.

Although many uncertainties remain, there is little room for serious doubt that measures to reduce CO_2 emissions are urgently required to minimize long-term climate change. While research and development efforts into low- or zero-carbon alternatives to the use of fossil fuels continues, the urgent need to move toward stabilization of [CO_2] means that measures such as the capture and storage of carbon that would otherwise be emitted can play an important role during the period of transition to low-carbon alternatives.

Within the field of carbon capture and storage (CCS), a diverse range of technologies is currently under research and development and a growing number of demonstration projects have been started or are planned. A few technologies have already reached the deployment stage, where local conditions or project specifics have made them economically viable, but for most technologies further development work is required to improve technical capabilities and reduce costs.

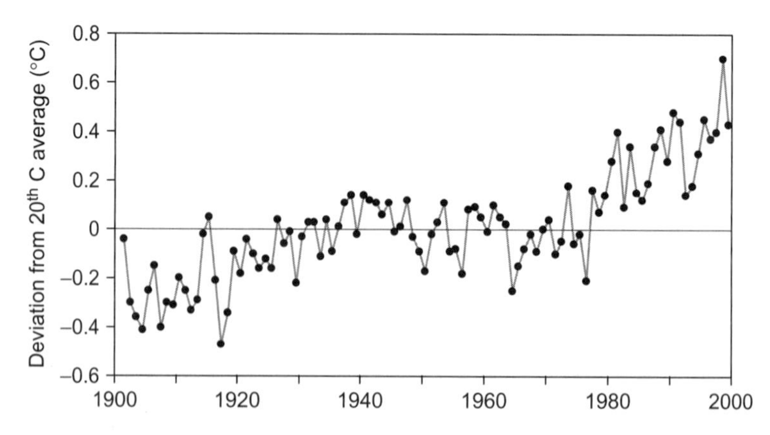

Figure 1.1 Variation of the earth's surface temperature during the 20th century (IPCC data)

Although it is possible with some confidence to identify the technologies that are most likely to yield to these efforts, it is also likely that the full long-term potential of CCS for emissions reduction will be achieved through the application of a broad portfolio of different technical solutions.

The remainder of this chapter provides the context for this challenge. Firstly, the inventories and fluxes that make up the global carbon cycle are discussed. While the current CCS frontrunners make a direct attack on anthropogenic emissions by capturing CO_2 from large sources before emission, reduction of the atmospheric carbon inventory can be achieved by any approach that can limit fluxes contributing to or enhance fluxes reducing this inventory. An understanding of these inventories and fluxes is therefore an essential grounding.

Finally, the process of technological innovation is described. The concepts and terminology introduced here will be used throughout the book to locate various technologies and projects within the life cycle of technology development from research to commercial deployment.

1.1 The carbon cycle

The carbon inventories in the atmosphere, biosphere, soils and rocks, and the oceans are linked by a complex set of natural and anthropogenic biogeochemical processes that are collectively known as the carbon cycle. Figure 1.2 illustrates the inventories (bold font, units of Gt-C) and fluxes (italic font, units of Gt-C per year) that make up this cycle.

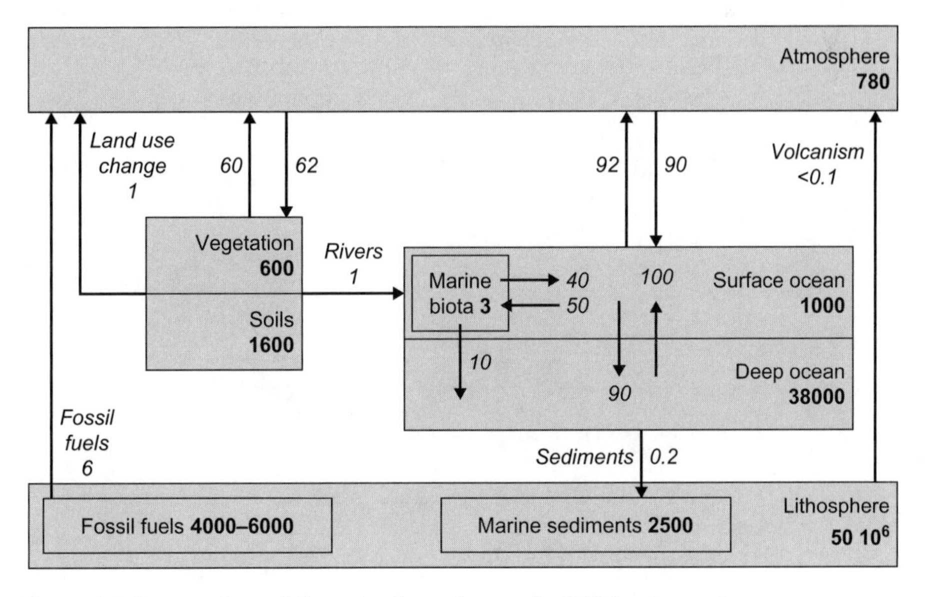

Figure 1.2 Inventories and fluxes in the carbon cycle (2008 estimates)

1.1.1 Carbon inventories

The main inventories relevant to the global carbon cycle are described in the following section.

Carbon inventory of the atmosphere

The atmospheric carbon inventory consists almost entirely of carbon dioxide, with a concentration [CO_2] of some 388 ppm (2010) or 0.04% by volume. As noted above, this inventory has risen by almost 40% since preindustrial times as a net result of emissions from fossil fuel combustion and changes in land-use and land-management practices. The remaining atmospheric carbon inventory consists of methane (CH_4) at ~1.8 ppm, with traces of carbon monoxide (CO) and anthropogenic chlorofluorocarbons (CFCs) also present.

Detailed measurements of [CO_2] were started by Charles Keeling at the National Oceanic and Atmospheric Administration (NOAA) Mauna Loa Observatory, Hawaii, in September 1957, establishing an average value of 315 ppm for the first full year of measurements. The curve of increasing [CO_2] established since that time is known as the Keeling curve, and the past two decades of data from Mauna Loa are shown in Figure 1.3.

The cyclical overprint on the continuously rising trend is shown in Figure 1.4 for each individual year from 2000 to 2008 and as an average over this period. The cycle is synchronized with the Northern Hemisphere seasons, where [CO_2] is drawn down ~3.5 ppm below the annual average trend by photosynthetic production from May to September and rebounds by a similar amount as a result of biomass decomposition from October to April.

The amplitude of this annual [CO_2] cycle at Mauna Loa has increased from ~5.7 ppm in the late 1950s to ~6.4 ppm over the 5 years to 2008. This is believed

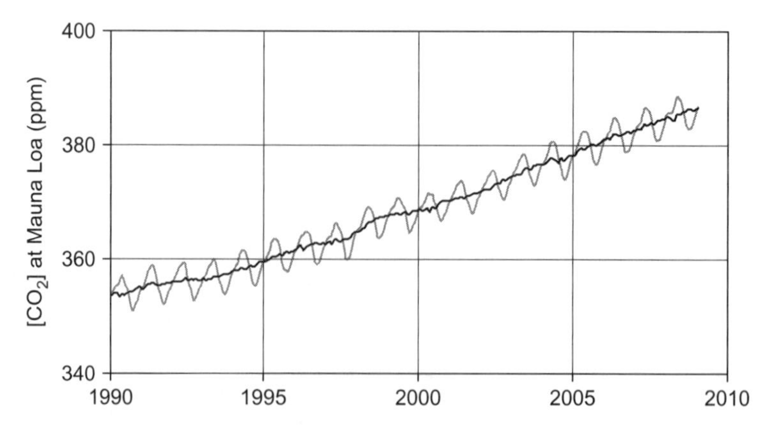

Figure 1.3 [CO_2] at Mauna Loa Observatory (Data courtesy NOAA, Earth System Research Laboratory, Global Monitoring Division)

to be a consequence of increased primary photosynthetic production in northern terrestrial ecosystems, as a result of increasing [CO_2] and temperatures.

The total inventory of carbon in the atmosphere with [CO_2] at 388 ppm is ~780 Gt-C, with an annual increase in [CO_2] of ~1.7 ppm corresponding to a net inventory increase of ~3.5 Gt-C per year.

Carbon inventory of the biosphere and soils

The terrestrial carbon inventory is estimated to hold 2200 Gt-C, of which 600 Gt-C is present as living biomass and 1600 Gt-C as organic carbon in soils and sediments. This inventory has declined by roughly 10% since preindustrial times, and predominantly since the mid-19th century, as a result of changes in land-use and land-management practices—deforestation, conversion of grasslands to agricultural use, and intensive agricultural practices being the main contributors.

The soil carbon inventory can be further classified according to the carbon residence time within soils, as shown in Table 1.1. This ranges from plant and animal detritus, which will be decomposed and emitted through respiration with a typical time scale of 1 to 10 years, to inert carbon, which is inaccessible to biological processes and will remain in the soil until physically removed by water or airborne transport. These processes are discussed further in Chapter 13.

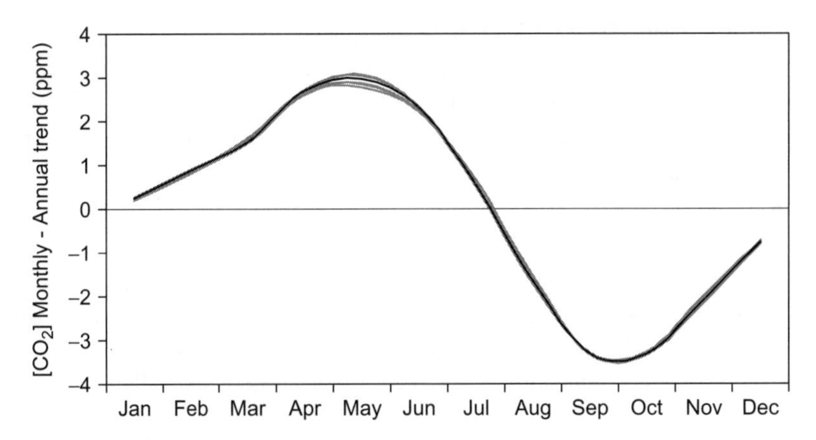

Figure 1.4 Annual [CO_2] cycle, 2000–2008 (Date source; as Figure 1.3)

Table 1.1 Soil carbon inventory, components, and lifetimes

Component	Lifetime	Inventory (Gt-C)
Plant and animal detritus	<10 years	350
Modified soil carbon	10 to 1000 years	1100
Inert carbwon	>1000 years	150
Total		1600

Carbon inventory of the oceans

The oceanic carbon inventory amounts to ~39,000 Gt-C, with more than 90% of this being present as bicarbonate ions (HCO_3^-), as shown in Table 1.2. In addition, some 2500 Gt-C is present in marine carbonate sediments, which are gradually transformed into sedimentary rock over geological time. Of the dissolved CO_2 inventory in the oceans, ~120 Gt-C (16%) is anthropogenic, with an estimated uptake rate of ~2 Gt-C per year.

Within the oceans, three key processes—the biological and solubility pumps and the thermohaline circulation—drive the distribution of carbon between organic and inorganic fractions, and its transport and eventual deposition in sediments. These processes are described in Chapter 12.

Carbon inventory of the lithosphere

The earth's crust, which represents the upper part of the lithosphere, is the final geological carbon sink and is estimated to hold $5 \cdot 10^7$ Gt-C in sedimentary rocks, ~20% of which is in the form of organic carbon and the remainder as limestone.

Fossil fuels—coal, oil, and gas—together account for between 4000 and 6000 Gt-C, or ~0.05% of the total organic carbon present in sedimentary rocks.

1.1.2 Carbon fluxes

These carbon inventories are subject to constant flux as a result of a web of interlinking natural processes. In addition, human activity has introduced new fluxes, and the effect of these has, in turn, modified some of the natural fluxes through various feedback mechanisms. To date, the net feedback has been negative, with the result that only ~45% of anthropogenic CO_2 emissions remain in the atmosphere.

Atmosphere ↔ ocean fluxes

The exchange of CO_2 between the atmosphere and the oceans occurs due to the difference in CO_2 partial pressure between the atmosphere and surface waters, with an estimated ~90 Gt-C being exchanged annually. This flux is

Table 1.2 Oceanic carbon inventory

Component	Inventory (Gt-C)
Bicarbonate ion	36,000
Carbonate ion	1300
Dissolved CO_2	740
Dissolved organic carbon	<700
Marine biomass	<10
Total	~39,000

controlled by two key processes: the global ocean circulation system, which exchanges surface and deep water on a 500- to 1000-year time scale, and the geochemistry of surface waters, in particular the removal of carbonate ions by ionic reactions and by precipitation (carbonate buffering; see Glossary).

An increase in the dissolved [CO_2] reduces the carbonate ion concentration ([CO_3^{2-}]) due to the ionic reaction that forms bicarbonate ions:

$$CO_2 + H_2O \leftrightarrow H_2CO_3 \leftrightarrow H^+ + HCO_3^- \quad (1.1)$$

and

$$H^+ + CO_3^{2-} \leftrightarrow HCO_3^- \quad (1.2)$$

The alkalinity of the ocean, measured as [HCO_3^-] + 2 × [CO_3^{2-}], is preserved in this reaction since one carbonate ion is converted into two bicarbonate ions. However, the decline in [CO_3^{2-}] means that the ion is less available to react with additional dissolved CO_2, reducing further uptake and resulting in an increase in acidity as a result of reaction 1.1.

The rise in atmospheric [CO_2] shown in Figure 1.3 has resulted in an increased rate of uptake of CO_2 by the oceans, reducing the atmospheric carbon inventory by an estimated ~2 Gt-C per year, or roughly one third of anthropogenic emissions, in the period from 1990 to 2005. This additional uptake has resulted in an increase in surface ocean acidity, as the carbonate buffer has been depleted in these waters. As noted above, this will limit the ability of the ocean to increase CO_2 uptake in response to future increases in atmospheric [CO_2].

The ocean provides a slow-acting buffer to stabilize atmospheric [CO_2], and any atmospheric perturbation will be dissipated by absorption into the ocean over a time scale of centuries. This is illustrated in Figure 1.5, which shows a simple model of the uptake by the ocean of a 100-year "pulse" of emissions at 6 Gt-C per year into the atmosphere, starting at year zero.

In this model the ocean has taken up 50% of the emitted CO_2 after 150 years and almost 75% after 1000 years. From an initial level of 350 ppm, [CO_2] peaks at 550 ppm at the end of the emission pulse, and declines to ~430 ppm over the final 100 years.

Atmosphere ↔ terrestrial biosphere and soil fluxes

Terrestrial photosynthesis removes an estimated 120 Gt-C per year from the atmosphere as gross primary production (GPP), of which 60 Gt-C per year is reemitted through plant (autotrophic) respiration and 60 Gt-C per year is retained as net primary production (NPP), resulting in biomass growth.

Soil respiration, primarily from the microbial communities that feed on plant detritus and root exudates (heterotrophic respiration), returns a further ~55 Gt-C per year to the atmosphere. Under steady state, the balance is made

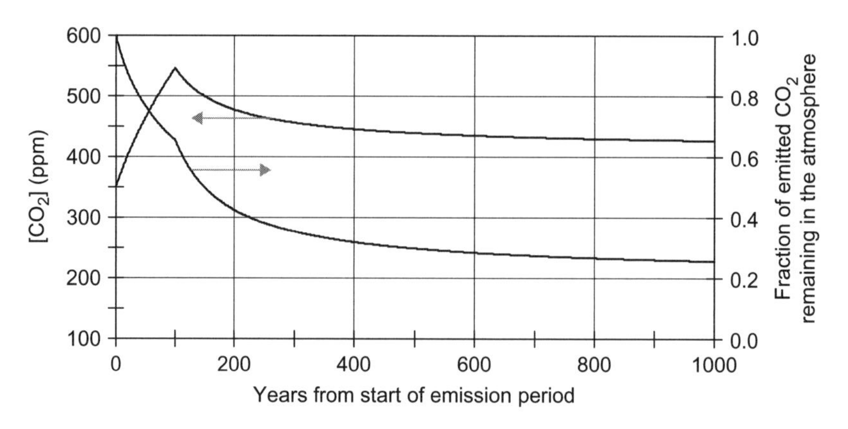

Figure 1.5 Simple model of oceanic uptake of atmospheric CO_2 emissions

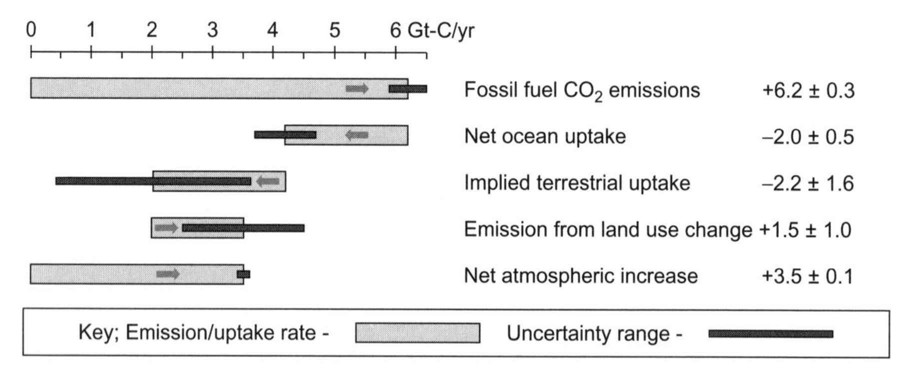

Figure 1.6 Implied terrestrial biosphere uptake from 1980 to 2005 carbon budget

up of emissions back to the atmosphere as a result of natural fires and dissolved organic carbon (DOC) export by rainwater runoff into rivers.

Over the 25-year period from 1980 to 2005, the net flux from the atmosphere to the terrestrial biosphere is estimated to have been 0.7 Gt-C per year, as shown in Figure 1.6. This figure is the net result of an estimated ~1.5 Gt-C per year of emissions resulting from land-use changes, balanced by an implied uptake of ~2.2 Gt-C per year into the terrestrial biosphere.

This net uptake is attributed to a CO_2 fertilization effect, which increases NPP with increasing [CO_2], as a result of increased photosynthetic efficiency and improved water-use efficiency in arid areas.

Atmosphere ↔ lithosphere fluxes

A carbon flux of ~0.5 Gt-C per year from the atmosphere occurs as a result of the carbonate–silicate cycle in which carbonic acid, formed by the dissolution of CO_2 in rainwater, causes the weathering of exposed silicate rocks in the reaction:

$$CaSiO_3 + H_2CO_3 \leftrightarrow CaCO_3 + SiO_2 + H_2O \quad (1.3)$$

The dissolved minerals are transported by rivers into the sea, adding to the ocean carbon inventory and carbonate buffer.

The carbonate–silicate cycle is eventually closed over geological time by the subduction of sedimentary rocks formed by the precipitation and sedimentation of the weathering products. Metamorphosis in subduction zones reforms the silicate minerals, while CO_2 is released through volcanoes. This CO_2 flux from volcanic venting is estimated to add on average <0.1 Gt-C per year to the atmospheric inventory.

1.2 Mitigating growth of the atmospheric carbon inventory

1.2.1 Anthropogenic emission scenarios

The future level of anthropogenic CO_2 emissions, both from fossil fuel combustion and from land-use changes, will be dictated by a wide range of demographic, socioeconomic, environmental, and technological factors, including:

- Population growth
- Economic growth and the globalization of trade
- Energy intensity of industrial production
- Fossil-fuel mix within total energy supply
- Technology development in primary energy production
- Environmental pressures and policy-driven incentives

Predicting any one of these factors over a 100-year time period carries a wide range of uncertainty, and the problem of combining multiple uncertainties is best handled by the creation of a number of scenarios, based on storylines that depict how these factors could play out in future.

The IPCC created a set of such scenarios in the *Special Report on Emissions Scenarios* (SRES), published in 2000, and Figure 1.7 illustrates the total CO_2 emissions, both fossil-fuel and land-use changes, generated for three of these scenarios as well as the maximum and minimum of the scenario range.

Although these scenarios do include technological developments such as advanced power-generation systems and decarbonization of transport fuels, the implementation of CCS is not considered. While the IPCC scenarios have been overtaken by actual data for the first decade of the scenario period, they still provide a broad indication of the magnitude of the challenge that CCS seeks to address.

Figure 1.8 illustrates the estimated $[CO_2]$ resulting from the SRES emissions scenarios depicted in the previous figure for a range of climate models, and shows $[CO_2]$ potentially rising to between 470 and 570 ppm by 2050 and into the 540- to 860-ppm range by 2100.

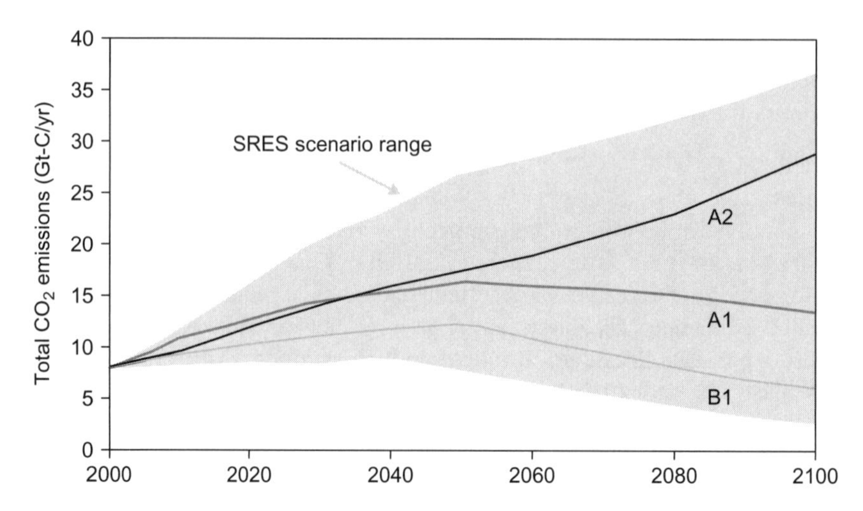

Figure 1.7 IPCC CO_2 emissions scenarios (Data courtesy IPCC, SRES 2000)

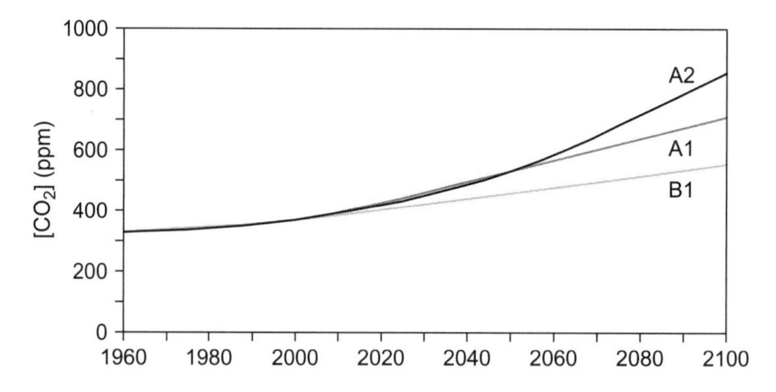

Figure 1.8 Estimated atmospheric [CO_2] resulting from SRES emission scenarios

1.2.2 CO_2 stabilization scenarios

The models used to predict [CO_2] for a given emissions scenario can also be run to establish the range of emissions scenarios that would result in stabilization of [CO_2] at a specific level. Figure 1.9 shows the emissions scenario data from Figure 1.7 together with the estimated range of emissions profiles that would permit [CO_2] stabilization at 450 ppm and 550 ppm.

The ranges of the two sets of scenarios are extremely broad and can give only a very rough indication of the emissions reductions required to achieve a specified [CO_2] target. SRES scenario B1, which is based on an environmentally conscious and resource-conservative storyline with technology development aimed at improving primary energy-conversion efficiency, is predicted to result in [CO_2] stabilization at around 550 ppm for many of the models,

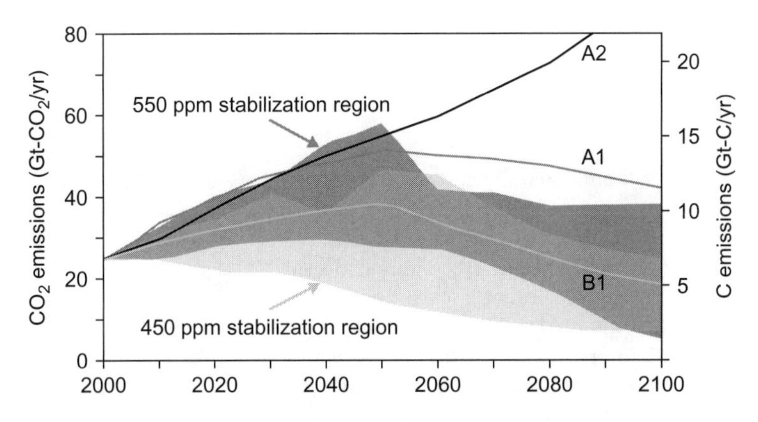

Figure 1.9 SRES emissions scenarios and [CO_2] target emissions ranges

while, under this emissions scenario, capture and storage of between 100 and 200 Gt-C (370–730 Gt-CO_2) would be required by 2050 to stabilize [CO_2] at 450 ppm. This reduction target would rise to between 200 and 300 Gt-C (730–1100 Gt-CO_2) by 2050 for stabilization at 450 ppm under the higher-emissions A1B SRES scenario. Climate sensitivity studies, over a range of climate models, suggest that the extra 100 ppm translates into an additional increase in global mean temperature of ~1°C.

This indicates the scale of the challenge that CCS addresses. In subsequent chapters a variety of ongoing and planned projects will be described, employing a range of technologies at various stages of development—from laboratory bench scale, to pilot and demonstration scale and finally up to full commercial scale—with throughputs ranging from a few kt-CO_2 per year up to 1 or 2 Mt-CO_2 per yr. To contribute significantly to meeting the challenge outlined above—to capture and store just 20 Gt-C by 2050—the largest of these current projects would need to be scaled by a factor of five and deployed 200 times in the next 10–20 years. The technology development process that will be followed in delivering this massive step change is outlined in the next section.

1.3 The process of technology innovation

The process of technology development generally goes by the acronym RDD&D—Research, Development, Demonstration, and Deployment, the four stages describing the route that most new technologies take in maturing from fundamental research to commercial application. Table 1.3 describes the characteristics of each of these stages.

An example of specific RD&D tasks at the first three stages is shown in Table 1.4 for the development of a membrane reactor based on technologies that will be discussed in Chapter 8.

Table 1.3 Stages in the technology development process

Stage	Description	Current CCS examples (2010)
Research	Fundamental research and experimental proof of concept, led by academic or industrial research organizations. Relatively low-cost exploration of a wide range of potential options. Definition of a "road map" detailing further fundamental and applied research requirements. Broad-brush estimate of eventual deployment costs and commercial viability.	Amine-facilitated transport membrane for postcombustion CO_2 separation from flue gas
Development	Progress along the development road map; applied research focusing on process engineering and system integration; laboratory- and pilot-scale demonstration of the process. Additional fundamental research may be spawned as further implementation issues are identified. Refined construction and operating cost estimates and indications of commercial viability.	Hybrid combustion–gasification chemical looping using calcium compounds
Demonstration	Initial industrial-scale implementation, often funded by government and industry partnerships. May involve the integration of existing, proven technologies in a new application. Evaluation and improvement of the design, construction, and operating processes. Budget-level definition of construction and operating costs.	Air-separation plant using ion transport membrane to supply oxyfuel combustion
Deployment	Progressive commercial implementation, which, in the early stages, may be accelerated by economic incentives in the form of capital grants or premium prices.	Transportation of CO_2 by pipeline; geological storage for enhanced oil recovery

Figure 1.10 shows the overall RDD&D timeline for the development of oxy-fuel combustion technology for a coal-fired power plant being undertaken by Vattenfall AB and partners, described further in Chapter 4. The 20-year time scale required to bring such technologies to the stage of readiness for commercial

Table 1.4 RD&D tasks for membrane reactor technology development

Stage	Example RD&D tasks
Research (bench-scale testing)	Basic membrane and module materials research, and mathematical modeling
	Validation of mathematical models in a bench-scale experimental set-up
	Build economic evaluation model
Development (pilot-scale testing)	Construct pilot testing set-up (membranes, module, and housing)
	Perform pilot-scale testing and technical evaluation
	Possible recycle to bench-scale testing to resolve design, fabrication, or performance issues
	Refine process simulations for field-scale reactor design and perform optimization and integration studies
	Perform preliminary economic and business assessment
Demonstration (field-scale testing)	Fabricate field-scale test reactor (membranes, modules, reactor, catalysts)
	Prepare field test site and install reactor
	Perform field testing and technical evaluation
	Possible recycle to bench- or pilot-scale testing to resolve design, fabrication, or performance issues
	Conduct system integration study for full-scale deployment
	Refine performance simulation to optimize process operating conditions
	Finalize economic and business assessment

deployment highlights the corporate vision and environmental commitment required to undertake such a project, especially when the work is being undertaken in parallel with the maturation of regulatory and fiscal frameworks that will ultimately determine the economic feasibility of the outcome.

The development of new technology is rarely as linear as that implied by this sequence of steps. In reality, backward loops frequently occur, particularly from development back to research, as obstacles encountered at the applied-research stage demand additional work to establish insights at the fundamental level. An example of this will be encountered in Chapter 10, where development efforts to improve the reaction rate of mineral carbonation reactions have spawned a number of lines of research aimed at understanding the impact of pretreatment on crystal and surface structure effects.

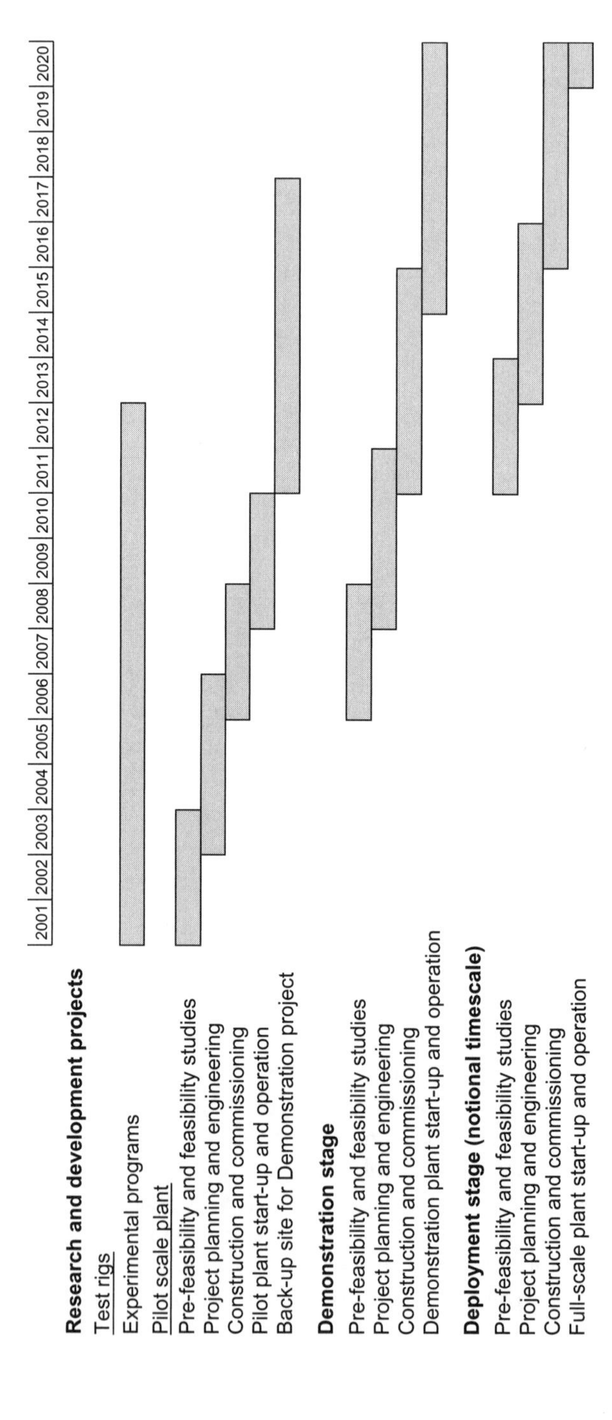

Figure 1.10 Vattenfall AB oxyfuel technology RDD&D timeline

A technology may span several stages at once, for example being proven in one application or industry but not yet in another, and the integration of various elements that may individually be considered proven technology typically requires an additional demonstration stage. While fundamental research and development work is generally directly transferable, the demonstration stage often needs to be repeated to address location- or industry-specific aspects.

Typically, the costs involved increase as a technology advances through successive stages in the process, and progress of any project to the next stage is controlled by the availability of and competition for funding. Technical viability and commercial viability or, in the early stages, the prospect of that viability, are the key drivers of funding decisions. Several other factors that characterize the process are shown in Table 1.5.

Governments often apply incentives to kick-start the demonstration and deployment phases by encouraging initial projects that may be marginal from a commercial perspective—for example, the premium pricing for early wind-power deployment in many countries. Continued deployment commonly results in reduced capital and operating costs as operating improvements and economies of scale are realized.

If there is a conclusion to be drawn from the following chapters it is that, given the many and varied technologies that are being developed to address the challenges of CCS and the global record of technical innovation and progress of the past century, pursuit of these technologies has a high likelihood of delivering solutions that can meet the challenge. However, success in this venture also requires a globally shared political will and sense of urgency to create the enabling conditions for the deployment of these technologies on a massive scale. Whether that consensus and urgency will develop at a pace equal to the challenge is perhaps a bigger uncertainty than whether the technologies can deliver.

Table 1.5 Characteristics of the technology development process

Factor	Research	Development	Demonstration	Deployment
Academic involvement	High	Moderate	Low	Low
Industry involvement	Variable	Moderate	High	High
Costs	Typically low	Moderate	Moderate to high	High
Diversity of options	Very broad	Broad	Narrower	Narrow
Government involvement	General or targeted research funding	Focused technology-development funding	Market incentives	Market incentives

1.4 References and Resources

1.4.1 Key References

The following key references, arranged in chronological order, provide a starting point for further study and access to a wider bibliography on the carbon cycle and climate change.

IPCC. (2000). *Special Report on Emission Scenarios*, Cambridge University Press, Cambridge, UK.

Wigley, T. M. L. and D. S. Schimel (eds.). (2000). *The Carbon Cycle*, Cambridge University Press, Cambridge, UK.

Stern, N. (2006). *Stern Review: The Economics of Climate Change*, Cambridge University Press, Cambridge, UK. Available at www.hm-treasury.gov.uk/stern_review_report.htm

Solomon, S., *et al.* (eds.). (2007). *Climate Change 2007— The Physical Science Basis*, Cambridge University Press, Cambridge, UK. Available at www.ipcc.ch/ipccreports/assessments-reports.htm

Pachauri, R. K. and A. Reisinger (eds.). (2008). *Climate Change 2007— Synthesis Report*, IPCC, Geneva, Switzerland. Available at www.ipcc.ch/ipccreports/assessments-reports.htm

1.4.2 Institutions and Organizations

Institutions and relevant web links relating to the carbon cycle and climate change:

Intergovernmental Panel on Climate Change: www.ipcc.ch/index.htm

North American Carbon Program (assessing the sources and sinks of CO_2, CH_4, and CO in North America and adjacent oceans): www.nacarbon.org

Ohio State University, Climate, Water, Carbon Program: http://cwc.osu.edu

U.S. Carbon Cycle Science Program: www.carboncyclescience.gov

1.4.3 Web Resources

Carbon Capture Journal: www.carboncapturejournal.com

Carbon Capture and Sequestration (educational website): www.ccs-education.net

First State of the Carbon Cycle Report (SOCCR): The North American Carbon Budget and Implications for the Global Carbon Cycle: www.climatescience.gov/Library/sap/sap2-2/final-report/default.htm

GRID-Arendal: United Nations Environment Programme (UNEP) Collaboration Center; access to IPCC Assessment and Special Reports (www.grida.no/publications), including IPCC Special Report on Emissions Scenarios (www.grida.no/publications/other/ipcc_sr/?src=/climate/ipcc/emission)

PointCarbon (carbon market information portal): www.pointcarbon.com

RealClimate (commentary on climate science): www.realclimate.org

U.S. Carbon Cycle Science Program: www.carboncyclescience.gov/about.php

U.S. Department of Commerce, National Oceanic and Atmospheric Administration (NOAA), Earth System Research Laboratory, Global Monitoring Division (latest trend in atmospheric CO_2): www.esrl.noaa.gov/gmd/ccgg/trends; data available at ftp://ftp.cmdl.noaa.gov/ccg/co2/trends/

U.S. Department of Energy (database of carbon sequestration RD&D projects): www.fossil.energy.gov/fred/feprograms.jsp?prog=Carbon+Sequestration

2 Overview of carbon capture and storage

The following sections provide an initial overview of the opportunities and technologies for CO_2 capture and storage, both currently available and under development, as an introduction to the more detailed discussion in Chapters 4–9.

2.1 Carbon capture

There are three main approaches to CO_2 capture (Figure 2.1):

- As a pure or near-pure CO_2 stream, either from an existing industrial process or by reengineering a process to generate such a stream (e.g., oxyfueling power-generation plant, precombustion fuel gasification)
- Concentration of the discharge from an industrial process into a pure or near-pure CO_2 stream (e.g., postcombustion separation from power plant or cement plant flue gases)
- Direct air capture into a pure CO_2 stream or into a chemically stable end product (e.g., mineralization of steel slag)

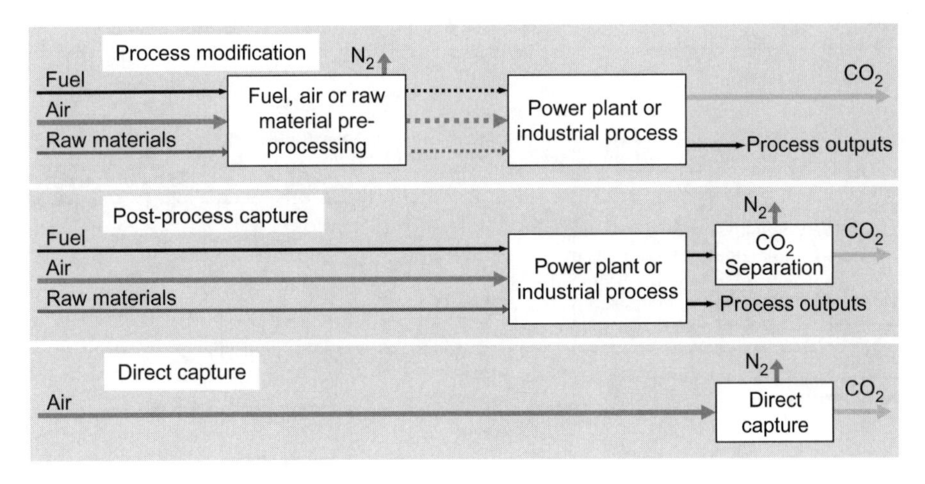

Figure 2.1 Main approaches to CO_2 capture

Doi: 10.1016/B978-1-85617-636-1.00002-X

2.1.1 Capture from power generation

IPCC analysis (See Key References, IPCC (2005)) shows that large fossil-fueled power plants account for almost half of the total CO_2 emissions from fossil fuel combustion. These large point sources—defined as emitting >0.1 Mt-CO_2 per year and summarized in Table 2.1—will be an essential area of application if CCS is to have a material impact on future CO_2 emissions.

Between 85–90% of global electrical power is generated from fossil fuel- and biomass-powered steam-driven turbines, a process illustrated schematically in Figure 2.2. The thermal efficiency of this type of plant is limited to ~40–45% by the achievable temperature of the working fluid (steam). This in turn is limited by the availability and cost of suitable materials that are required to withstand the high temperatures and pressures. Current technology limits steam temperatures to ~650°C, although technologies under development are expected to raise this to 700°C by 2020.

Generation efficiency is significant improved in a combined cycle power plant, which uses hot combustion gases to directly drive a gas turbine and subsequently to generate steam in a heat-recovery steam generator, which then drives a steam turbine. A thermal efficiency of 60% can be achieved in current combined cycle power plants, and this can be increased to >80% if low-temperature waste heat

Table 2.1 CO_2 emissions from fossil-fueled power plant: large point sources (IPCC data)

Power plant fuel	Emissions from identified sources (Mt-CO_2 per year)	Number of large point sources	Average annual emissions per source (Mt-CO_2)
Coal	7984	2025	3.9
Natural gas	1511	1728	0.9
Oil	980	1108	0.9

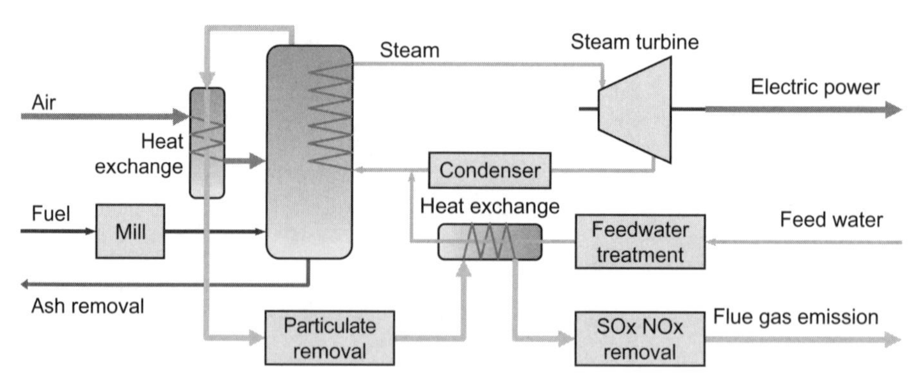

Figure 2.2 Schematic fossil-fueled electric-generation plant

is also recovered for residential or industrial heating in a combined heat and power (CHP) application.

The typical characteristics of flue gas from fossil fuel combustion are summarized in Table 2.2. Carbon monoxide (CO) is not shown in the table but may also be present in the case of incomplete combustion as a result of a lack of oxygen in the combustion chamber.

The CO_2 content ranges from 3% to 15%, the lower end of this range (3–5%) being typical for gas-fired plants and the upper end (12–15%) for coal-fired plants. Three alternative approaches to CO_2 capture from power generation are at various stages of development and are illustrated schematically in Figure 2.3.

CO_2 capture from flue gas is known as postcombustion capture, and techniques have been developed and implemented in the natural gas processing industry that are directly applicable to existing power plants. These technologies, including chemical and physical sorbents and membranes, are further described in Chapter 4 and Chapters 6–9.

The two alternatives to postcombustion capture aim to modify the combustion process so that a pure or high-concentration CO_2 stream is generated. In the first of these (oxyfueling), combustion of the fuel in pure oxygen rather than air results in a near-pure CO_2 combustion gas stream that may require only minimal further processing before being compressed for transportation and storage. Oxyfueling is further discussed in Chapter 4.

Table 2.2 Typical fossil fuel combustion flue gas characteristics

Parameter	Typical range of values
Pressure	At or slightly above atmospheric pressure
Temperature	30–80°C or higher, depending on the degree of heat recovery
CO_2	Coal-fired, 14%
	Natural gas-fired, 4%
O_2	Coal-fired, 5%
	Natural gas-fired, 15%
N_2	~81%
SO_x	Coal-fired, 500–5000 ppm
	Natural gas-fired, <1 ppm
NO_x	Coal-fired, 100–1000 ppm
	Natural gas-fired, 100–500 ppm
Particulates	Coal-fired, 1000–10,000 mg per m^3
	Natural gas-fired, 10 mg per m^3

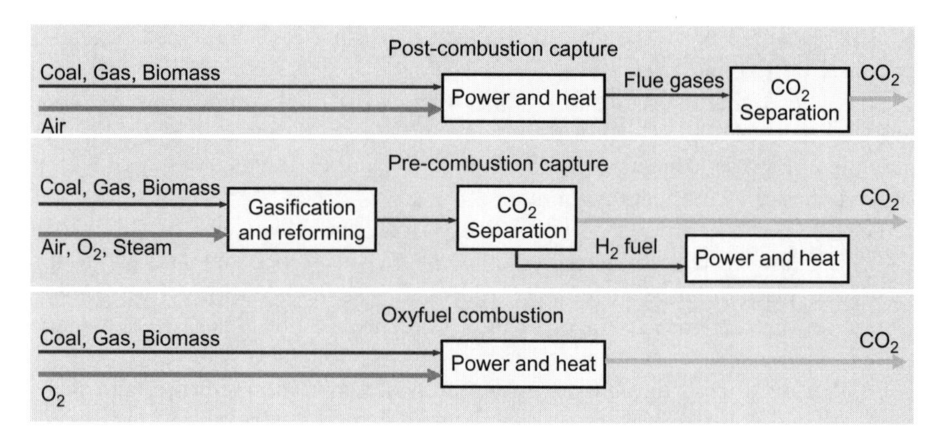

Figure 2.3 Approaches to CO_2 capture from power generation plants

In the second alternative (precombustion capture), the fuel is partially oxidized and reacted with steam to form a CO_2 and H_2 mixture containing 15–60% CO_2, from which the CO_2 can be separated by a range of techniques similar to those used for postcombustion capture. The resulting hydrogen fuel stream could then be combusted conventionally in a boiler or gas turbine, or put to a variety of other fuel uses. Precombustion capture is discussed further in Chapter 4.

2.1.2 Capture from other industrial processes

As well as power generation, a number of other industrial processes contribute a significant fraction of the total CO_2 emissions from large (>0.1 Mt-CO_2 per year) stationary sources. The most significant of these are summarized in Table 2.3 (IPCC (2005)).

Capture from cement and steel production is briefly introduced below and is further discussed, along with other industrial processes, in Chapter 5.

Cement production

Total global CO_2 emissions from cement production are roughly double the volume identified in Table 2.3 from large point sources, and amounted to ~2 Gt-CO_2 in 2008.

The main raw material for cement production is calcium carbonate ($CaCO_3$), derived from crushed limestone, chalk, marl, or shells. Small amounts of clay, shale, blast furnace slag, or ash are added during the production process to provide controlled quantities of aluminum, iron, and silicon. Cement is produced by heating the raw materials to produce slaked lime (CaO) and other compounds in the chemical process of calcination. Large roller kilns, operating at 1450–1500°C, fire the input slurry to produce marble-sized cement clinker, which is then crushed to yield the final fine-powder product. The energy input to the kiln is provided by combustion of coal, oil, or natural gas and, when

Table 2.3 CO_2 emissions from industrial processes: large point sources (IPCC data)

Process	Emissions from identified sources ($Mt\text{-}CO_2$ per year)	Number of large point sources	Average annual emissions per source ($Mt\text{-}CO_2$)
Cement production	930	1175	0.8
Integrated steel mills	630	180	3.5
Oil refineries	800	638	1.3

combined with the gas released from calcination, yields a flue gas with [CO_2] of typically 14–33%. Global average CO_2 emissions from cement production are ~0.8 t-CO_2 per t-cement, and are split roughly 50:50 between emissions resulting from the calcination reaction and those resulting from fuel combustion for kiln firing and other process requirements.

The capture of CO_2 from cement production is thus similar to postcombustion capture applied to power generation, with the key difference being the higher [CO_2] compared to normal power-generation applications. Alternatively, by drafting the burners in the kiln with oxygen rather than air (analogous to oxyfuel combustion in power generation), an offgas stream can be produced that would be suitable for compression and storage after minimal further treatment. These options are described in Chapter 5.

A potentially material CO_2 storage opportunity also arises where cement finds its end use in the construction industry, since CO_2 can be absorbed in significant quantities by precast concrete products, such as masonry blocks, during the curing process. This opportunity is discussed in Chapter 14.

Integrated steel mills

CO_2 emission from steel production typically amounts to ~1.4 t-CO_2 per t-steel produced for integrated steel mills using basic oxygen steelmaking, reducing to roughly a quarter of this (~0.3 t-CO_2 per t-steel) in the case of a steel recycling plant.

At the end of the basic oxygen steelmaking process, described in Chapter 5, blast furnace gases contain ~30% CO_2, and the concentration in the overall flue gas stream from an integrated steel mill is ~15%. This is in the same range as for coal-fired power plant, and the capture options introduced above for power generation are equally applicable for steel production, namely:

- "Postcombustion" capture from the overall flue gas stream
- Firing the blast furnace with oxygen rather than air to give a CO plus CO_2 offgas stream
- Capturing CO_2 in a precombustion step and using hydrogen to reduce iron oxide in the blast furnace rather than carbon monoxide

These options are further described in Chapter 5.

2.1.3 Other capture options

Direct air capture

At the opposite end of the [CO_2] spectrum from capture at ~30% from a cement plant, the possibility of capture directly from ambient air at ~0.04% has also been proposed, using various chemical absorption approaches.

Although at first glance it would seem that capture from a flue gas stream with a high [CO_2] would require far less energy input than capture from the air, the theoretical difference is surprisingly little: just 15 kJ per mol (equivalent to ~95 kWhr per t-CO_2). This is some 4–6% of the overall energy requirement of current amine-based absorption processes for flue gas CO_2 removal (1600–2400 kWhr per t-CO_2).

A sodium hydroxide (NaOH)-based spray-tower system capable of capturing 15 t-CO_2 per year per square meter of contactor area has been demonstrated at laboratory scale and is described in Chapter 6. Scale-up of such a system to capture 1 Mt-CO_2 per year would require an absorber wall ~10–20 m tall and 3–6 km long.

Unlike other capture systems, which will reduce future emissions and limit the resulting increase in atmospheric [CO_2], direct air capture also has the potential to accelerate the natural decline in atmospheric [CO_2] below a future peak level.

2.2 Carbon storage

2.2.1 Geological storage

Injection into oil-, gas-, and water-bearing geological formations is widely regarded as the front-running option for CO_2 storage and is the only option that has so far been applied on a commercial scale. The readiness of this option for commercial deployment is due to the use of site characterization, injection, and monitoring technologies that have been developed and widely deployed in the oil and gas industry.

Two main storage options are available: storage in formations containing non-potable water (saline aquifers) or in oil and gas reservoirs. The use of oil or gas reservoirs, whether producing or depleted, has a potential economic advantage if injection can enhance hydrocarbon recovery, as well as a risk-management advantage since the occurrence of hydrocarbons already demonstrates the presence of a sealing cap rock that has remained competent on a geological time scale. For saline-aquifer storage, this sealing capacity needs to be demonstrated by initial studies and must be more carefully monitored throughout the life of the storage project.

CO_2 injection into oil reservoirs is widely practiced as an enhanced oil recovery (EOR) technique, particularly in the Permian Basin oilfields in the United States, using CO_2 primarily sourced from naturally occurring CO_2 reservoirs.

EOR has also been applied using anthropogenic CO_2 on a 2 Mt-CO_2 per year scale at the EnCana-operated Weyburn and Apache Canada-operated Midale in Saskatchewan, and a number of projects on a similar scale are being planned for start-up before 2015. Injection into depleted gas reservoirs has also been demonstrated, for example in the Gaz de France K12-B field in the North Sea, although enhanced gas recovery (EGR) benefits remain uncertain.

While lacking the direct economic incentives, saline-aquifer storage has the potential advantage of being more geographically accessible to capture sites; global storage potential in saline aquifers is estimated to be one to two orders of magnitude greater than in oil and gas reservoirs. Saline-aquifer storage of 1 Mt-CO_2 per year has been demonstrated and is continuing at the StatoilHydro-operated Sleipner field in the North Sea, driven by the need to avoid venting of CO_2 removed from natural gas production. A number of projects are currently being planned worldwide storing volumes of up to 1.5 Mt-CO_2 per year, and with start-up in the 2011–2017 period.

The technologies involved in site characterization, injection, and monitoring for geological storage are described in Chapter 11.

2.2.2 Ocean storage

With a carbon inventory some 50 times greater than the atmosphere, the ocean is a prime candidate for storage of captured CO_2, and several options have been investigated. Since surface waters exchange CO_2 with the atmosphere on a time scale of months to years, storage must be at depth in order to prevent rapid release back to the atmosphere.

Long-term storage by direct dissolution into deep waters could be achieved by venting gaseous CO_2 or supercritical fluid at sufficient depth to ensure dispersal of the rising buoyant plume before it reaches surface waters. Venting could be either from a fixed pipeline or from a riser trailed behind a moving ship (Figure 2.4).

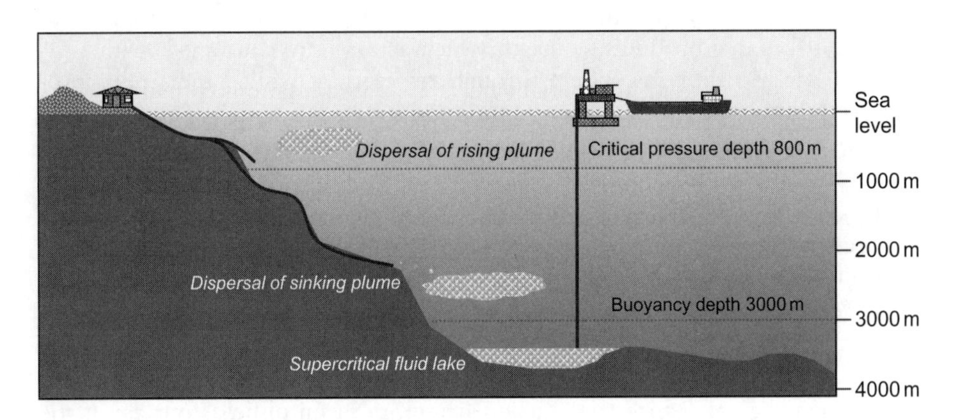

Figure 2.4 Options for CO_2 storage in the oceans

Alternatively, CO_2 could be stored as a lake of supercritical fluid if injected below the depth at which it becomes negatively buoyant in seawater (~3000 m) and at a location where the seabed topology provides lateral containment. Pools of this type have been observed in the vicinity of deepwater hydrothermal vents as a result of the separation of CO_2 from vented gases.

Experimental and small-scale in situ trials have been conducted to investigate the behavior and local environmental impact of CO_2 injected into the sea. However, attempts to conduct larger trials have met with severe environmental opposition and have not gone ahead. Although the ocean will be the ultimate sink for all CO_2 released into the atmosphere, over a time scale of centuries, the acceptability of the ocean as a direct storage site for CO_2—the perception of solving one environmental problem by creating another—will be a major hurdle for further RD&D progress.

These and other ocean-storage options, including increasing photosynthetic production in nutrient-depleted surface waters, either by direct fertilization or by wave-assisted upwelling of nutrient-rich deep waters, are discussed further in Chapter 12.

Processes that mimic the natural weathering of carbonate and other alkaline rocks have also been proposed that would, in some cases, use the ocean as both a source of water and a storage site for reaction products. This could potentially have a beneficial effect on ocean chemistry, for example by replacing depleted carbonate ions, although dispersal over a wide area would be essential to avoid adverse local environmental impact. These and other mineral carbonation storage options are discussed in Chapter 10.

2.2.3 Storage in terrestrial ecosystems

Unlike geological or ocean storage, in which the captured carbon is injected directly into the storage site, storage in terrestrial ecosystems is less direct and relies on the identification and control of the biogeochemical processes and conditions that determine the fate of carbon in these ecosystems.

From the point of "capture" in photosynthesis, carbon is partitioned into a range of organic products, some of which are rapidly consumed, with CO_2 released back to the atmosphere through respiration, while others have progressively longer residence times in the biomass and soil carbon inventories. Net storage in terrestrial ecosystems requires an enhancement of the processes that move carbon into longer-lived pools by controlling factors such as:

- Land use (e.g., preserving or reestablishing forests and wetlands)
- Land-management practices, particularly relating to soil disturbance
- Plant types and cropping systems: debris retention, water-use efficiency, and lignin production
- Microbial community make-up: balance of fungal versus bacterial populations
- Soil fertility and irrigation, including wetting–drying cycles

Although the total terrestrial carbon inventory is only some four times the atmospheric inventory (roughly three times in soils and one in biomass), active control of these factors could nevertheless have a material and low-cost impact on carbon uptake.

At the same time, anthropogenic carbon emissions are being partially offset by a natural increase in the atmosphere-to-terrestrial carbon flux that is occurring through enhanced photosynthetic production as a result of increasing [CO_2] and related climate changes. Interventions to increase carbon storage in terrestrial ecosystems will also enhance this natural feedback mechanism.

Processes and approaches to CO_2 storage in terrestrial ecosystems are described in Chapter 13.

2.2.4 Storage by mineral carbonation

Mineral carbonation is a potential storage method that accelerates the geological weathering process. It involves the formation of stable carbonates by the reaction of CO_2 with naturally occurring oxides or silicates of magnesium, iron, and calcium. In particular, igneous silicate rocks are globally abundant and contain important silicate minerals such as olivine ((Fe_x, $Mg_{1-x})_2SiO_4$), woolastonite ($CaSiO_3$), and serpentine ($Mg_3Si_2O_5(OH)_4$) which are potential feedstocks for mineral carbonation.

The carbonation reactions typically require ~2 tons of silicate mineral per t-CO_2 captured, so application of mineral carbonation would entail very-large-scale mining and disposal operations. For example, a 100 kt per day mining operation would be able to support capture of ~18 Mt-CO_2 per year and could serve about five 500 MW_e coal-fired power stations. In addition, backfilling operations would need to accommodate an excess of 50–100 kt per day of carbonation products.

Apart from this type of large-scale application, the alkaline waste from many industrial processes is also suitable as feedstock for mineral carbonation, providing the opportunity for smaller-scale application. Wastes such as ash from municipal waste incineration, coal combustion, and cement production, as well as slag from steel making and asbestos mine tailings, are potential feedstocks. Some of these wastes have a resale value into other industries, but the end products after mineral carbonation generally have a higher value, offering an economic incentive to plant owners.

Mineral carbonation options for carbon capture and storage are described in Chapter 10.

2.2.5 Other storage and use options

The use of CO_2 an industrial feedstock is dominated by the production of urea (NH_2CONH_2) as a nitrogen fertilizer, which currently consumes ~65 Mt of industrially produced CO_2 per year. Other uses include the production of methanol (CH_3OH), polyurethanes, and the food industry, and total industrial use is estimated at ~120 Mt-CO_2 per year.

Although this is a significant quantity against the scale of current capture and storage projects, the scope to increase this usage is limited by the demand for the end products. Also, the retention time of carbon in these products is very limited; it is less than a year for urea, which quickly hydrolyzes to ammonia and CO_2 when applied. The relevance of these uses for material long-term storage is therefore very limited.

Two potential applications that could have a material impact are the production of precipitated calcium carbonate (PCC) for use in the paper and cement industries, and the direct use of cooled flue gases as a CO_2 source for microalgal photosynthesis, generating biomass for biofuel production. The latter application would be carbon neutral if the biofuel is subsequently burned without capture, or would reduce net emissions if CO_2 is also captured in the biofuel combustion process. These applications are described in Chapter 14.

2.3 References and Resources

2.3.1 Key References

The following key references, arranged in chronological order, provide a starting point for further study and access to a wider bibliography on carbon capture and storage.

OECD/IEA. (2004). *Prospects for CO_2 Capture and Storage*, OECD, Paris, France.

VGB. (2004). *CO_2 Capture and Storage: VGB Report on the State of the Art*, VGB PowerTech e.V. Available at www.vgb.org/en/CO2cace.html.

European Carbon Dioxide Network. (2005). *Report on the Current State and the Need for Further Research on CO_2 Capture and Storage*, EU CO2Net, August 2005 Rev 6. Available at www.co2net.eu/public/reports/CCSRTDStrategyRev6.pdf

IPCC. (2005). *Special Report on Carbon Dioxide Capture and Storage*, Cambridge University Press, Cambridge, UK.

U.S. Department of Energy, National Energy Technology Laboratory. (2007). *Carbon Sequestration Technology Roadmap and Project Plan 2007*. Available at www.netl.doe.gov/technologies/carbon_seq/refshelf/project%20portfolio/2007/2007Roadmap.pdf

Gibbins, J. and H. Chalmers. (2008). Carbon capture and storage, *Energy Policy*, 36 (4317–4322).

2.3.2 Web Resources

NatCarb (U.S. Department of Energy, National Energy Technology Laboratory, information on carbon sequestration): www.narcarb.org

U.S. Department of Energy National Energy Technology Laboratory Carbon Sequestration Newsletter (an email newsletter produced by the National Energy Technology Laboratory presenting CCS news summaries and newly published literature): http://listserv.netl.doe.gov/mailman/listinfo/sequestration

U.S. Department of Energy National Energy Technology Laboratory Carbon Sequestration Reference Shelf (links to National Energy Technology Laboratory Carbon Sequestration Program documents and reference materials): www.netl.doe.gov/technologies/carbon_seq/refshelf/refshelf.html

3 Power generation fundamentals

Carbon capture from fossil fuel-burning power-generation plant will be a necessity if CCS is to make a material impact on total anthropogenic emissions. It is also an area where the opportunity exists for a rapid reduction of emissions, since some key technologies have been developed and deployed in other industries. As a precursor to the discussion of capture technologies in Part II, the fundamentals of fossil-fueled power generation are described in this chapter.

This begins with a review of the chemistry of combustion and gasification, the physics of the thermodynamic cycles used in standard and combined cycle power plants (the Rankine steam cycle and the Brayton gas turbine cycle), and a discussion of some relevant aspects of the metallurgy of steel. The components and processes of a typical fossil-fueled power plant and of a combined cycle plant are then described.

3.1 Physical and chemical fundamentals

3.1.1 Fossil fuel combustion

The combustion of fossil fuels remains the primary energy source for power generation worldwide, representing ~85% of the global primary energy supply in 2008. The typical characteristics and compositions of fossil fuels, in the order of their importance for power generation, are shown in Table 3.1, while those for some biomass fuels are shown in Table 3.2.

Combustion converts the chemical energy of the fuel to heat through a series of exothermic reactions. The general form of the reaction for the complete combustion of a hydrocarbon is:

$$C_xH_y + (x + y/4)O_2 \rightarrow xCO_2 + (y/2)H_2O + \text{heat} \quad (3.1)$$

For example:

$$C + O_2 \rightarrow CO_2 + \text{heat} \quad \Delta H = -383.8\,\text{kJ/mol C} \quad (3.2)$$

$$CH_4 + 2O_2 \rightarrow CO_2 + 2H_2O + \text{heat} \; \Delta H = -800\,\text{kJ/mol}\,CH_4 \quad (3.3)$$

Doi:10.1016/BB978-1-85617-636-1.00003.

Table 3.1 Typical heating value and composition of fossil fuels

Fuel	Heating value (LHV kJ per kg)	Composition (wt% dry)					
		C	H	O	S	N	Ash
Coal (lignite)	17,700	49.8	3.3	27.7	0.5	1.0	17.7
Coal (anthracite)	33,860	88.9	3.4	2.3	0.8	1.6	2.9
Natural gas (Groningen)	38,050	58	18.7	1.5	0	21.6	0
Crude oil	18,500	88	8.2	0.5	3	0.1	0.2
Wood	18,700	50.7	6.1	42.8	0.1	0.4	2.2
Peat	22,600	56.7	6.0	35.9	0.2	1.7	4.9

Table 3.2 Average heating value and composition of biomass fuels

Fuel	Heating value (LHV MJ per kg)	Moisture content (wt%)	Composition (%)					
			C	H	O	S	N	Ash
Wood	18.7	18.6	50.7	6.1	42.8	0.1	0.4	2.2
Grass/plants	18.3	29.8	49.2	6.0	43.4	0.2	1.2	6.9
Manure	18.9	44.0	47.2	6.5	36.4	0.7	5.0	28.5
Straw	18.0	14.6	48.7	6.0	43.3	0.2	0.9	7.5
Peat	22.6	20.8	56.7	6.0	35.9	0.2	1.7	4.9
Algae	24.8	31.9	53.8	7.4	30.9	0.5	7.5	6.1

The enthalpy of combustion (ΔH; see Glossary)—essentially the exothermic energy released—measures the enthalpy change when one mole of the fuel is fully oxidized, and is quoted at 25°C (298°K) unless otherwise stated. For combustion of carbon in Equation 3.2, this translates to release of 1 MWhr of thermal energy for the combustion of 9400 moles (113 kg) of carbon.

As well as the oxidation of the carbon or hydrocarbon, other components of the fuel will also be oxidized during combustion:

$$S + O_2 \rightarrow SO_2 + \text{heat} \quad \Delta H = -296.8\,\text{kJ/mol} \tag{3.4}$$

$$N + O_2 + \text{heat} \rightarrow NO_2 \quad \Delta H = +33.1\,\text{kJ/mol} \tag{3.5}$$

$$2N + O_2 + \text{heat} \rightarrow 2NO \quad \Delta H = +90.3\,\text{kJ/mol N} \tag{3.6}$$

The latter two reactions are endothermic (heat input is required).

Partial oxidation

Equation 3.1 expresses the stoichiometric requirement for complete combustion of the fuel. Partial combustion, or partial oxidation (POX), occurs if oxygen availability in the reaction zone is less than the stoichiometric requirement for full oxidation, and is important in gasification, discussed in the next section. The general equation is:

$$C_xH_y + (x/2 + y/4)O_2 \rightarrow xCO + (y/2)H_2O + \text{heat} \tag{3.7}$$

For example:

$$C + \tfrac{1}{2}O_2 \rightarrow CO + \text{heat}\ \Delta H = -123.1\,\text{kJ/mol}\,C \tag{3.8}$$

$$CH_4 + \tfrac{1}{2}O_2 \rightarrow CO + 2H_2 + \text{heat} \quad \Delta H = -38.0\,\text{kJ/mol}\,CH_4 \tag{3.9}$$

Heating value of a fuel

The heating value of a fuel can be expressed either as a higher heating value (HHV) or a lower heating value (LHV). This measures the amount of heat released when a quantity of fuel, initially at 25°C, is combusted and cooled either to 25°C in the case of the HHV or to 150°C in the case of the LHV. The difference between the two values is therefore due to the difference in heat content of the combustion products between the two end temperatures including the latent heat of vaporization of any steam in the combustion products, resulting from moisture in the fuel and from water formed by the combustion of hydrogen in the fuel. The thermal efficiency of a plant is typically referenced to the LHV of the fuel, except where waste heat is recovered down to low temperatures, such as in combined heat and power (CHP) plants.

Oxyfueling

The reaction products shown in Equations 3.1–3.9 reflect combustion in oxygen, whereas in traditional power plants oxygen is just one component of the air blown into the furnace. A more accurate representation of Equation 3.2 would therefore be:

$$C + O_2 + 3.73N_2 \rightarrow CO_2 + 3.73N_2 + \text{heat} \tag{3.10}$$

Separation of CO_2 from this reaction product stream is the essence of the postcombustion carbon capture problem. Oxyfuel combustion, or oxyfueling, sidesteps this separation problem by burning the fuel using pure oxygen rather than air, thereby adhering strictly to Equation 3.2. The combustion gases then comprise CO_2, steam, and possibly SO_x and NO_x, depending on the S and N content of the fuel. Steam can be separated by condensation, with recovery of the low-grade heat, and the remaining dried gas can be further treated if required and compressed for transportation and storage.

Technologies for air separation to generate oxygen for oxyfueling are described in Part II and include membrane-based (Section 8.5) and cryogenic systems (Section 9.3).

3.1.2 Gasification of fossil and other fuels

An alternative to fully releasing the chemical energy of the fuel through complete combustion is to gasify the fuel to produce an intermediate synthesis gas product (syngas), which can then be used as a feedstock for the production of liquid fuel, hydrogen, or methanol. Syngas is a mixture of CO and H_2 and results from the partial combustion of carbon and its high-temperature reaction with steam, according to the reactions shown in Table 3.3.

Table 3.3 Carbon-based feedstock gasification processes and reactions

Process	Reaction
Partial oxidation	$C + {}^1/_2\, O_2 \rightarrow CO$
Carbon–steam reaction	$C + H_2O \rightarrow CO + H_2$ $C + 2H_2O \rightarrow CO_2 + 2H_2$
Water–gas shift reaction	$CO + H_2O \leftrightarrow CO_2 + H_2$
Boudouard reaction	$C + CO_2 \leftrightarrow 2CO$

Any CO_2 present under partial oxidation conditions will be converted to CO via the Boudouard reaction, shown in the table, particularly at high temperatures, which shift the equilibrium of this reaction to the right.

3.1.3 Syngas production from methane

Syngas can also be produced from methane either by partial oxidation (Equation 3.9) or by steam reforming according to the reactions shown in Table 3.4.

The partial oxidation reaction takes place at temperatures $>1200°C$ to drive the reaction equilibrium toward the partial oxidation products and requires a high-temperature catalyst in order to achieve a high reaction rate. The 2:1 H_2:CO ratio of the partial oxidation reaction products provides the optimal feed for gas-to-liquids (GTL) conversion such as the Fischer–Tropsch process.

Catalytic steam reforming of methane is the most important process used for syngas production, the reaction taking place in catalyst-packed reactor tubes

Table 3.4 Syngas production by steam reforming of methane

Process	Reaction
Partial oxidation	$CH_4 + {}^1/_2\, O_2 \rightarrow CO + 2H_2$
Steam methane reforming	$CH_4 + H_2O \rightarrow CO + 3H_2$
Water–gas shift reaction	$CO + H_2O \leftrightarrow CO_2 + H_2$

operating at pressures of 1.5–4.0 MPa and temperatures of 700–950°C. Steam reforming of methane produces a 3:1 H_2:CO ratio in the product stream, which is too hydrogen rich for liquid synthesis processes. This can be modified using the water–gas shift (WGS) in a separate reactor at temperatures of 300–500°C, reducing the H_2:CO ratio as the WGS equilibrium shifts to the left. Alternatively, at higher temperatures, as the equilibrium shifts to the right, a $H_2 + CO_2$ reaction product stream results, from which the CO_2 can be removed to generate hydrogen. This process is used for hydrogen production in refineries as a hydrocracker feed, with amine absorption to remove CO_2 from the product stream, and is also used for the generation of hydrogen from natural gas as a carbon-free automotive transport fuel.

3.1.4 Thermodynamic cycles

A thermodynamic cycle is a sequence of changes in the state of a thermodynamic system such that at the end of the sequence all properties of the system are returned to their initial values. Examples of a change would be the addition of heat to a working fluid such as water in a boiler, or allowing a working fluid such as steam to expand through a turbine. If the cycle results in the transfer of energy from a hot source to a cold sink and conversion of part of that energy to work, the system is termed a heat engine.

Thermodynamic cycles are described and analyzed using the temperature–entropy (T-S) or pressure–volume (P-V) diagrams of the working fluid, as shown in Figure 3.1 for water.

Above the freezing temperature of water, three regions are shown on the T-S diagram:

- The liquid region, to the left of the bubble point line. Points on the T-S diagram in this region represent 100% liquid water at varying temperature (T) and pressure (P).
- The vapor region to the right of the dew point line. In this region water exists as 100% vapor (steam) at varying T and P.
- The intervening liquid–vapor region, where points represent varying water–steam mixtures, from 100% liquid at the bubble point line to 100% vapor at the dew point line.

The line ABCD represents a thermodynamic process in which, at constant pressure P_1, a quantity of water is heated to the boiling point (T_1) at this pressure

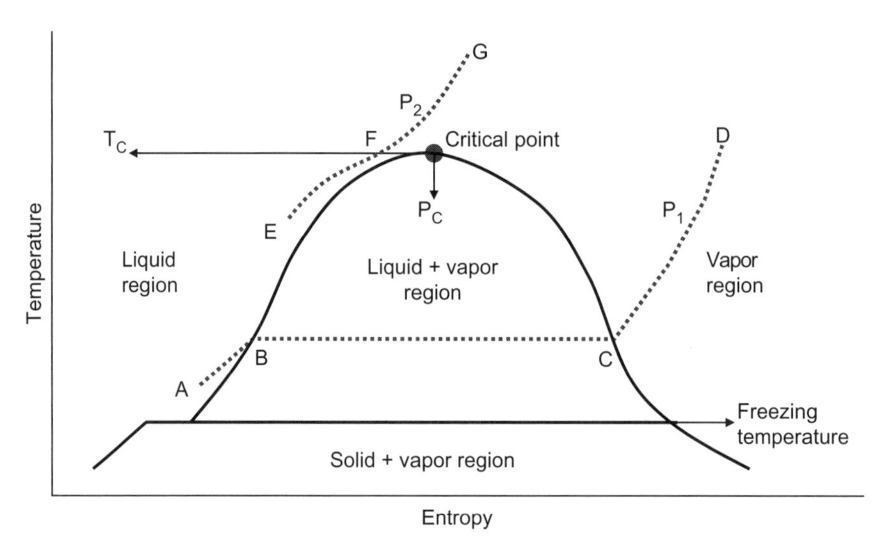

Figure 3.1 Temperature–entropy (T-S) diagram for water

(A to B), is fully vaporized (B to C), and is then superheated above the boiling temperature (C to D).

The point where the bubble point and dew point lines meet is called the critical point, and represents the temperature (T_c) and pressure (P_c) above which the transition from liquid to vapor occurs without a discernible change of state and without any intervening mixed liquid-plus-vapor state. A thermodynamic process operating at temperatures and pressures above the critical point is termed supercritical. Line EFG in the figure shows the T-S curve for a quantity of fluid at a pressure P_2, greater than P_c, being heated to the critical temperature (E to F) and then superheated (F to G).

Steam generators operating at supercritical (SC) and ultrasupercritical (USC) (see Glossary) conditions are an important area of technology development to improve the thermal efficiency of fossil-fueled power plants (Section 3.2.3).

The simplest ideal thermodynamic cycle, the Carnot cycle, is illustrated in Figure 3.2 on T-S and P-V diagrams, and consists of four thermodynamic processes:

- From A to B: a reversible isothermal expansion of the working fluid, with heat being drawn from the hot source at temperature T_H (°K)
- From B to C: an isentropic (reversible adiabatic) expansion of the working fluid, delivering work, for example, to a turbine
- From C to D: a reversible isothermal compression of the working fluid, with heat transfer out of the system to the low temperature sink at temperature T_C
- From D to A: an isentropic compression of the working fluid. Work is done to compress the working fluid, resulting in a rise in temperature from T_C to T_H, at which point the ideal system has returned to its initial state.

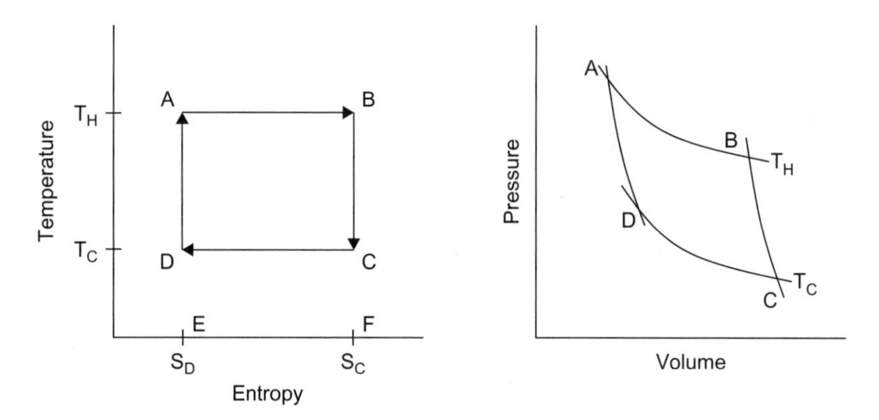

Figure 3.2 The Carnot thermodynamic cycle

In the cycle shown in Figure 3.2, the area CDEF, equal to $T_C(S_C-S_D)$, is the amount of heat exchanged with the cold sink (Q_C) while the area ABFE, equal to $T_H(S_C-S_D)$, is the heat absorbed from the hot sink (Q_H). The difference between these is the energy transferred from the system as work (W):

$$W = (T_H - T_C)(S_C - S_D) \tag{3.11}$$

and the Carnot efficiency (η_{Carnot}), defined as the amount of work done divided by the heat exchanged with the hot source, is:

$$\eta_{Carnot} = W/Q_H = (T_H - T_C)/T_H = 1 - T_C/T_H \tag{3.12}$$

temperatures here being measured in °K. The Carnot efficiency represents the theoretical limit for perfectly reversible systems where all heat input and output are done at the hot and cold reservoir temperatures. In practice, processes such as compression, expansion, and heat transfer are irreversible, and various other energy losses through friction, radiation, and leakage prevent the Carnot efficiency from being achieved in practice.

Directionally, the Carnot efficiency is increased as the ratio T_H/T_C is increased, which is also true for practical cycles, a simple example of this being the roughly 2-percentage-point-higher efficiency of fossil-fueled power plants on coastal sites, where seawater is available for cooling, compared to those inland. Several further examples will be discussed in the following section, including combined cycles, which reduce the final temperature T_C of heat rejection, as well as SC and USC steam cycles, which increase T_H.

3.1.5 Rankine steam cycle

The Rankine cycle is the ideal cycle that describes a heat engine using water and steam as the working fluid, and is the cycle that applies to power generation using a steam turbine. Figure 3.3 illustrates schematically the physical components of a Rankine heat engine and the T-S diagram of the cycle.

The four thermodynamic processes that make up the cycle are as follows:

- From A to B: High-pressure feed water is heated in a boiler by the combustion of fuel. At point B, on the dew point line, the working fluid has become dry saturated steam.
- From B to C: The saturated steam is expanded through a steam turbine. Work is delivered to the turbine shaft, with the drop in pressure and temperature likely bringing the working fluid inside the dew point line, resulting in some condensation.
- From C to D: The wet steam is cooled in a condenser, and heat is transferred to a cooling medium. The working fluid is returned fully to the liquid state.
- From D to A: Water is pressurized by a feed pump before reentering the boiler.

In Figure 3.3, H_A, H_B, H_C, and H_D measure the enthalpy ($H = TS + PV$) carried by the working medium into the boiler, turbine, condenser, and feed pump respectively, W_T is the net shaft work delivered from the turbine and W_{FP} is the work delivered to the feed pump, Q_B is the external heat delivered to the boiler from the furnace, and Q_C is the heat rejected to the condenser cooling medium.

The efficiency of the Rankine cycle can be derived as follows:

$$\eta_{Rankine} = (W_T - W_{FP})/Q_B \tag{3.13}$$

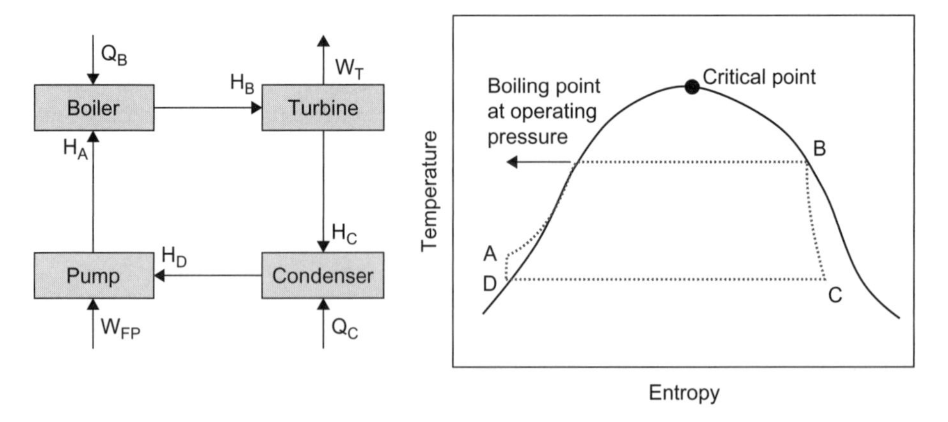

Figure 3.3 Rankine cycle heat engine and T-S diagram

Since $W_{FP}/Q_B \approx 0$, this reduces to:

$$\eta_{Rankine} = W_T / Q_B = (H_B - H_D)/(H_B - H_A) \tag{3.14}$$

The overall efficiency can be improved by reducing losses and by recovery of low-grade heat from flue gases and from the condenser cooling medium, for example to preheat feed water.

The efficiency is also improved by increasing both the temperature and pressure of steam driving the turbine, since either of these will increase the enthalpy at the turbine inlet (H_B). This leads to the use of superheated, reheated, and supercritical steam cycles, as illustrated in Figure 3.4.

Superheating (Figure 3.4a) increases the temperature of dry steam leaving the boiler, resulting in an increase in turbine work (W_T) and also has the practical advantage of reducing condensation within the turbine, since point C is moved closer to the dew point line. Cycle efficiency is also raised since the additional heat input to the working fluid is occurring at a higher temperature. Reheating (Figure 3.4b) adds additional heat from the boiler to steam between the high-pressure and low-pressure sections of the turbine, with similar improvement in turbine work but at a lower incremental efficiency.

A supercritical steam cycle (Figure 3.4c) operates at a steam pressure and temperature above the critical point (22.1 MPa, 374°C) with higher thermal efficiency. To achieve this, special high-cost materials are required to withstand the extreme conditions, particularly in the furnace wall, as described in Section 3.1.7.

3.1.6 Brayton gas turbine cycle

The Brayton cycle is the ideal thermodynamic cycle that describes the operation of a closed-loop system comprising a compressor, a combustion chamber, and a turbine, as illustrated in Figure 3.5.

The operation of the cycle involves four process steps:

- From A to B: Air is drawn into a compressor and undergoes isentropic compression.
- From B to C: The air is mixed with fuel and combusted in a combustion chamber at constant pressure.
- From C to D: The combustion products are expanded through a turbine and work is extracted via the turbine shaft.
- From D to A: The loop is closed by the rejection of heat from the turbine exhaust gas to an ambient temperature cold sink.

The efficiency of the ideal Brayton cycle is given by:

$$\eta_{Brayton} = 1 - T_A / T_B = 1 - r_p^{(\gamma-1)/\gamma} \tag{3.15}$$

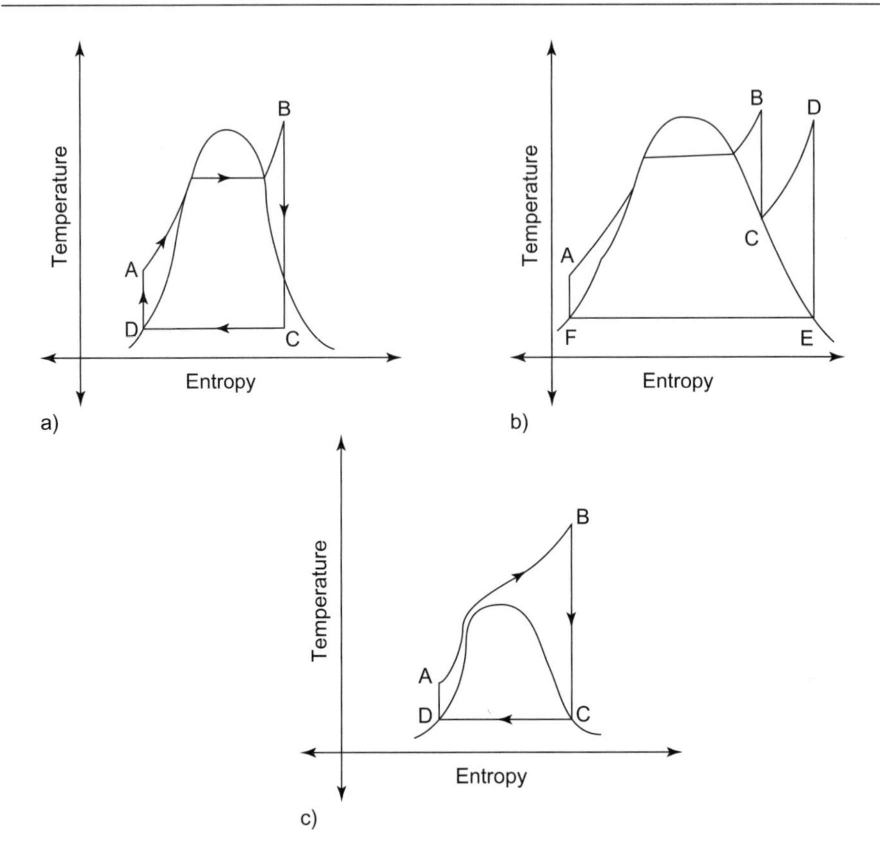

Figure 3.4 T-S diagrams for a) superheated, b) reheated, and c) supercritical steam cycles

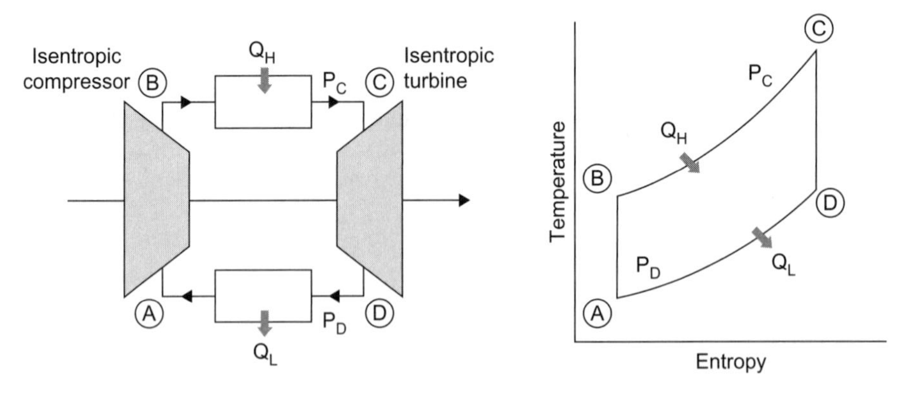

Figure 3.5 Brayton cycle heat engine and T-S diagram

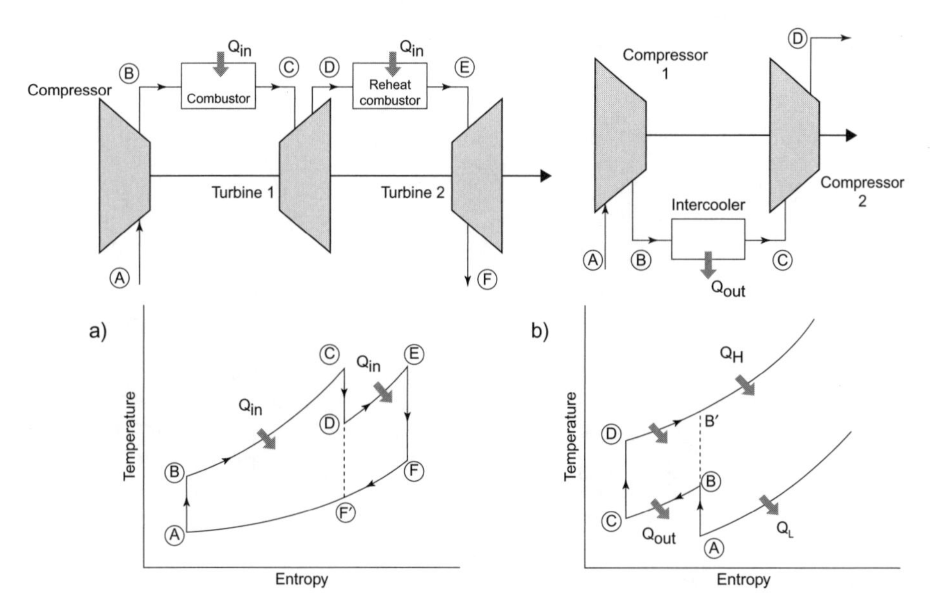

Figure 3.6 Process schematics and T-S diagrams for a) reheated and b) intercooled Brayton cycles

where r_p is the ratio of the inlet and outlet pressures of the turbine (P_C :P_D) and γ is the ratio of the specific heats of the gas for constant pressure and constant volume processes ($\gamma = C_p$:C_v).

Similar to the use of superheating and reheating in the Rankine cycle, the work output from a Brayton cycle can be increased by reheating, in which the exhaust from the first gas turbine is reheated in a second combustion chamber and drives a second turbine (Figure 3.6).

Reheating allows additional heat input to the turbine without exceeding the maximum temperature limit determined by the rotor material, but the overall cycle efficiency is reduced because, as can be seen from the T-S diagram, heat rejection for the second part of the cycle (F–F′ in Figure 3.6a) occurs at a higher temperature than the original cycle.

Intercooling splits compression into two stages, cooling the working fluid between stages. An increase in cycle efficiency can be achieved since a higher pressure can be achieved with the same compressor outlet temperature.

Regeneration (or recuperation) recovers heat from the turbine exhaust to preheat gas entering the combustion chamber. The amount of heat recovered can be increased by increasing the turbine exhaust temperature and by reducing the compressor outlet temperature. The impact of regeneration on overall thermal efficiency is therefore maximized when applied in combination with reheating and intercooling. Figure 3.7 illustrates a Brayton cycle process with reheat, intercooling, and regeneration.

Other approaches to increasing the power output level of a Brayton cycle, generally at higher thermal efficiency than the standard cycle, include injecting steam into the gas turbine combustion chamber (STIG) and the use of humid air as the turbine working fluid (HAT).

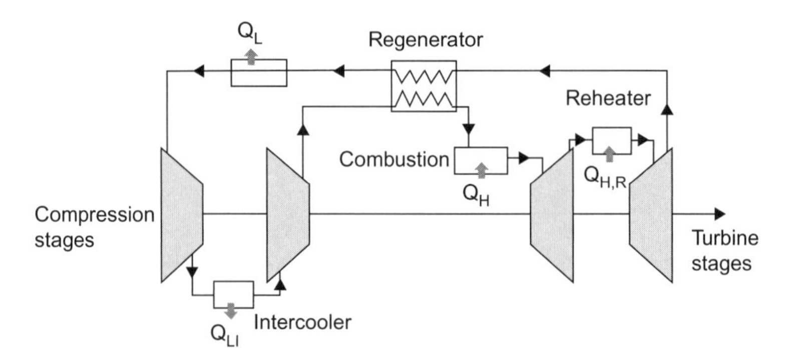

Figure 3.7 Brayton process schematic with reheating, intercooling, and regeneration

3.1.7 Aspects of steel metallurgy for fossil-fueled power plants

Since the live steam temperature is the key factor that determines the thermal efficiency of the Rankine steam cycle (Equation 3.14), the quest for higher cycle efficiency boils down to a quest for materials capable of operating under ever-higher temperatures and pressures, and under the corrosive conditions of a steam boiler.

Stainless steel is the principal construction material for boiler and turbine components, and the three main types of steel—ferritic, austenitic, and martensitic (see Glossary)—each has advantages in particular power plant applications (Table 3.5). The two key properties of steel that are essential in high-temperature, high-pressure, corrosive environments are strength, particularly creep strength, and corrosion resistance. Slow thermally activated creep of components under load can eventually lead to rupture or excessive plastic deformation. A creep strain limit of $\sim 3 \cdot 10^{-11}$/s is not uncommon and would result in a dimensional elongation of ~2% after a typical 30-year service life. Corrosion resistance must address both the fire-side and steam-side environments and, on the fire side, the impact of different fuels used in the furnace.

Corrosion resistance

The oxidation resistance of steel is primarily a result of the chromium content, and to a lesser extent, in austenitic steels, the high nickel content. Chromium on the steel surface is oxidized to form an invisibly thin layer of chromium oxide (Cr_2O_3), and this so-called passivation layer will be quickly reformed if damage occurs, provided the bulk chromium content remains above a threshold value of ~7%.

Table 3.5 Advantages and disadvantages of various steel types

Steel type	Composition	Advantages	Disadvantages
Ferritic steels	Fe-Cr alloy steels with 10–25% Cr, 2–4% Mo, <1% Ni, and <0.75% C	Lower thermal expansion coefficient resulting in lower thermal stresses under cyclic operation, particularly for thick-walled components	Alloying (e.g., tungsten and vanadium) required to overcome low creep strength of conventional ferritic steels
Austenitic steels	Fe-Cr-Ni alloys with 16–25% Cr, 1–37% Ni, and <0.24% C	High corrosion resistance—important in corrosive service such as refuse incineration boilers	Higher thermal expansion and low creep strength for high-temperature service
Martensitic steels	Fe-Cr alloys with 12–18% Cr and <1% C; may also include up to 2% Ni, Mo, V, or W	High creep resistance as a result of the microcrystalline structure	Poorer corrosion resistance, which can be improved by precipitation hardening; more brittle

Cracking of the protective layer can occur as a result of creep or thermal cycling, and the presence of water vapor then leads to the formation of volatile chromium-containing hydroxyl species and a loss of chromium from the steel through evaporation. A chromium content well above the 7% threshold is therefore essential for long-term service under corrosive conditions. Further alloying can improve corrosion resistance; the addition of silicon aids in the process of crack healing in the passivation layer, manganese slows chromium evaporation loss as a result of the formation of a hard manganese–chromium ($MnCr_2O_4$) spinel layer (see Glossary), while tungsten and other metals increase creep strength as a result of carbide formation.

Carbides, creep, hardening, and embrittlement

A key process that determines the strength and thermal creep resistance of steel is the precipitation of carbides, and in some cases nitrides or borides, that occurs during the initial cooling or subsequent heat-treating of steel, the latter process also known as precipitation hardening.

The solubility of carbon in an Fe-Cr-Ni austenitic steel declines from 0.15% by weight to <10 ppm as the steel is cooled to ~600°C. As a result, carbon comes out of solution during cooling, resulting in the formation of carbides of Fe, Cr, Mo, Mn, Nd, etc., depending on the alloying metals present in the steel. Steels containing nitrogen or boron will also precipitate nitrides and borides. Some common precipitates are summarized in the Table 3.6, where M represents the metal alloy atom.

Table 3.6 Carbide, nitride, and boride precipitates

Precipitate	Composition
Carbides	
MC	(Ti,Nb,V)C
M_3C	$(Fe,Mn,Cr)_3C$
M_7C_3	$(Fe,Cr)_7C_3$
$M_{23}C_6$	$(Fe,Cr,Mo)_{23}C_6$
M_6C	$(Fe,W,Mo,Nb,V)_6C$
Nitrides	
MN	(Ti,Nb)N
M_2N	$(Fe,Cr)_2N$
Borides	
M_2B	$(Fe,Cr)_2B$
M_3B_2	$(Fe,Cr,Mo)_3B_2$

As well as primary precipitation, which occurs during solidification and typically produces precipitate particles in the 1- to 10-μm size range, secondary precipitation also occurs during heat-treating or as a result of cycling in high-temperature service and results in smaller precipitate particles in the 5- to 50-nm range.

The most stable MC carbides nucleate predominantly on dislocations and stacking faults within crystal grains and are beneficial because they prevent creep by locking and preventing the movement and growth of these crystal defects that lead to creep. In contrast, $M_{23}C_6$ carbides tend to nucleate on grain boundaries and result in chromium depletion and increased susceptibility to intergranular corrosion. This detrimental formation of $M_{23}C_6$ can be reduced by the addition of Ti or Nb, which encourage MC precipitation, or by the addition of nitrogen, which promotes the precipitation of M_6C-type carbides. Similar to the MC carbides, secondary intragranular precipitation of very fine complexes such as NiAl, Ni_3Al, Ni_3Ti, Ni_3Nb, and Ni_3Cu during heat treatment can also lead to increased strength, a process known as precipitation hardening.

In high-temperature service, steel components experience repeated temperature cycling through the temperature range where these precipitation processes are active. As a result the crystal structure of a steel component is continually changing throughout its service life due to dissolution, transformation, and reprecipitation. For example, continued growth of the intermetallic NiAl-type complexes that result from precipitation hardening can occur in high-temperature service and can lead to a decline in the strength of the material since dislocations are no longer locked once these particles exceed a certain size relative to the lattice spacing. Carbide precipitates will also coarsen during service, and more rapidly at high temperatures when the diffusivity of impurities is increased, providing sites for crack formation. This process is known as grain boundary embrittlement.

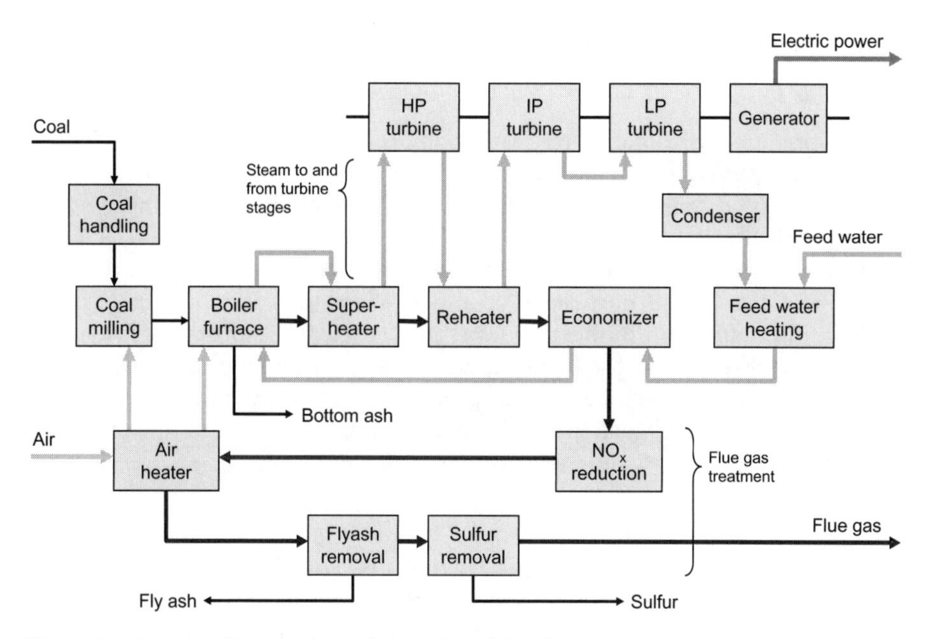

Figure 3.8 Process diagram for pulverized coal-fired power plant

3.2 Fossil-fueled power plant

3.2.1 Introduction

The process diagram of a typical pulverized coal power plant is shown in simplified form in Figure 3.8.

The pulverized fuel is blown into the furnace by a draft of air, which is preheated in a heat exchanger by flue gas exiting the boiler. Initial boiling takes place in the water-cooled wall of the furnace, and the final steam temperature to drive the high-pressure (HP) steam turbine stage is achieved in a superheater.

HP turbine exhaust steam is reheated (first reheat) to drive the intermediate-pressure (IP) turbine stage, and may be reheated again (second reheat) to drive the low-pressure (LP) stage. The three turbine stages are linked by a single shaft to the generator.

Exhaust steam from the LP stage is condensed using a cooling water supply, pressurized, and preheated in a heat exchanger before reentering the boiler. Following a NO_x reduction step, the final stage of heat recovery from the flue gas is used to heat the boiler air supply. Fly ash removal and desulfurization complete the flue gas treatment before emission.

The following sections describe each of the main process steps. Although the emphasis is on pulverized coal firing, which dominates recently installed plant capacity, variations to cater for other fuels, particularly syngas and biomass, are also described.

3.2.2 Fuels and fuel handling

Coal firing

In coal-fired power plants, the traditional mechanically stoked furnace, in which a layer of chunk coal burns on a grate, has been largely superseded by the combustion of pulverized coal (PC), which is blown into the furnace by the preheated air draft.

The fuel is prepared by milling in a ball mill to produce a fine powder with 70–75% of the fuel in particles $<75\ \mu m$ in diameter and $<2\%$ in particles $>300\ \mu m$ in diameter. Combustion takes place at 1300–1600°C for low-grade coals and 1500–1700°C for higher grades, and particles typically spend only 2–5 seconds in the boiler before being deposited as bottom ash or, predominantly, carried out as fly ash in the flue gas stream.

Natural gas firing

For fossil fuels other that coal, combustion of natural gas results in ~50% less carbon emission per unit of heat generated as a result of the higher H:C ratio of methane (CH_4) compared with the "average" hydrocarbon in fuel oil. Methane produces $^1/_2$ CO_2 per H_2, while saturated hydrocarbons (C_nH_{2n+2}) produce $2n/(2n + 2)$ or roughly one CO_2 per H_2. Although natural gas can be used to fire a boiler for Rankine cycle steam generation, it is more commonly used as a fuel for gas turbine-driven generation, either in a simple Brayton cycle or more commonly in a variety of combined cycle configurations (see Section 3.3).

For most natural gas compositions, no pretreatment is required before combustion, although removal of acid gas (H_2S and CO_2), as well as water and hydrocarbon dew point control, may be required at the point of production if the gas is to be transported in a pipeline or liquefied.

Gasification

The gasification of coal or other carbonaceous feedstock, described above in Section 3.1.2, provides a flexible approach to fuel usage that has a number of advantages, including:

1. Wide range of feedstocks, including heavy oils and hydrocarbon residues, biomass, municipal solid wastes (MSW), or other refuse-derived fuel (RDF)
2. Adaptable for carbon capture
3. Production of a flexible intermediate fuel that can be used for direct firing as well as hydrogen or synthetic liquid production

As a fuel for direct firing, syngas can be used on its own to fire steam boilers or gas turbines and can also be combined in co-firing applications with natural gas or fuel oil.

Biomass co-firing

In the co-firing of biomass with coal, biomass may be introduced at the start of the fuel processing process so that the mixed coal plus biomass fuel is handled using the existing processing and injection systems, or by the direct injection into the furnace of a separately processed biomass feed, either through the existing coal burners or through additionally installed biomass burners.

The premixing option has the advantage of rapid implementation and low capital cost and is now well established at many large coal-fired plants, although it is generally limited to 5–10% biomass on a heat input basis due primarily to limitation at the co-milling stage. This limitation can be overcome by direct injection co-firing, where the biomass is processed separately. Although the modification of existing coal burners for dedicated biomass injection has been successfully applied, the direct injection into existing coal firing systems has generally been preferred as a simple and lower-cost option.

The main technical issues arising from biomass co-firing are increased fireside corrosion of boiler materials, particularly with biomass feeds containing chlorine; reduction of the ash fusion temperature with addition of biomass ash (Section 3.2.5); and impact on ash handling as well as NO_x and particulate control systems.

3.2.3 Steam generation

The steam generator consists of a furnace, in which heat is released from the fuel by combustion, and a boiler, in which the heat is transferred to water to create steam.

Boiler technology

In modern plants the functions of furnace and boiler are combined into a single unit in which a major stage of water heating is achieved by water-cooling the walls of the furnace, a so-called water wall or membrane wall.

Feedwater processing

The purity and properties of the make-up feedwater are tightly specified in order to prevent scale deposition and corrosion, which increase maintenance costs and reduce boiler life. Raw feedwater is treated with chemical coagulants in settling tanks, followed by filtration to remove suspended solids. Dissolved salts are removed using chemical softeners and ion-exchange demineralizers (clays). Deaeration is achieved through multistage feedwater heaters, with residual oxygen removed to a few parts per billion (ppb) using oxygen scavengers such as hydrazine. Finally, the pH is controlled by chemical dosing to reduce acidity and avoid corrosion.

Evaporator design

A key component of the subcritical boiler is the steam drum, in which steam and water exiting the water walls are separated, with water being recycled through the boiler together with make-up feed water, while steam is further heated and piped to the HP turbine stage. In a boiler operating above the critical pressure, no phase change occurs as the feedwater is heated, so separation and recycling is not required, resulting is a so-called once-through design.

The water wall in the combustion zone—the lower part of the boiler—consists either of vertical or spiral wound tubing, while vertical tubing is standard in the upper section where the heat flux is much lower. Vertical evaporator tubing is standard in drum boilers, but spiral wound tubing has the advantage that fewer parallel paths are required to cover the furnace wall, increasing the water mass flow rate through each tube, which ensures adequate cooling. In contrast, vertical tubes are internally ribbed to improve heat transfer.

While spiral wound designs have become the standard in supercritical boilers, recent development of advanced tubing designs with internal rifling allows the advantages of vertical tubing (primarily lower manufacturing, installation, and maintenance costs) to be combined with the technical advantages of spiral configurations.

The design that dominates the supercritical boiler market today, the Benson boiler, was patented in 1925 and is owned by Siemens AG (Figure 3.9).

The operating conditions of a boiler are commonly expressed as (Live steam pressure/Live stream temperature/first recycle steam temperature/second recycle steam temperature); for example, (31 MPa/610°C/565°C/540°C).

Superheating, reheating, and steam temperature control

Superheating raises the temperature of steam exiting the evaporator to the operating temperature of the HP turbine stage. This increase in temperature is desirable as it increases the overall steam cycle efficiency (Section 3.1.5) and because the thermal energy delivered to the steam turbine by a given quantity of steam is also increased. Superheaters are heat exchangers located in the upper part of the furnace that bring the steam to its live operating temperature with heat transfer occurring both by convection and radiation. Reheaters, located downstream of the superheaters, reheat the exhaust steam from one turbine stage to provide additional energy to the next stage.

Maintaining high overall plant efficiency under varying load conditions requires stream temperatures to be maintained within a narrow operating range. This is typically achieved using a spray attemperator, in which water is sprayed into the superheated steam to control the steam temperature, combined with either flue gas bypass or flue gas recirculation, which reduces either the quantity or the temperature of flue gas directed at the superheater. In once-through boiler, the feedwater flow and firing rates are also coordinated to control steam temperature.

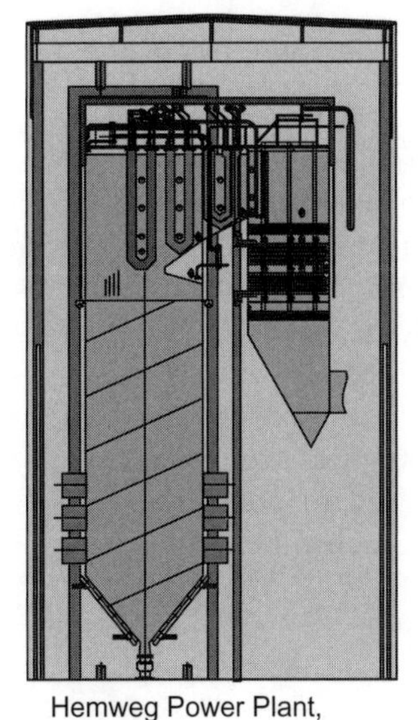

Hemweg Power Plant, Netherlands (261 bar / 540°C / 540°C)

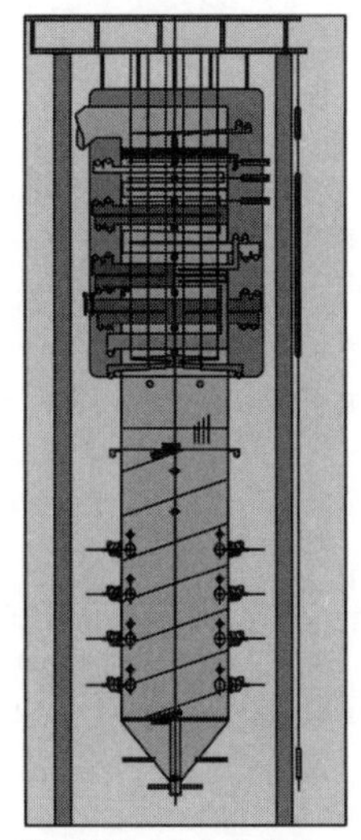

Nordjyllandsvaerket Power Plant, Denmark (310 bar / 582°C / 580°C)

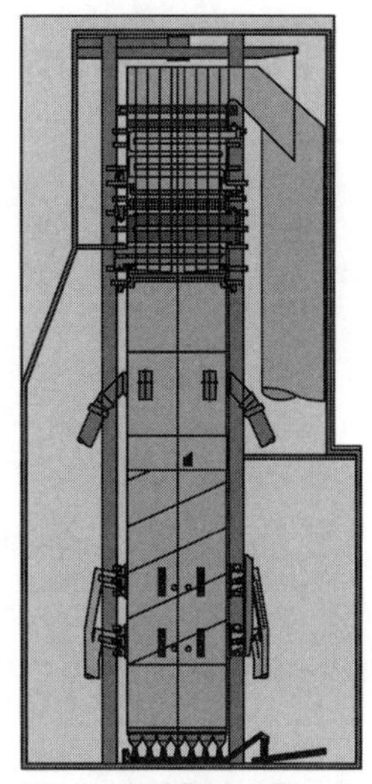

Lippendorf Power Plant, Germany (285 bar / 554°C / 583°C)

Figure 3.9 Schematic of Benson boilers – two pass and tower designs (Courtesy Siemens AG)

Condenser and heat recovery

The thermal efficiency of the boiler (the fraction of energy released in combustion that is transferred to the steam) is a key factor determining the overall plant efficiency and is influenced by the temperature of gas exiting the heat recovery area (HRA), downstream of the furnace, and by the operating pressure of the condenser.

As well as superheaters and reheaters, the HRA includes one or more heat exchangers, known as economizers, which preheat the condensed steam plus make-up feedwater between the condenser and the boiler. Further heat is recovered using a heat exchanger to preheat the air that will draft the furnace. In practice, reducing the temperature of the flue gas at the air heater exit to below the dew point temperature is not desirable to avoid acid corrosion resulting from the condensation of water, and with conventional materials a boiler exit temperature of ~130°C is typical.

Combustion technology

Combustion chambers and burners

A variety of combustion chamber and burner configurations are used for PC combustion, as shown in Figure 3.10. The choice is dictated by the fuel type and the attendant residence time required to achieve full oxidation of carbon in the fuel (burn-out).

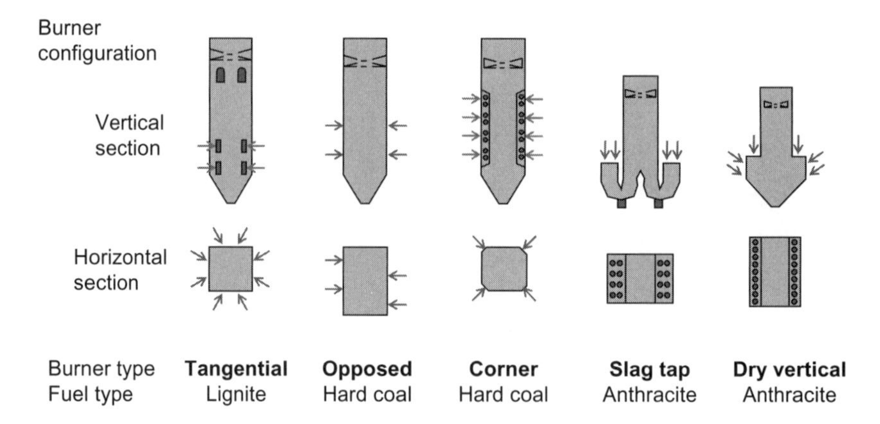

Figure 3.10 Combustion chamber configurations

Fluidized bed combustion

In fluidized bed combustion (FBC) a bed of coarse fuel particles, typically milled to ~5–10 mm, is fluidized by an upward-flowing stream of combustion air. If the air flow rate is sufficiently high the particles will be suspended in the circulating gas flow, resulting in a circulating fluidized bed (CFB). At lower air rates the bed will resemble a bubbling fluid—a bubbling fluidized bed (BFB). Figure 3.11 illustrates the configuration of BFB and CFB combustors.

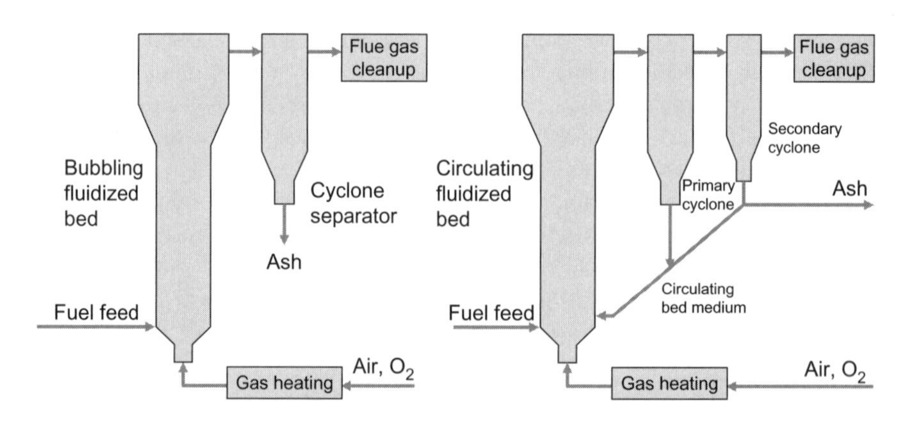

Figure 3.11 Fluidized bed combustion configurations

Fluidized beds have a number of advantages and disadvantages when compared to traditional pulverized coal combustion, the main ones being summarized in Table 3.7.

Fluidized bed combustors cover a wide range of capacities: bubbling beds typically range up to 35 MW_e, while circulating beds are common in the 100 to

Table 3.7 Main advantages and disadvantages of atmospheric fluidized bed combustion

FBC advantages	FBC disadvantages
• Combustion at relatively low temperatures (800–900°C) with low NO_x formation • Efficient heat transfer from the bed to immersed heat transfer surfaces • SO_x can be removed within the combustor by adding limestone particles to the bed. • FBC combustors can be designed to cater for a wide range of fuel types, including low-grade coals, biomass, and wastes. • High combustion efficiency due to turbulent mixing and longer fuel particle residence time	• Thermal efficiencies are ~2–4% lower than conventional PCC due to higher fan power and possible heat losses from the cyclones (CFBCs), as well as heat loss from the removal of ash and sorbent (e.g., if used for SO_x control). • In-bed heat transfer surfaces, primarily in BFBs, are subject to erosion. • N_2O formation is higher than in conventional PCC. Higher unburned fuel losses in CFBs, depending on fuel type.

460 MW_e range, with some designs up to 800 MW_e. CFB circulating velocities are typically 4–5.5 m per second, and 50–100 kg of solids are recycled through the bed for each kg of fuel burned.

Pressurized bubbling and circulating fluidized beds, operating in the 1.0 to 1.5 MPa range, have also been developed and are similar to atmospheric systems except that the combustor and cyclones are housed in a pressure vessel. The advantage of pressurizing the bed is that the high-pressure combustion gas can be used directly to drive a gas turbine, significantly improving overall thermal efficiency. However, the system also becomes much more complicated due to the need to pressurize the fuel, and air feeds, to depressurize the ash for removal, and possibly to filter the hot combustion gases to avoid erosion and deposition in the gas turbine system.

Advanced PFBC combined cycle systems under development aim to raise thermal efficiency from the current (2010) 40–45% to >50%.

SC and USC steam operation

Until the 1960s, power-plant steam generators traditionally operated at subcritical conditions, with live steam at 540°C and 18.5 MPa—below the 22.1 MPa critical pressure—and used a single reheat stage, commonly at the same temperature. The overall thermal efficiency that could be achieved at these conditions was ~36%, although state-of-the-art subcritical plants can now reach 39% (LHV basis). By moving to higher-pressure SC conditions with live steam at 600°C and 30 MPa, an increase in efficiency to ~45% could be achieved, approaching ~50% with steam temperature >650°C. The first units operating supercritical steam cycles were commissioned in the late 1950s and early 1960s, with live steam at 605–610°C, two reheat cycles at 540–565°C, and operating at pressures of 31–32 MPa.

USC operation is defined as a live steam temperature >600°C and pressure of 30 MPa and above, and efficiencies of 45–50% are being targeted by ongoing development programs. The maximum live steam temperature that can be achieved at a given operating pressure is determined by the availability of materials with the strength and gas-side corrosion steam-side oxidation resistance required to construct the final superheater and reheater.

3.2.4 Steam turbine technology

The steam turbine is the site of expansion and cooling in the Rankine steam cycle, where the energy contained in the steam is converted into mechanical work to drive an electrical generator. As shown in Figure 3.8, a large power plant will use one or more multistage steam turbines, with HP, IP, and LP stages.

The LP stage in this conventional power application exhausts to a condenser at or slightly below atmospheric pressure, and this type of turbine is called a condensing turbine. In CHP or other applications where process steam is required, the LP stage may exhaust at a higher pressure (noncondensing turbine) or steam may be extracted from the turbine casing at an intermediate point (extraction turbine).

Figure 3.12 shows the internal structure of a three-stage steam turbine. Each stage consists of one or more sets of fixed nozzles and of moving rotor blades, with mechanical force being delivered to the moving rotors either by impulse or by reaction.

In an impulse stage, commonly used in the HP and IP sections, the fixed nozzles direct jets of steam onto the bucket-shaped rotor blades. Expansion through the nozzles increases the steam velocity, and momentum is transferred to the turbine shaft as the impact of the jets on the rotor blades causes a change in direction of the steam jets. No change in pressure occurs across the moving rotor blades.

The LP section is typically a reaction stage, in which expansion through the fixed nozzles again results in acceleration of the steam. In this case, however, the rotor blades are also shaped like nozzles, and the transfer of momentum to the shaft is the result of both a change of direction of the steam and also its acceleration relative to the rotor, due to expansion through the rotor nozzles. Roughly half of the overall pressure drop across a reaction stage is due to expansion over the moving rotor. Impulse stages tend to be more efficient in delivering work to the turbine shaft at higher pressures, while the two designs have similar efficiency at lower pressures.

High-capacity steam turbines rated >500 MW_e often have dual parallel flows in the IP section and may have two, four, or six flows in the LP section.

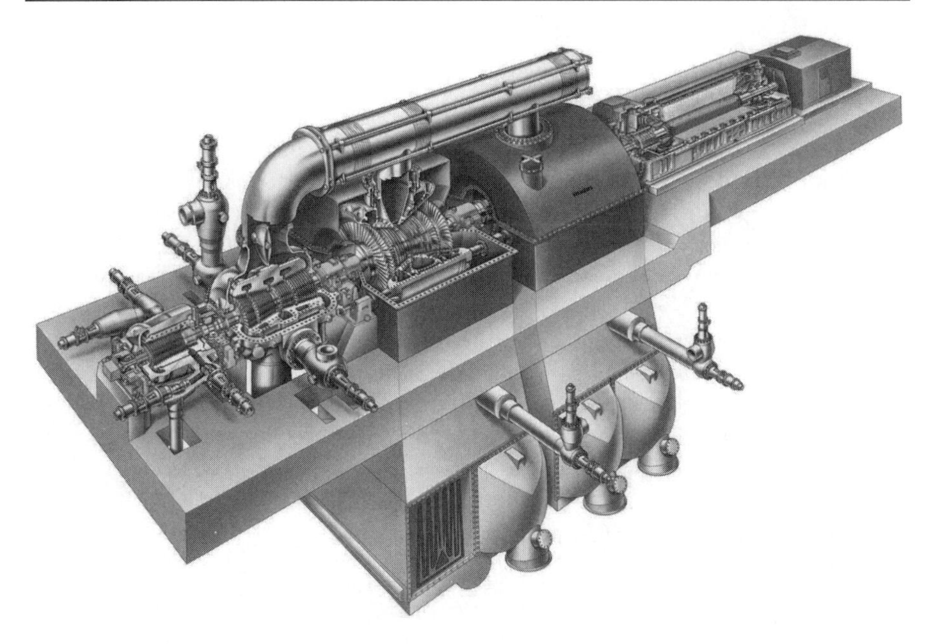

Figure 3.12 Three-stage steam turbine internal structure with dual flow IP and four flow LP (Courtesy Siemens AG)

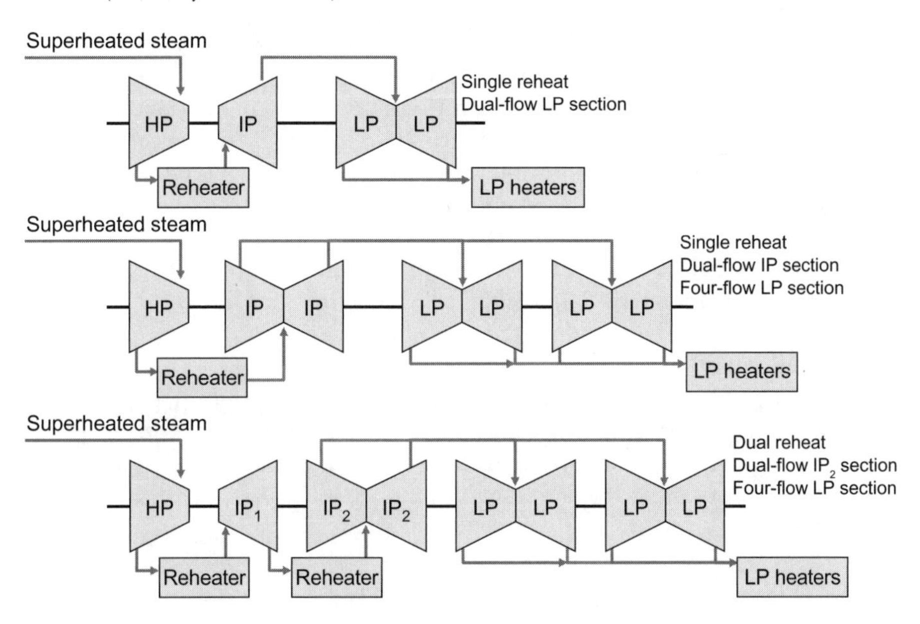

Figure 3.13 High-capacity steam turbine configurations

These configurations are shown schematically in Figure 3.13. For advanced USC steam conditions, double reheat is used and a second IP stage, then called a reheat stage, is introduced, the second reheat section also being commonly dual flow.

3.2.5 Flue gas clean-up

Ash and particulate removal

In PC-fired furnaces, the temperature at which bottom ash fuses into a hard slag is an important design factor, since the buildup of slag leads to increased maintenance costs. While this can be avoided by operating the furnace at a temperature above the ash melting point (a so-called wet-bottom furnace), this leads to high NO_x formation and is at odds with the now-common use of low-NO_x burners. Modern PC-fired furnaces maintain the ash in a dry state (dry-bottom) and bottom ash is removed through a hopper at the base of the furnace.

Typically 70–80% of the ash from PC combustion is fly ash, which leaves the furnace entrained in the flue gas. Electrostatic precipitators or bag filters (baghouses) are used to remove this and can achieve 99.9% ash removal down to levels of 1–2 mg per Nm^3.

Flue gas desulfurization (FGD) systems

If small amounts of sulfur are present in the fuel, sulfur dioxide (SO_2) will be formed during combustion and will therefore be present in the flue gas, with typical concentrations in the range of 400–2500 ppm, depending on the fuel. Sulfur trioxide (SO_3) will also be present if the combustion temperature exceeds 800°C.

The permitted emission levels of SO_x are strictly controlled to prevent acid rain, which causes damage to terrestrial and aquatic ecosystems and corrodes building materials. Within the European Union, the SO_2 emission limit for new liquid- or solid-fueled plants with capacities $>300\,MW_{th}$ was established by the 2001 Revised Large Combustion Plant Directive at 200 mg per Nm^3, while in the United States the New Source Performance Standard (NSPS), established by the Environmental Protection Agency in 2006, specifies an SO_2 emission limit of 0.64 kg per MWh for newly built plant, equivalent to 550 mg per Nm^3 for a coal-fired station emitting flue gas at $\sim 320\,Nm^3$ per GJ.

Wet scrubbing FGD process

The most common method for SO_2 removal is a wet limestone process in which a $CaCO_3$ slurry is sprayed into a countercurrent flow of hot flue gas in an absorber tower. The alkaline solid reacts with the acid gas to produce calcium sulfite according to the reaction:

Absorption:

$$CaCO_{3(s)} + SO_{2(g)} + \tfrac{1}{2}H_2O \rightarrow CaSO_3 \cdot \tfrac{1}{2}H_2O_{(s)} + CO_{2(g)} \qquad (3.16)$$

Air can also be blown into the absorber to oxidize the sulfite to sulfate:

Oxidation:

$$2CaSO_3 \cdot \tfrac{1}{2}H_2O_{(s)} + O_{2(g)} + 3H_2O \rightarrow 2CaSO_4 \cdot 2H_2O_{(s)} \qquad (3.17)$$

This results in the unwanted pollutant being converted into a useful product, gypsum, which is marketed for use in the building industry. Other alkaline sorbents, such as magnesium hydroxide ($Mg(OH)_2$), can also be used. In this case the resulting $MgSO_4$ is regenerated by heating and sulfuric acid is produced as a marketable end product:

Absorption:

$$Mg(OH)_{2(aq)} + SO_{2(g)} \rightarrow MgSO_{3(aq)} + H_2O \quad (3.18)$$

Oxidation:

$$2MgSO_{3(aq)} + O_{2(g)} \rightarrow 2MgSO_{4(aq)} \quad (3.19)$$

Precipitation:

$$MgSO_{4(aq)} + 7H_2O \rightarrow MgSO_4 \cdot 7H_2O_{(s)} \quad (3.20)$$

Regeneration:

$$MgSO_4 \cdot 7H_2O_{(s)} + heat \rightarrow MgO_{(s)} + SO_{3(g)} + 7H_2O \quad (3.21)$$

$$SO_{3(g)} + H_2O \rightarrow H_2SO_{4(aq)} \quad (3.22)$$

Slaking:

$$MgO_{(s)} + H_2O \rightarrow Mg(OH)_{2(aq)} \quad (3.23)$$

Wet scrubbing FGD can achieve 99% SO_x removal, and the wet limestone process, producing gypsum, is expected to be the dominant technology for newly built plants in view of the ease of disposal of the end product.

Electron beam flue gas treatment

An FGD approach that has also been applied uses a high-powered electron beam to ionize the main components of the flue gas stream (N_2, O_2, H_2O, CO_2) in an irradiation chamber. The resulting ionic species (N, O, OH, and HO_2) react with SO_2 to produce sulfuric acid:

$$SO_2 + OH + M \rightarrow HSO_3 + M \quad (3.24)$$

$$HSO_3 + O_2 \rightarrow SO_3 + HO_2 \quad (3.25)$$

$$SO_3 + H_2O \rightarrow H_2SO_4 \quad (3.26)$$

where M is any inert molecule. Ammonia is sprayed into the flue gas upstream of the chamber and reacts with sulfuric acid to form ammonium sulfate:

$$H_2SO_4 + 2NH_3 \rightarrow (NH_4)_2SO_4 \tag{3.27}$$

The resulting solid product is collected using electrostatic precipitation and is also marketable, in this case as a fertilizer.

An electron beam FGD installation on the 200 MW_{th} EPS Pomorzany power plant at Szczecin in Poland uses electrons accelerated to an energy of 700 keV and a total beam power of 1.04 MW to treat 135,000 Nm^3 per hour of flue gas, with an SO_2 removal efficiency of 90–95%.

NO_x control and removal

The NO_x emitted in flue gases reacts with volatile organic compounds (VOCs) in the presence of sunlight to form ozone, which can adversely affect human health and causes damage to terrestrial ecosystems. NO_x emissions limits currently (2010) in force in the European Union range from 200 to 600 mg per Nm^3 (~166 ppm to ~500 ppm), depending on fuel type and plant capacity, with the lower level applying to plant >500 MW_{th}. Untreated flue gases from a modern PC-powered plant would be typically 20–50 ppm, reducing to 2–5 ppm after treatment.

NO_x control during combustion

During combustion, nitrogen from the air or the fuel can be oxidized to form NO_x in parts of the combustion zone that are at high temperatures and in which oxygen is present above the stoichiometric requirement for combustion (i.e., a "fuel-lean" environment).

Methods to reduce NO_x emissions start with low-NO_x burners, which are designed to produce large swirling or branched flames with reduced flame temperature. Some burner designs also internally mix recirculated combustion gas with the incoming air plus fuel mixture to further reduce flame temperature, and a similar result can be achieved by recirculating flue gas into the burner or furnace.

Air staging or two-stage combustion is also used to control NO_x formation. The primary combustion zone is maintained in an oxygen-deficient state by mixing only 70–90% of the stoichiometric requirement with the fuel at the burner. The remaining 10–30% secondary or overfire air is injected above the burners in a lower-temperature combustion zone. Thus, oxygen-rich conditions can occur only at lower temperatures, limiting NO_x formation.

These primary measures can reduce NO_x in the combustion gas from a typical 300 ppm for conventional coal combustion to <100 ppm.

NO_x removal by selective reduction

The most common method to further reduce NO_x in flue gas is selective catalytic reduction (SCR), in which ammonia or urea (NH_2CONH_2) is injected into the flue gas stream over a catalyst. The NO_x is catalytically reduced to form nitrogen and water, the primary reactions being:

$$NH_2CONH_2 + H_2O \rightarrow 2NH_3 + CO_2 \tag{3.28}$$

$$4NH_3 + 2NO_2 + O_2 \rightarrow 3N_2 + 6H_2O \tag{3.29}$$

$$4NH_3 + 4NO + O_2 \rightarrow 4N_2 + 6H_2O \tag{3.30}$$

$$2NH_3 + NO_2 + NO \rightarrow 2N_2 + 3H_2O \tag{3.31}$$

The most common catalysts are oxides of vanadium and tungsten, configured in a ceramic-supported flat plate or honeycomb structure. Titanium and iron oxides, activated carbon, and zeolite catalysts are also used, depending on the proportions of NO and NO_2 present.

The optimal temperature for the catalyzed reaction is in the range of 300–400°C, and removal efficiencies of 90–95% can be achieved. Selective reduction can also proceed without a catalyst, at temperatures in the range of 900–1100°C. However, this selective noncatalytic reduction (SNCR) process can achieve NO_x removal performance comparable to SCR only if reagent mixing and distribution within the reaction zone are carefully controlled to match the temperature profile. This is problematic under varying load conditions, particularly since feedback control from flue gas NO_x levels, which provides efficient control for SCR, is ineffective with SNCR due to the complexity of the reagent injection system. The typical NO_x removal efficiency for SNCR in practice is in the range of 30–50%. Where this is sufficient to meet emissions standards, removal of catalyst costs gives SNCR an economic advantage over SCR.

Electron beam flue gas treatment

As for FGD, electron beam flue gas treatment can also be used to remove NO_x. In reactions similar to Equations 3.24–3.26, NO is either reduced to and released as nitrogen, or oxidized to NO_2. NO_2 is converted to HNO_3 (c.f. Equation 3.24), which reacts with ammonia to produce ammonium nitrate:

$$HNO_3 + NH_3 \rightarrow NH_4NO_3 \tag{3.32}$$

which is collected and sold as a fertilizer. NO_x removal efficiency of up to 70% was achieved using electron beam treatment in the EPS Pomorzany installation described earlier.

3.2.6 *Thermal efficiency of conventional power plants*

The boiler efficiency of modern pulverized fuel, hard coal-fired units is 94–95% (LHV basis), reducing slightly to 90–91% for lignite-fired units due to the higher moisture content. When combined with the other elements of the Rankine cycle, the overall efficiency currently achievable in conventional plants is in the 40–45% range.

Supercritical plants are more efficient and also sustain a higher efficiency under reduced load conditions. Table 3.8 illustrates the typical effect of turn-down on plant efficiency for subcritical and supercritical units.

Table 3.8 Power plant thermal efficiency penalty under part load conditions

	η_{th} (%) 100% load	η_{th} (%) 75% load	η_{th} (%) 50% load
Supercritical	45	43	37–39
Subcritical	39	35	28–29

3.3 Combined cycle power generation

As discussed earlier, the thermal efficiency of a Carnot cycle, and of other practical thermodynamic cycles, is increased if the ratio T_H:T_C is increased. A combined cycle plant combines two or more thermodynamic cycles to exploit a wider temperature range and reach a higher thermal efficiency than would be possible with a single cycle. This is achieved by using the relatively high-temperature reject heat from the first cycle to drive a second cycle, effectively reducing the overall T_C. The concept can be simply illustrated in terms of two Carnot cycles, as shown in Figure 3.14.

In power-generation applications, the most common combined cycle plant comprises two stages:

- A gas turbine operating a Brayton cycle (Section 3.1.6) and fired either by natural gas or integrated with a gasification plant, followed by
- A steam turbine operating a Rankine cycle (Section 3.1.5), driven by a heat recovery steam generator that recovers heat from the high-temperature exhaust gases exiting the gas turbine

A natural gas-fired combined cycle power plant is illustrated schematically in Figure 3.15. In the first-generation cycle natural gas is burned in a combustion turbine, which directly drives an electrical generator. Combustion temperatures are typically in the range of 1000–1400°C, depending on the fuel, while final exhaust exit temperatures following additional turbine stages are in the range of 450–650°C.

The steam cycle operates on similar principles to a normal fossil fuel-fired power plant, with the boiler replaced by a heat-recovery steam generator.

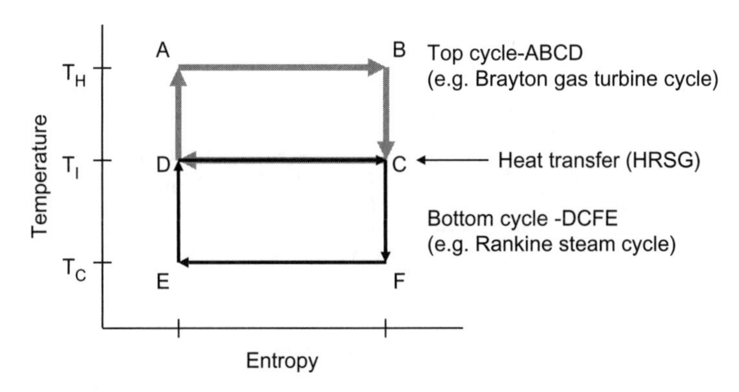

Figure 3.14 Combined cycle concept based on ideal Carnot cycles

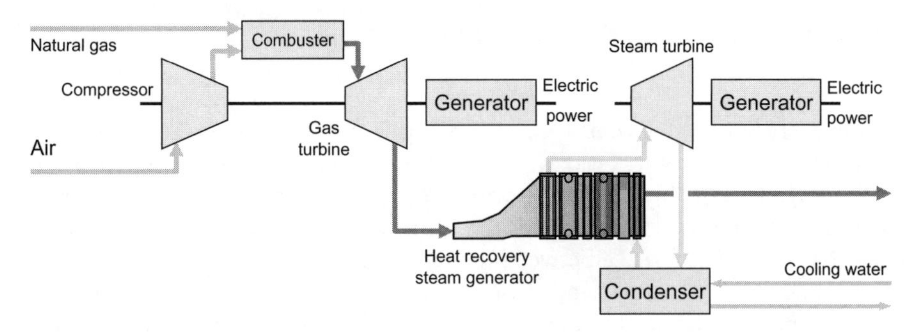

Figure 3.15 Process schematic of a natural gas-fired combined cycle power plant

Other combined cycle variants use pressurized fluidized bed combustion (PFBC) in a number of configurations:

- High-pressure PFBC combustion gas is used to drive a gas turbine.
- Natural gas plus PFBC combustion gas is used to co-fire a combustion turbine.
- Fuel is gasified in a PFBC and syngas is used to fire a gas turbine in a gasification fluidized-bed combined cycle (GFBCC).

In each case, steam is raised in the FBC, and additional heat may also be recovered from gas turbine exhaust.

3.3.1 Heat recovery steam generation

The heat recovery steam generator (HRSG) recovers heat from the gas turbine exhaust to generate steam at temperatures up to ~650°C and pressures of 13–20 MPa. Although supercritical systems have been developed, heat recovery steam generation is most commonly applied under subcritical conditions.

The components of the HRSG—evaporator, superheater, and economizer—are functionally equivalent to those in a conventional steam boiler. Figure 3.16 illustrates schematically the heat transfer in the three sections of

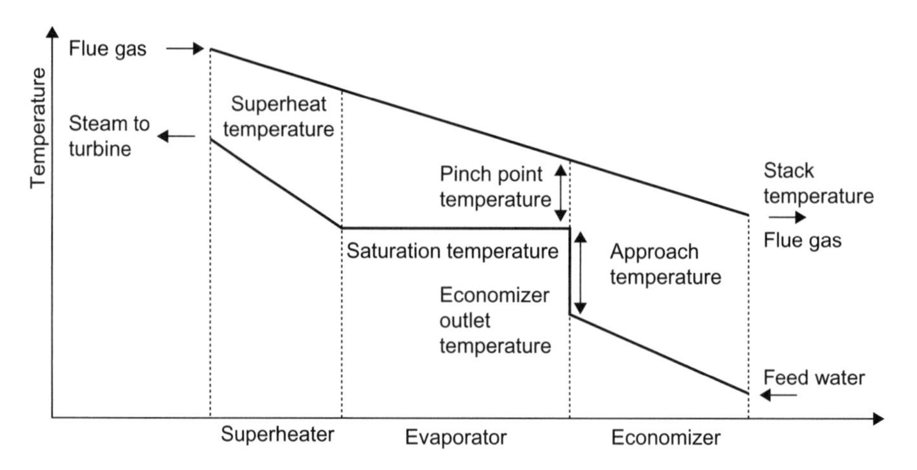

Figure 3.16 Heat transfer in heat recovery steam generator

a simple HRSG, with exhaust gases moving from right to left and water and steam from left to right. Feed water entering the economizer is heated by the gas exiting the HRSG to a temperature close to the saturation temperature of steam at the operating pressure of the unit. The so-called approach temperature (ΔT_A) is the difference between the temperature of water leaving the economizer and the saturation temperature, and is kept at 10°C or more to prevent steam formation in the economizer. The gas exit temperature from the HRSG is held >100°C if it is necessary to prevent condensation, although lower temperatures are commonly used with suitable materials in order to maximize heat recovery.

In the evaporator section, flue gas is cooled to close to the saturation temperature, providing the latent heat of evaporation to the hot water exiting the economiser. Heat recovery is maximized if the gas exiting the evaporator section is at the saturation temperature (i.e., the pinch temperature—ΔT_P—is zero). In practice this requires very large heat-exchange surfaces, leading to increased capital costs, and in practice a ΔT_P of 10–20°C is typical.

After evaporation, steam is superheated to close to the gas entry temperature. Additional burners (duct firing) may be included to increase the gas temperature and steam production capacity of an HRSG to meet peak demand, although this is not generally used to increase the baseload capacity in view of the relatively low efficiency of this supplemental firing.

More advanced multipressure HRSGs improve overall thermal efficiency and achieve higher work output by including two or three parallel steam flows to drive HP, IP, and LP stages of the steam turbine, although this comes at the cost of a more complex plant. Figure 3.17 illustrates a state-of-the-art triple-pressure HRSG, with multistage economizers for the IP and HP flows.

The key operating parameters of this HRSG design are summarized in Table 3.9.

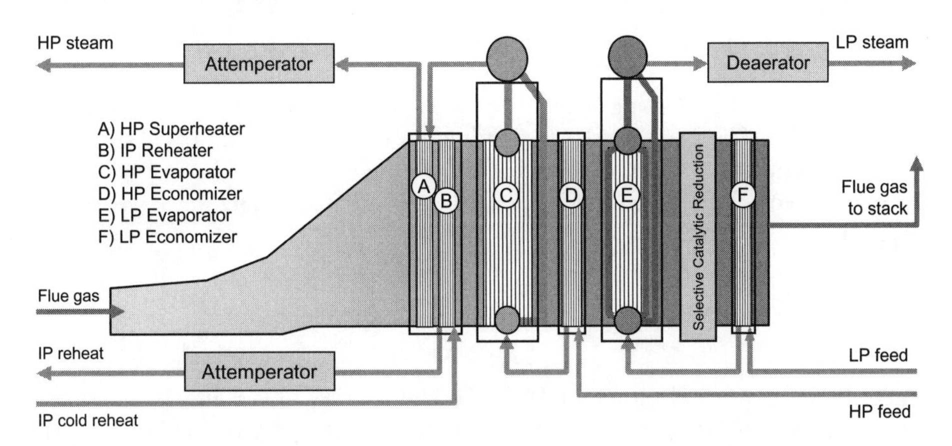

Figure 3.17 Process schematic for triple-pressure HRSG

Table 3.9 Process parameters for triple-pressure HRSG

Subsystem	Parameter	Value
Gas turbine	Exhaust temperature	570°C
HP steam system	Live steam temperature	535°C
	Live steam pressure	16.9 MPa
IP steam system	Live steam temperature	535°C
	Live steam pressure	2.9 MPa
LP steam system	Live steam temperature	200°C
	Live steam pressure	0.4 MPa
	Pinch point temperature	11°C
Flue gas	Outlet temperature	81°C

Modifications to the standard subcritical systems to optimize performance in HRSG applications include replacement of the high-pressure steam drum by a thin-walled steam–water separator, allowing rapid start-up to match that of the gas turbine and therefore improving operational flexibility.

3.3.2 Combined cycle thermal efficiency

Considering the ideal Carnot cycle, the theoretical maximum efficiency is given by:

$$\eta_{max} = 1 - T_C / T_H \tag{3.33}$$

where for a combined cycle power plant the hot source corresponds to the combustion turbine, with T_H typically 1000–1400°C (1270–1670°K), and the cold sink is the cooling medium of the Rankine cycle condenser, with T_C typically 10–25°C (280–300°K), depending on the plant location. These values yield theoretical maximum efficiencies in the range of 75–85%.

In practice, the thermal efficiency of a combined cycle power plant is given by:

$$\eta_{th} = \text{Net power output of the plant} / \text{Heating value of the fuel} \qquad (3.34)$$

and can be expressed either in LHV or HHV terms. Current (2010) state-of-the-art combined cycle plants have efficiencies of up to 60% (LHV base), which are still significantly lower than η_{max}, largely due to low-grade heat loss. If low-grade heat is fully used down to the cold sink temperature, for example in a combined heat and power plant, the overall energy utilization efficiency can approach the Carnot cycle efficiency.

3.3.3 Integrated gasification combined cycle power generation

In an integrated gasification combined cycle (IGCC) power-generation system, the fuel for the combined cycle plant is syngas, generated from a gasification system. This approach, which was first demonstrated in 1984 at Southern California Edison's 100 MW_e Cool Water coal gasification plant, results in a system with high efficiency, flexibility to use a wide range of feedstocks (including biomass and waste), and low or potentially zero carbon emissions.

Figure 3.18 illustrates a simplified process scheme for an IGCC power plant, at the heart of which is the gasification process. This operates, as discussed in Section 3.1, by partial oxidation and steam reaction of carbon in the fuel to produce a syngas of CO and H_2, which is combusted in the gas turbine.

Although the gasifier can be either air- or oxygen-blown, the latter system avoids syngas dilution with nitrogen, resulting in a product with typically double the heating value when compared to an air-blown system. Gasifier and turbine operability and carbon conversion efficiency also improve with oxygen-blown gasification. Cryogenic air separation (Chapter 9) has been the technology most commonly used to date to produce oxygen for gasification, although a number of other technologies are also under development, including membrane systems (Chapter 8).

The integration of steam use between the gasification plant and the HRSG is a key element in the efficiency of IGCC. Equally important is the relative ease with which precombustion carbon capture can be integrated into the IGCC scheme, as discussed in the following chapter.

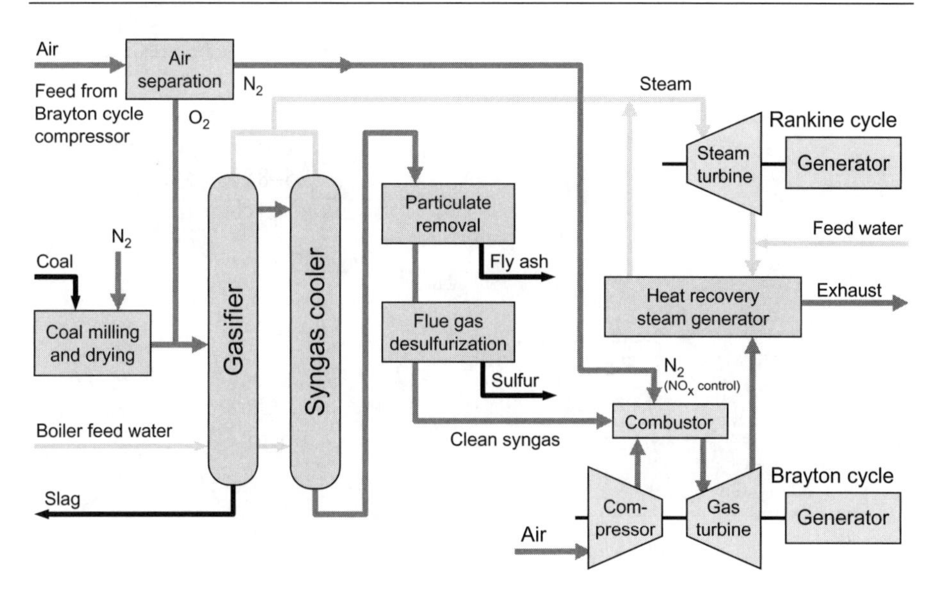

Figure 3.18 Simplified IGCC power plant process scheme

3.4 Future developments in power-generation technology

Since steam temperature is the fundamental determinant of thermal efficiency in a steam cycle, the development of materials capable of operating at ever-higher temperatures is the main focus of RD&D work aiming to achieve higher efficiencies.

3.4.1 Steel development for SC and USC boilers

The progressive development of steels for more advanced steam conditions and the resulting trend in plant efficiency is illustrated in Figure 3.19.

A key design parameter in assessing material suitability is the creep rupture stress that can be maintained for the life of the plant. Creep strength is commonly assessed for a period of 10^5 hours at the operating temperature of the component, with an ability to withstand 100 MPa being a typical requirement. Table 3.10 summarizes the characteristics of a number of advanced steels and superalloys that have been developed to address the challenge of SC and USC operation.

Advances in welding processes have also been made in order to ensure tight control of the weld deposit composition, as well as pre- and postweld heat treatment (PWHT), which are important to ensure long life at high temperatures and pressures.

RD&D work with the aim of raising steam temperature to 700°C or above has progressed under the European Union-funded Thermie AD700 project (now part of the VGB E_{max} Power Plant Initiative), targeting 700°C, and the

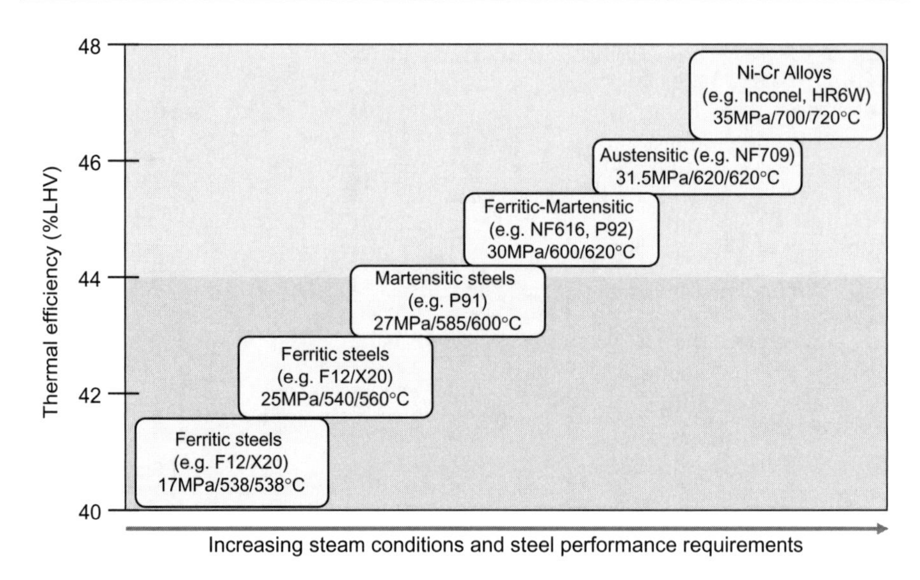

Figure 3.19 Development of steels for SC and USC boilers

Table 3.10 Steel and superalloys developed for SC and USC boiler fabrication

Designation	Composition	Description
P91	9Cr-1Mo-Nb-V	High-strength martensitic steel used for operating temperatures up to 600°C and pressures to 27 MPa
NF616	9Cr-0.5Mo-1.8W-V-Nb	A ferritic–martensitic steel developed for use in boiler tube and pipework to overcome high expansion of austenitic alloys
NF12	12Cr-2.6W-2.5Co-0.5Ni-V-Nb	Developed by Nippon Steel with the design aim of achieving a 30% improvement in creep rupture strength compared to NF616 (P92) steel
NF709	20Cr-25Ni	A novel austenitic steel alloy suited to high temperatures due to low creep and high corrosion resistance
HR6W	23Cr-45Ni-6W-Nb-Ti-B	A high Ni-Cr superalloy developed for 700°C service; high creep rupture strength due to the strengthening effect of W and Nb
Inconel	50-70Ni, 15-25Cr, 3-10Mo, 3-6Nb	A Ni-Cr superalloy with <10% Fe

U.S. Department of Energy-funded Vision 21 project, targeting 732°C. As part of these projects nickel–chromium superalloys (such as NF709, HR6W, and Inconel) are being developed for the fabrication of boiler components exposed to the highest temperatures, while high-strength 9-12Cr martensitic steels are being further developed for components operating at temperatures up to 650°C.

3.4.2 Other materials developments

For USC service, the HP sections of steam turbines generally use a triple-shell construction to distribute the thermal and burst stresses. Similar to boiler service, nickel superalloys are required for turbine components operating at the highest temperatures and pressure (HP nozzles and rotors), while advanced high-strength austenitic steels are used for the second line of temperatures, and with inner and outer shells of progressively lower alloy steels.

As noted earlier, reducing the flue gas exit temperature to below the dew point is desirable to maximize boiler efficiency but runs the risk of acid corrosion to heat exchangers and ducting due to condensation of water and the formation of sulfuric acid. The use of plastic, glass-fiber, and Teflon materials for these cold end surfaces can allow exit temperatures to be reduced to below 100°C, leading to an increase in overall cycle efficiency. In early 2009, Foster Wheeler commissioned the world's first SC CFB boiler, the Lagisza 460 MW_e unit at the Poludniowy Koncern Energetyczny SA (PKE) plant in Poland. This unit has a plastic heat exchanger as the final heat recovery stage, cooling flue gas to 82°C.

3.5 References and Resources

3.5.1 Key References

The following key references, arranged in chronological order, provide a starting point for further study.

Drbal, L., K. Westra, and P. Boston (eds.). (1996). *Power Plant Engineering*, Springer Science + Business Media, New York.

Department of Trade and Industry. (1999). *Cleaner Coal Technologies*, DTI, UK. Available at www.berr.gov.uk/files/file22078.pdf

Pansini, A. J. and K. D. Smalling. (2006). *Guide to Electric Power Generation*, The Fairmount Press, Lilburn, GA.

Stiegel, G. J. and M. Ramezan. (2006). Hydrogen from coal gasification: An economical pathway to a sustainable energy future, *International Journal of Coal Geology*, 65 (173–190).

IEA. (2007). *Fossil Fuel-Fired Power Generation; Case Studies of Recently Constructed Coal- and Gas-Fired Power Plants*, OECD Publishing, Paris, France.

3.5.2 Institutions and Organizations

Institutions and organizations that are active in the development of advanced power-generation systems relevant for CCS application are listed below.

Gasification Technologies Council: www.gasification.org
IEA Clean Coal Centre: www.iea-coal.org.uk/site/ieacoal/home
University of Utah, Utah Clean Coal Program: www.uc3.utah.edu
U.S. Department of Energy National Energy Technology Laboratory, RD&D in coal and power systems: www.netl.doe.gov/technologies/coalpower
U.S. Department of Energy Vision 21 project: www.fossil.energy.gov/programs/powersystems/vision21
VGB E_{max} (formerly the Thermie AD700 project): www.vgb.org/en/emax.html

3.5.3 Web Resources

Clean-energy.uc (news and information about coal gasification): http://www.cleanenergy.us/index.phpwww.clean-energy.us/index.php
Clean Coal Today (National Energy Technology Laboratory quarterly newsletter): www.netl.doe.gov/technologies/coalpower/cctc/newsletter/newsletter.html
Cooperative Research Centre for Coal in Sustainable Development (CCSD), Power Station Emissions Handbook: www.ccsd.biz/PSE_Handbook
Future of Coal (MIT interdisciplinary study): http://web.mit.edu/coal
Heat recovery steam generator design: www.hrsgdesign.com
IEA Clean Coal Centre (clean coal technologies database): www.iea-coal.org.uk/site/ieacoal/databases/clean-coal-technologies
Phyllis (database of biomass and waste composition): www.ecn.nl/Phyllis

Part II

Carbon capture technologies

4 Carbon capture from power generation

4.1 Introduction

The fundamental chemical process involved in the generation of power from carbon-based fuel is the exothermic oxidation of carbon, as described in Section 3.1, and since CO_2 is the lowest-energy endpoint of the oxidative reaction chain, its production is unavoidable. The elimination of carbon from power plant emissions therefore requires either:

- Decarbonation of the fuel prior to combustion (precombustion capture)
- Separation of CO_2 from the products of combustion (postcombustion capture)
- Reengineering the combustion process to produce CO_2 as a pure combustion product, obviating the need for its separation (oxyfueling or oxyfuel combustion)

These approaches are illustrated schematically in Figure 4.1.

As shown in the figure, the pre- and postcombustion approaches both require technologies to separate CO_2 from a gas mixture comprising $CO_2 + H_2$ or $CO_2 + N_2$ respectively. For oxyfueling, the oxygen supply can be achieved either through a separation of O_2 from air ($O_2 + N_2$ + trace gases) or by the

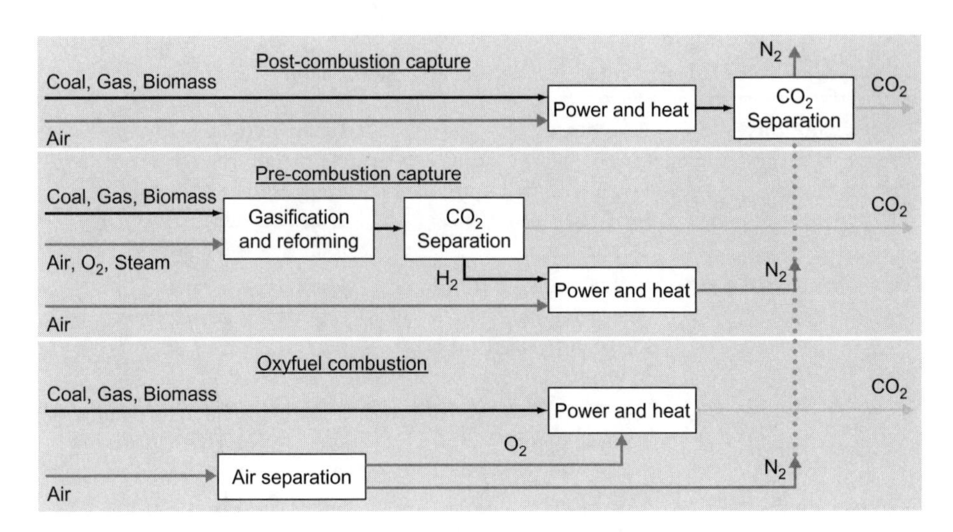

Figure 4.1 Options for CO_2 capture from power generation

Doi: 10.1016/B978-1-85617-636-1.00004-3

delivery of oxygen to the combustion process in the form of a solid oxide. Some advantages and disadvantages of these capture options are summarized in Table 4.1.

Four fundamental technology areas are in use or under development to address these gas separation challenges—absorption, adsorption, membranes, and cryogenic separation systems. The application of these technologies to pre- and postcombustion capture and to oxyfuel combustion is outlined in the next three sections, and the technologies are described in detail in Chapters 6–9.

4.2 Precombustion capture

Precombustion capture involves decarbonation by gasification of the primary fuel, commonly coal or biomass, to produce hydrogen through a combination of partial combustion, reforming and water–gas shifting (Section 3.1), and the separation of CO_2 from the resulting reaction product stream.

The current development and demonstration focus of precombustion capture is on IGCC plants, using the process shown schematically in Figure 4.2, although in principle the approach is equally applicable to all integrated gasification systems where hydrogen is the final syngas product, such as integrated gasification fuel cell (IGFC) systems.

The separation of CO_2 and H_2 can be achieved using a number of technologies, as shown in Table 4.2, of which the use of physical solvents (such as the Selexol and Fluor processes, described in Section 6.2) is currently the most commercially developed. In this and similar tables below, underlined technologies are those that are currently deployed or close to being deployed in CCS projects.

Table 4.1 Advantages and disadvantages of capture options

Capture option	Advantages	Disadvantages
Precombustion	Lower energy requirements for CO_2 capture and compression	Temperature and efficiency issues associated with hydrogen-rich gas turbine fuel
Postcombustion	Fully developed technology, commercially deployed at the required scale in other industrial sectors	High parasitic power requirement for solvent regeneration
	Opportunity for retrofit to existing plant	High capital and operating costs for current absorption systems
Oxyfuel combustion	Mature air separation technologies available	Significant plant impact makes retrofit less attractive

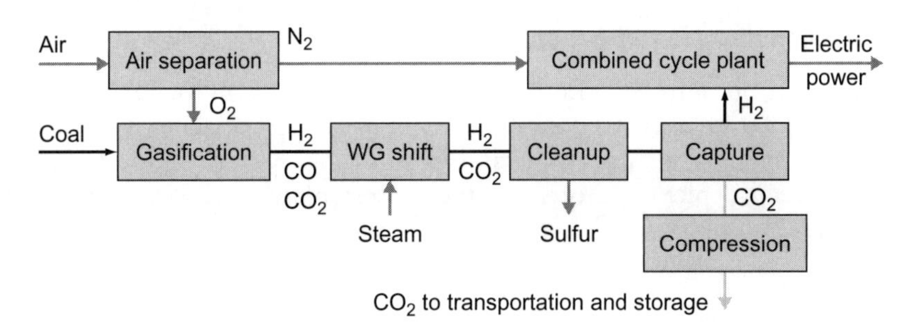

Figure 4.2 Precombustion capture IGCC process schematic

Table 4.2 CO_2:H_2 separation technologies for precombustion capture

Technology area	Currently developed technologies	Example technologies under development
Absorption-based separation (Chapter 6)	Physical solvents (e.g., Selexol, Fluor processes), chemical solvents	Novel solvents to improve performance; improved design of processes and equipment
Adsorption-based separation (Chapter 7)		Zeolite, activated carbon, carbonates, hydrotalcites and silicates
Membrane separation (Chapter 8)		Metal membrane WGS reactors; ion transport membranes
Cryogenic separation (Chapter 9)	CO_2 liquefaction	Hybrid cryogenic + membrane processes

CO_2:H_2 separation is somewhat easier than the postcombustion separation of CO_2 and N_2 due to the greater difference in molecular weights and molecular kinetic diameters for CO_2 versus H_2 than for CO_2 versus N_2.

In addition to the further development of technologies related to CO_2:H_2 separation, other aspects of precombustion capture processes that are being addressed by RD&D efforts include:

- Development of H_2-fueled gas turbines (addressing combustion processes, including flameless combustion, burner design, heat transfer and cooling, materials impact, and operational aspects)
- Physical, energetic, and operational integration of precombustion capture process into an IGCC plant
- Optimization of gasification process—catalysts, heat integration, and adaptation for CO_2 capture, purification, and transmission

4.2.1 Precombustion RD&D projects

The European Union-funded European Technology Platform for Zero Emission Fossil Fuel Power Plants (ETP ZEP) exemplifies RD&D activity in the area of precombustion capture and was established in 2005 with the aim of enabling the commercial deployment of fossil-fuel power plants with zero CO_2 emissions by 2012.

Research and development

A number of European Union-funded projects have been in progress since 2004 addressing some of the key technologies required to enable precombustion capture from IGCC plants, as summarized in Table 4.3.

The ENCAP SP2 project concluded that the technologies for zero emission IGCC (ZEIGCC) can be considered largely ready for full-scale demonstration. Exceptions are the need for further design optimization and testing of burners for H_2-rich combustion, and the optimization of turbine blade material for higher turbine inlet temperatures.

Table 4.3 European Union-funded R&D projects addressing IGCC precombustion

Project	R&D objectives
ENCAP (SP2) (leading partner, Vattenfall AB)	Optimization of CO-shift conversion
	Modeling of H_2-rich combustion and experimental validation
	Development and testing of burners for H_2-rich combustion in gas turbines
	Hydrogen compatibility of turbine components
	Development of precombustion capture plant specifications
COORIVA (leading partner, TU Frieberg)	Review of IGCC operational experience, focusing on operational problems, corrosion and plugging, and plant integration
	Analysis of further optimization potential
	Identification of technology development requirements for gasification and syngas cleaning
	Feasibility study for optimized zero-emission IGCC (ZEIGCC)

Demonstration

Building on the ENCAP and COORIVA R&D results, the German energy company RWE Power is planning to construct a 450 MW_e IGCC plant with CCS at its Goldenbergwerk site near Cologne in Germany. The gasification plant will be lignite-fueled, with a gross electrical output of 450 MW_e, reducing to 360 MW_e net after deducting the plant and CCS power utilization, and is expected to be commissioned at the end of 2014.

A capture efficiency of 90% is targeted, and storage of the ~300 t-CO_2 per hour (2.6 Mt-CO_2 per year) captured in the plant is planned to be in a saline aquifer or depleted gas reservoir, with evaluation of potential storage sites expected to be completed in 2010. Figure 4.3 shows the overall project plan.

A major challenge faced by the project is the lack of a European Union-wide legal and regulatory framework for CO_2 transport and storage; because of this, the Goldenbergwerk plant will include an option to run without capture.

Several other demonstration and commercial-scale precombustion capture projects are planned to be operational in the 2010–2015 time scale, as shown in Table 4.4.

4.3 Postcombustion capture

In postcombustion capture, CO_2 is removed from the combustion reaction product steam—the flue gases—before emission to the atmosphere. Postcombustion capture is thus an extension of the flue gas treatment for NO_x and SO_x removal, made more challenging by the relatively higher quantities of CO_2 in the gas stream (typically 5–15%, depending on the fuel being used).

A similarly wide range of technologies is available or under development to address the postcombustion CO_2:N_2 separation problem, as shown in Table 4.5. The use of chemical solvents, such as MEA (described in Section 6.1), is the most mature and is widely deployed for natural gas treatment, although many of the currently planned demonstration projects are expected to use the chilled ammonia process (CAP; Section 6.2.1).

As well as technologies to address the gas separation challenge, several other aspects of the postcombustion process are also the focus of ongoing RD&D, notably:

- Integration and optimization of the postcombustion process within the power plant
- Environmental impact of the overall postcombustion capture process and of specific solvent use
- Procedures for optimal postcombustion process operation under varying plant conditions

An alternative postcombustion capture approach that has been proposed is to use cooled, CO_2-rich flue gases to feed bioreactors producing microalgal biomass that would be used as a biofuel, either by direct combustion or gasification with subsequent liquid fuel production (biodiesel). This is discussed together with other industrial use options in Chapter 14.

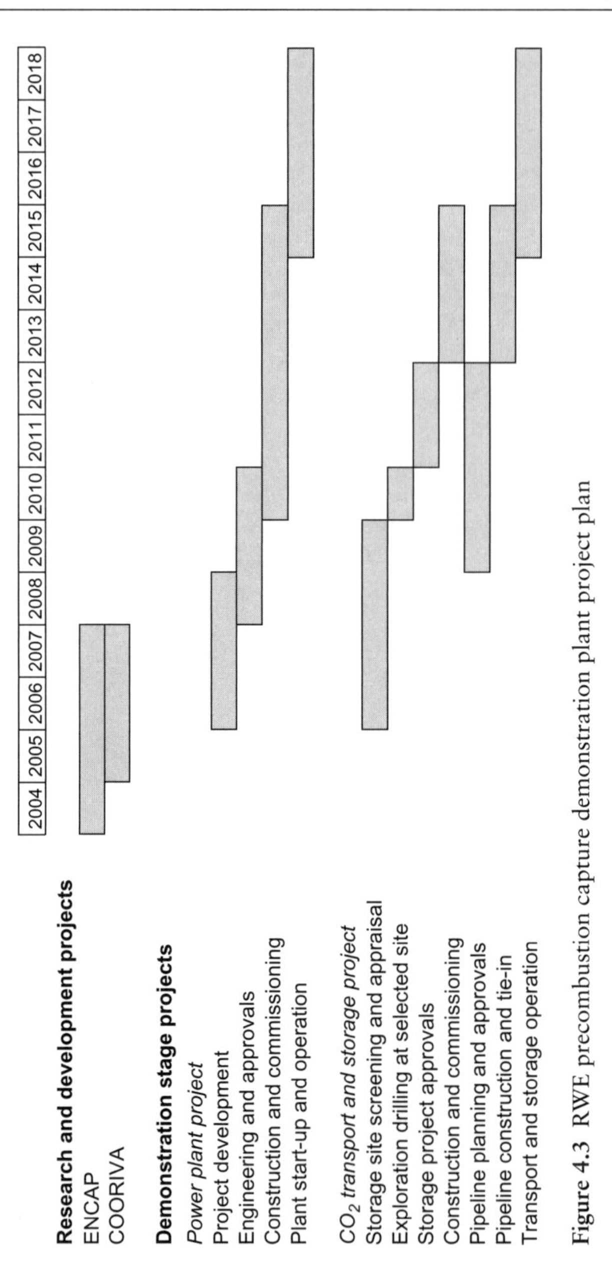

Figure 4.3 RWE precombustion capture demonstration plant project plan

4.3.1 Postcombustion RD&D projects

A number of postcombustion capture systems using monoethylene amine stripping (Section 6.2.1) have been in commercial operation since the late 1970s, providing CO_2 for refrigeration and food-processing purposes. These installations, of which three are summarized in Table 4.6, provide pilot-scale proof of

Table 4.4 Precombustion demonstration and early deployment projects

Project; operator and partners	Project description	Planned start-up
Edwardsport IGCC; Duke Energy, USA	630 MW IGCC with capture using physical absorption process	2012
Callide A; ZeroGen, Australia	100 MW IGCC with saline aquifer storage	2012
Teesside IGCC; Coastal Energy (Centrica and Progressive Energy), UK	800 MW coal-fired IGCC with 5 Mt-CO_2 per year capture and North Sea EOR use	2012
Hatfield IGCC; Powerfuel plc, UK	740 MW coal-fired IGCC with North Sea storage	2013
GreenGen; Peabody Energy, China	Staged coal-fired IGCC project with CCS development from 100 MW scale in 2015 to 400–650 MW by 2020	2015 on

Table 4.5 CO_2:N_2 separation technologies for postcombustion capture

Technology area	Currently developed technologies	Example technologies under development
Absorption-based separation (Chapter 6)	Chemical solvents (e.g., MEA, chilled ammonia)	Novel solvents to improve performance; improved design of processes and equipment
Adsorption-based separation (Chapter 7)	Zeolite and activated carbon molecular sieves	Carbonate sorbents; chemical looping
Membrane separation (Chapter 8)	Polymeric membranes	Immobilized liquid membranes; molten carbonate membranes
Cryogenic separation (Chapter 9)	CO_2 liquefaction	Hybrid cryogenic + membrane processes

this absorption technology. However, these installations have not been used as pilots to investigate scale-up to capture the full plant emissions, due to the limited quantity of CO_2 required for the specific end users served by each plant.

R&D and pilot-scale testing

The European Union CASTOR project, established as part of the EC's Sixth Framework Program, included among its aims the development of new sorbents

Table 4.6 Postcombustion capture plants in commercial operation

Operator	IMC Global	AES Power	AES Power
Plant and location	Trona, California, USA	Warrior Run, Maryland, USA	Shady Point, Oklahoma, USA
Commissioned	1978	2000	2001
Plant rating and type	Coal-fired boiler	180 MW_e coal-fired FBC	320 MW_e coal-fired FBC
CO_2 production	800 t-CO_2 per day	200 t-CO_2 per day, 5% slip stream	800 t-CO_2 per day, 3% slip stream
Capture process	Absorption (MEA)	Absorption (MEA)	Absorption (MEA)
CO_2 use	Brine carbonation	Refrigeration and fire extinguishers	Food processing

for postcombustion capture with an energy consumption target of <2 GJ per t-CO_2 at 90% recovery efficiency and a cost of €20–30 per t-CO_2 avoided. The project scope included completion of pilot-plant testing to demonstrate reliable and efficient operation of postcombustion capture. The pilot plant was constructed and commissioned during 2005 and started operation in March 2006 at the 400 MW_e pulverized bituminous coal-fired Esbjerg power station, operated by Danish energy company DONG Energy. The plant captures 90% of the CO_2 from a 5000 Nm^3 per hour, 0.5% flue gas slip stream taken after wet FGD, yielding 1 t-CO_2 per hour.

A number of solvents were pilot-tested, including 30%-wt MEA and two proprietary solvents, CASTOR-1 and CASTOR-2. The pilot testing validated the postcombustion capture process and resulted in energy consumption of 3.5 GJ per t-CO_2 for the CASTOR-2 solvent, with the potential to reduce this to 3.2 GJ per t-CO_2 with further plant heat integration, and indicated capture costs in the range of €35–37 per t-CO_2 avoided.

Planned demonstration plant

Building on this pilot-scale experience, DONG Energy is constructing a 280 MW_e plus 350 MW_{th} natural gas-fired combined heat and power (CHP) plant at StatoilHydro's Mongstad refinery in Bergen, Norway. From start-up of the CHP plant in 2010, a 100 kt-CO_2 per year postcombustion pilot will go into operation, establishing the European CO_2 Test Centre Mongstad (TCM).

Both amine and chilled ammonia processes will be tested in the facility, as well as capture from both the CHP plant and from other refinery offgas (Figure 4.4). Full capture of 1.1 Mt-CO_2 per year from the CHP plant is expected to follow from 2014. Other partners in the TCM are Vattenfall AG, Gassnova SF, and A/S Norske Shell.

A number of other demonstration-scale postcombustion capture projects are planned to be operational in the 2010–2015 time scale, as shown in Table 4.7.

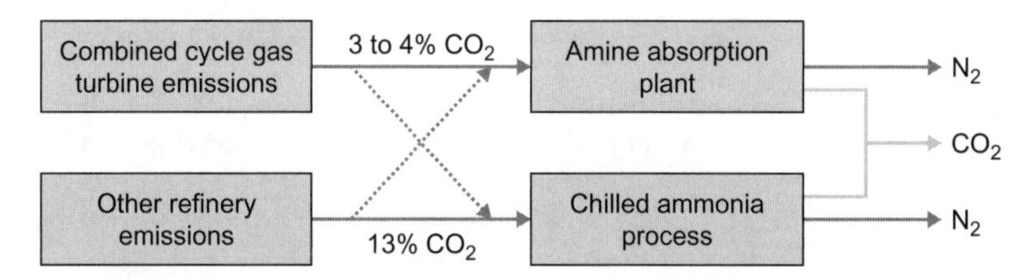

Figure 4.4 European CO_2 Test Centre Mongstad configuration

Table 4.7 Postcombustion demonstration and early deployment projects

Project; operator and partners	Project description	Planned start-up
Mountaineer; American Electric Power	20 MW demonstration plant, 110 kt-CO_2 per year capture using Alstom chilled ammonia process and geological storage	2009
Antelope Valley Station; Basin Electric Power Cooperative	120 MW plant, capture using PowerSpan ammonia-based ECO_2 technology, delivery into existing pipeline for EOR	2012
Northeastern; America Electric Power, Public Service of Oklahoma	200 MW plant, full-scale 1 Mt-CO_2 per year test using Alstom chilled ammonia process, with CO_2 sale for EOR	2012
Project Pioneer; TransAlta, TransCanada Pipelines, Alstom	200 MW plant, 1 Mt-CO_2 per year, Alstom chilled ammonia process with possible geological storage or EOR use	2012
New build; RWE, nPower, Peel Energy, DONG Energy	400 MW supercritical pulverized coal using absorption technology developed by RWE, BASF, and Linde, and storage in a depleted North Sea gas field	2016

The last of these projects will bring the demonstration of postcombustion capture up to the full scale of commercial deployment.

4.4 Oxyfuel combustion capture

Oxyfuel combustion requires the delivery of oxygen rather than air to the combustion chamber, so that the gaseous combustion reaction product is near-pure CO_2 rather than a mixture from which CO_2 needs to be separated. Oxygen may be delivered either as a gas stream, produced by the separation of O_2 from

Table 4.8 O_2:N_2 separation technologies for oxyfuel combustion capture

Technology area	Currently developed technologies	Example technologies under development
Adsorption-based separation (Chapter 7)	Zeolite and activated carbon molecular sieves	Perovskites and chemical looping technology
Membrane separation (Chapter 8)	Polymeric membranes	Ion transport membranes; carbon molecular sieves
Cryogenic separation (Chapter 9)	Distillation	Improvements in distillation processes

air (effectively an $O_2 + N_2$ binary mixture), or as a solid oxide in a chemical looping process. Table 4.8 summarizes the technologies that can be applied to oxyfuel combustion.

In addition to RD&D focused on improvements in air separation for oxygen supply, other aspects of the oxyfuel process that are the subject of current research include:

- Characteristics of oxyfuel combustion processes (heat transfer, ash properties and composition, fouling) for a variety of solid fuel types
- Impact on boiler and combustion system (e.g., fluidized bed) design, construction materials, and operation, including flexibility for air firing
- Optimization of existing flue-gas treatment technologies for specific oxyfuel conditions (CO_2-rich flue gas stream)
- Optimization of air separation units (ASU) to reduce energy requirement, improve efficiency, and optimize overall system integration

4.4.1 Oxyfuel RD&D projects

In 2001 the Swedish power company Vattenfall AB, Europe's third-largest power company, began a comprehensive RD&D program into oxyfuel combustion, which it had identified as the preferred option for lignite-fueled plants. Vattenfall generates >40% of its power from fossil fuels, and the aim of the program is to develop oxyfuel technology for full commercial deployment by 2015 at a target capture cost of less than €20 per t-CO_2 avoided.

Research and development

Between 2001 and 2007, five laboratory-scale test rigs were commissioned at universities in Sweden and Germany to investigate oxyfuel processes and performance, as summarized in Table 4.9.

Following this laboratory-scale program, a 30 MW pilot plant was put into operation in September 2008 at the Schwarze Pumpe power plant in Germany as an intermediate step-up toward a full demonstration plant. This is the first

Table 4.9 Vattenfall AB-commissioned laboratory-scale test rigs for oxyfuel R&D

Institute	Test rig	R&D objectives
Technical University of Hamburg-Harburg, Germany	20 kW combustion test rig with nonrecirculating oxyfuel combustion	Thermodynamics of oxyfuel process
Technical University of Dresden, Germany	50 kW recirculating oxyfuel test rig	FGD operation on oxyfuel flue gas stream
		Combustor dynamics when shifting between oxygen and air firing
		Novel FGD processes
Chalmers University of Technology, Gothenburg, Sweden	100 kW recirculating oxyfuel test rig	Oxyfuel combustion research with wet and dry flue gas recirculation, assessing the changes in radiation and reaction kinetics under oxyfuel conditions
Brandenburg Technical University (BTU), Germany	500 kW recirculating oxyfuel test rig	Drying and oxyfuel combustion of lignite
Stuttgart University Institute of Process Engineering and Power Plant Technology (IVD), Germany	20 kW combustion test rig with nonrecirculating oxyfuel combustion	Impact of recirculation conditions and staged combustion
	500 kW recirculating oxyfuel combustion test rig, with all power plant systems	Power plant system integration and combustion flame behavior

oxyfuel pilot plant in the world, and the main objectives of the initial 3-year pilot program are:

- Validation of laboratory-scale results and engineering work for lignite and hard coal combustion
- Optimizing the integration of oxyfuel requirements into power plants
- Improving knowledge and gaining experience of oxyfuel combustion dynamics and operational issues
- Demonstration of capture technology and possible underground storage, pending suitable site identification and permitting

Planned demonstration plant

A 250 MW_e demonstration plant is currently envisaged as the final step-up before full commercial deployment. This is planned at the 3 GW_e Jänschwalde power plant in Germany, which currently consists of six 500 MW_e generation blocks, each comprising two 650 MW_{th} (250 MW_e) conventional lignite-fired boilers driving a single steam turbine. It is planned to add one 250 MW_e oxyfuel boiler to one of the generation blocks, allowing demonstration of oxyfuel technology alongside conventional combustion. Storage options for the site are still under investigation.

Feasibility studies began in mid-2008, with construction planned to commence in 2011 and start-up scheduled for 2015. The overall RDD&D timeline is shown in Figure 4.5 and clearly illustrates the 20-year time scale required to bring these technologies to the stage of full commercial deployment.

Fewer demonstration-scale and early deployment oxyfuel capture projects are being planned compared to pre- and postcombustion capture; a few that have been announced are summarized in Table 4.10.

4.5 Chemical looping capture systems

Chemical looping is a form of oxyfueling in which oxygen is introduced to the combustion reactor via a metal oxide carrier (Me_xO_y), which is either fully or partially reduced in the combustion reaction. The chemical looping concept is very flexible, and options have been demonstrated for full combustion, reforming to produce syngas, and for hydrogen production, either by reforming, WGS and CO_2 removal, or indirectly during carrier regeneration.

Chemical looping can also be applied to capture CO_2, either in a postcombustion application or in gasification (precombustion) systems, where it can be used in combination with a chemical loop to provide oxygen to the gasifier.

4.5.1 Chemical looping combustion

Considering methane as the fuel, chemical looping combustion proceeds according to the reaction:

$$CH_{4(g)} + 4Me_xO_{y(s)} \rightarrow 4Me_xO_{y-1(s)} + CO_{2(g)} + 2H_2O_{(g)} \quad (4.1)$$

This reaction can be applied either at high pressure in a gas turbine cycle, or at atmospheric pressure in a steam cycle. After heat recovery from the offgas, steam can be condensed out, leaving a high-purity CO_2 stream for storage.

Following this fuel oxidation step, the spent carrier is circulated out of the combustion reactor into a carrier reoxidation reactor and regenerated in the reaction:

$$4Me_xO_{y-1(s)} + 2O_{2(g)} \leftrightarrow 4Me_xO_{y(s)} \quad (4.2)$$

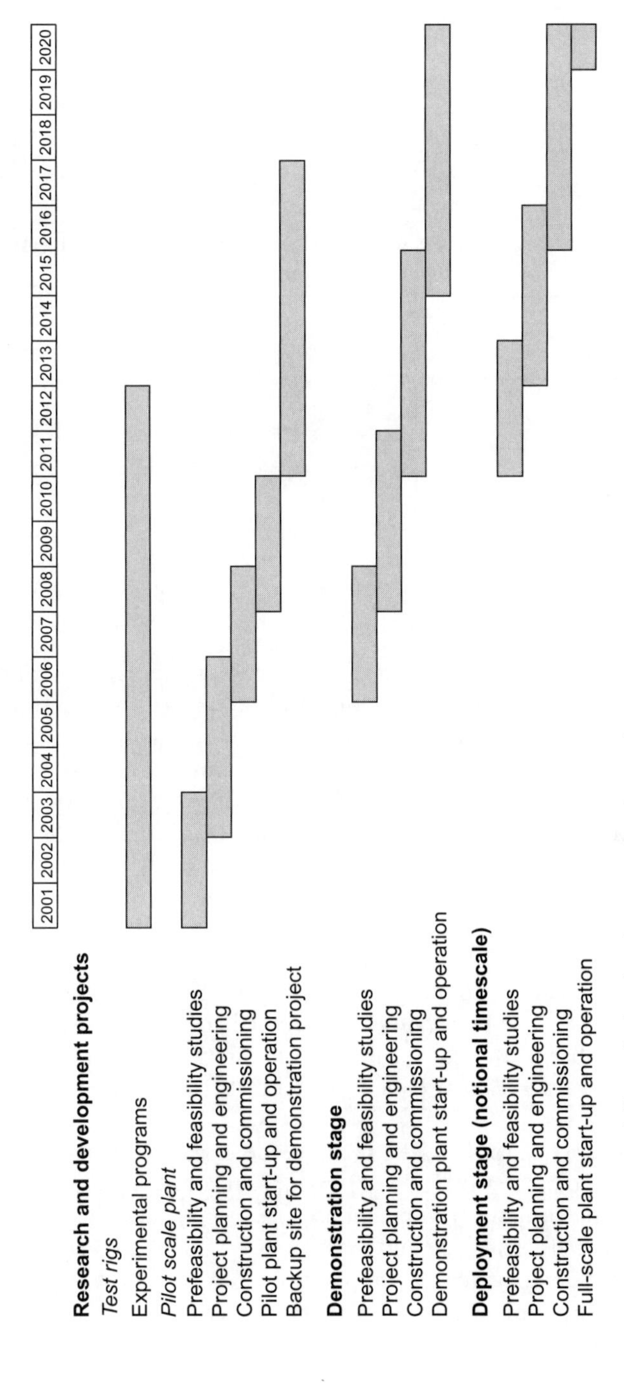

Figure 4.5 Vattenfall oxyfuel technology RDD&D timeline

Table 4.10 Oxyfuel demonstration and early deployment projects

Project; operator and partners	Project description	Planned start-up
Renfrew; Doosan Babcock	40 MW oxyfuel boiler demonstration test	2009
Shand 2, SaskPower, Canada	300 MW_e oxyfuel plant with EOR or saline aquifer storage	2012 but now on hold

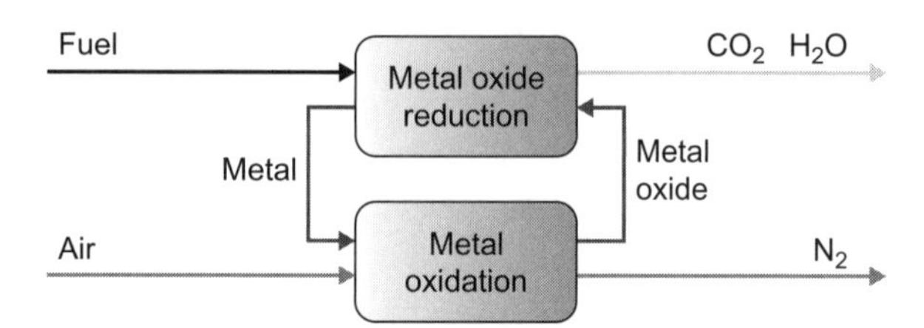

Figure 4.6 Chemical looping combustion process

This reoxidation captures oxygen directly from a supply of compressed air, eliminating the need for a separate air separation unit for oxygen supply.

The overall chemical looping process is illustrated schematically in Figure 4.6.

Oxides of the common transition metals (Fe, Cu, Ni, Mn, Ba) are possible carriers, with Fe_2O_3/Fe_2O_2, NiO/Ni, and BaO_2/BaO being extensively studied. The fuel oxidation step may be exothermic (for Cu, Mn, or Ba carriers) or endothermic (for Fe and Ni carriers), while the carrier reoxidation step is always exothermic, the total energy released in the overall process being the same as for the direct combustion of the fuel. Table 4.11 shows the reactions and energetics with NiO as the carrier.

A schematic of chemical looping combustion applied in a gas turbine cycle is shown in Figure 4.7. The fuel oxidation reactor here replaces the conventional combustion chamber of the gas turbine, while the CO_2 plus steam offgas from the reoxidation reactor can also be expanded through a turbine to generate additional power.

In development work to date, the metal oxide carrier is typically in the form of particles with a diameter of 100–500 μm, and in some cases (e.g., Ni) it is carried on an alumina support. The reactions take place in two fluidized bed reactors, which achieve efficient heat and mass transfer. The recycling of solid carrier between the two reactors is both an important control on the operating temperatures and overall heat balance between the two reactors as well as being a major technological challenge. For example, the BaO_2/BaO chemical loop requires 40 kg BaO_2 per kg CH_4, and applying this carrier to generate supercritical steam for a 500 MW_e power plant requires a carrier transport rate of 1 kt per second (3.6 Mt per hour), while also avoiding gas leakage between the reactors.

Table 4.11 Chemical looping combustion using NiO carrier

Process	Reaction	ΔH
Fuel oxidation	$CH_4 + 4NiO \rightarrow CO_2 + 2H_2O + 4Ni$	175 kJ per mol CH_4
Regeneration	$Ni + \frac{1}{2}O_2 \rightarrow NiO$	−489 kJ per mol O_2

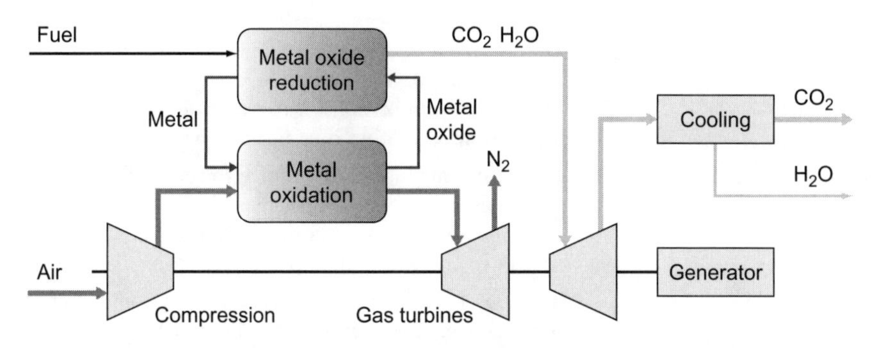

Figure 4.7 Chemical looping combustion applied in a gas turbine cycle

Chemical looping combustion is still at a very early stage of development and has been the subject of laboratory-scale experiments, at scales up to $120kW_{th}$ using a NiO/Ni carrier, as well as process simulation studies. Current areas of research are summarized in Table 4.12.

4.5.2 Chemical looping reforming

A number of chemical looping configurations are possible that can generate syngas for liquids production or hydrogen for fuel cell and other uses. If the supply of metal oxide to the fuel oxidation reaction in Equation 4.1 is limited, partial oxidation of the fuel can be achieved, yielding a $CO + H_2$ syngas:

$$CH_{4(g)} + Me_xO_{y(s)} \leftrightarrow Me_xO_{y-1(s)} + CO_{(g)} + 2H_{2(g)} \quad (4.3)$$

For example, using NiO as carrier, the reactions would be as follows:

Fuel partial oxidation

$$CH_4 + NiO \rightarrow CO + 2H_2 + Ni \quad (4.4)$$

Steam reforming

$$CH_4 + H_2O \rightarrow CO + 3H_2 \quad (4.5)$$

Table 4.12 Current RD&D areas in chemical looping combustion

Research area	Description
Performance of mixed oxide carriers	Assessment of mixed carriers, such as Fe_2O_3-CuO, Ni-Ba, and Ni-La, with the aim of maximizing reactivity and stability under cyclic operations. Adding a second metal can aid in the formation of easily reducible oxides of the primary carrier while improving carrier–support interaction in the case of supported oxides.
Carrier material handling	Development of systems for high-rate carrier transport, including interconnected high-pressure fluidized beds, and optimization for carrier rates, carrier–ash separation, and other operating parameters
Supported carrier particles	Exploiting the high-temperature stability of ceramic materials to improve carrier particles' sinter resistance, by embedding metal nanoparticles into a ceramic matrix to create nanocomposite carrier particles
Sulfur resistance of carrier and support	The presence of sulfur, mainly as H_2S in fuel streams derived from coal, can result in degradation of carrier performance due to the production of sulfides, both of the carrier metal and of the support, if present.

Regeneration

$$\mathrm{Ni} + \frac{1}{2}\mathrm{O_2} \rightarrow \mathrm{NiO} \tag{4.6}$$

4.5.3 Chemical looping hydrogen production

Several chemical looping variants have been demonstrated for pure hydrogen production. The example illustrated in Figure 4.8 shows hydrogen production in the carrier regeneration step, using $Fe/Fe_3O_4/Fe_2O_3$ as the carrier and water as the source of oxygen for regeneration.

The process reactions for this concept are as follows:

Fuel partial oxidation

$$\mathrm{C} + \frac{1}{2}\mathrm{O_2} \rightarrow \mathrm{CO} \tag{4.7}$$

Syngas oxidation

$$\mathrm{CO} + 3\mathrm{Fe_2O_3} \rightarrow \mathrm{CO_2} + 2\mathrm{Fe_3O_4} \tag{4.8}$$

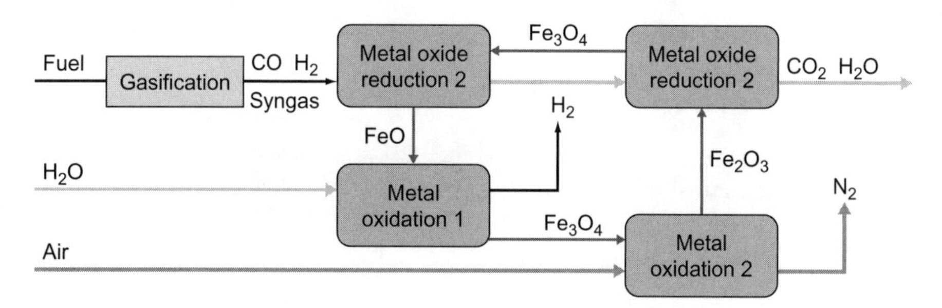

Figure 4.8 Chemical looping hydrogen production using FeO carrier

$$CO + Fe_3O_4 \rightarrow CO_2 + 3FeO \tag{4.9}$$

Carrier oxidation

$$3FeO + H_2O \rightarrow Fe_3O_4 + H_2 \tag{4.10}$$

Carrier combustion

$$4Fe_3O_4 + O_2 \rightarrow 6Fe_2O_3 \tag{4.11}$$

The first gasification stage may be applied to the full range of carbon-based fuels.

An advanced chemical looping system that uses two loops to generate H_2 and CO_2 streams in a coal gasification process has been demonstrated by Alstom Power. The chemical loops employed in the process are a $CaS/CaSO_4$ loop to provide oxygen for syngas production from partial combustion of coal:

Fuel partial oxidation

$$4C + CaSO_4 \rightarrow 4CO + CaS \tag{4.12}$$

Carrier A oxidation

$$CaS + 2O_2 \rightarrow CaSO_4 \tag{4.13}$$

and a $CaO/CaCO_3$ loop to remove CO_2 after water shifting CO:

Water-gas shift

$$CO + H_2O \leftrightarrow CO_2 + H_2 \tag{4.14}$$

Carrier B carbonation

$$CaO + CO_2 \rightarrow CaCO_3 \tag{4.15}$$

Carrier B calcination

$$CaCO_3 + \text{heat} \rightarrow CaO + CO_2 \tag{4.16}$$

This calcium looping process is shown schematically in Figure 4.9.

4.6 Capture-ready and retrofit power plant

During the period of further development and demonstration of CCS technologies, and while the framework of regulations and incentives remains uncertain, there will be continued demand for the construction of new power-generation plants, both to provide new capacity and to replace retiring units. If new-generation plants are constructed without a capability to retrofit carbon capture, the operator may eventually need to either buy carbon credits to offset emissions (so-called carbon lock-in) or shut down the plant if the cost of carbon credits were to exceed the marginal cash flow per unit carbon emission.

This risk can be mitigated by ensuring that the requirements for retrofitting carbon capture are considered in the plant design and construction, resulting in a capture-ready plant. In Europe, the package of energy measures agreed by the European Union Heads of Government at the 2007 Spring Energy Council recognized that CCS will be required on coal- and gas-fired generation plants to meet emissions reduction targets for 2020 and beyond. The European Union Commission has therefore proposed that all planning consents from 2010 onward will require that plants should be capture-ready, while CCS installation will be a requirement from 2020. The intention of the capture-ready requirement is to ensure that once they have been successfully demonstrated, capture technologies can be rapidly adopted to maximize the impact on cumulative emissions.

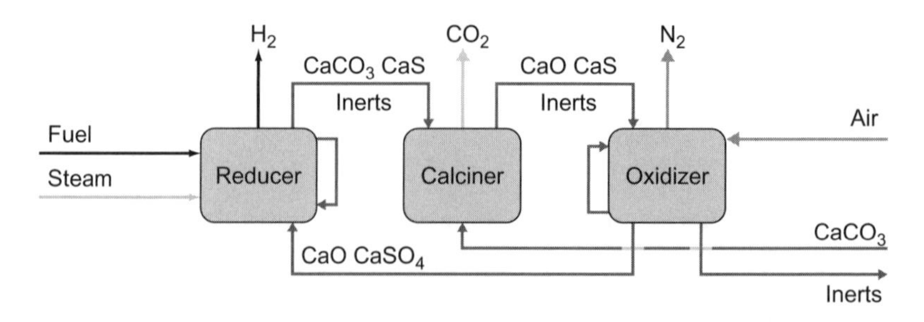

Figure 4.9 Calcium chemical looping hydrogen production process

4.6.1 Capture-ready power plants

To be considered as capture-ready, the retrofitting of capture systems should be both technically and economically feasible, although the latter requirement may be problematic to demonstrate at the planning stage in view of uncertainties in the future carbon market and cost of installation. The main technical factors to be considered in designing a capture-ready plant are summarized in Table 4.13, including specific considerations for precombustion, oxyfuel, and postcombustion capture.

As noted in the table, each of the three capture options—pre- and postcombustion or oxyfueling—has specific requirements to ensure capture readiness, and the risk associated with preinvestment for capture readiness therefore differs between the options. These considerations are summarized in Table 4.14.

In the case of oxyfueling, the risk that prohibitive additional requirements may emerge as the technology reaches full deployment can be mitigated by also considering the requirements for postcombustion capture when designing the plant for capture readiness, and carrying this option as a fallback.

Table 4.15 summarizes a number of new power projects that will be capture-ready. It is notable that, in line with the risks summarized in Table 4.14, all of these projects are proposed to be capture-ready for postcombustion capture.

4.6.2 Retrofitting capture capability

In retrofitting carbon capture to existing power plants that have not been designed as capture-ready, all of the considerations in the previous section will have a bearing on the retrofit, and the feasibility and cost of installation will be specific to the design, age, and location of each individual plant.

A key decision will be whether to accept the derating of the net power output of the plant, as a result of the energy cost of capture, or to add additional generation capacity to sustain the previous plant rating. Plant space and layout will be a limiting factor in some cases. The power derating impact of retrofitting postcombustion capture to a 500 MW_e power plant is illustrated in Table 4.16 for a range of existing and future power plant efficiencies. The analysis assumes 96% capture efficiency using an amine stripping system with an energy penalty of 15% of the gross thermal power rating of the plant.

Clearly, for less-efficient plant, the capital cost as well as the increased operating and maintenance cost of the post-retrofit plant, coupled with a ~40% reduction in net output if make-up power is not added, will result in a very substantial increase in the cost of electricity supply.

The location of an existing plant in relation to potential storage sites will also be a major factor in determining the economic feasibility of retrofitting, in view of the cost of construction and operation of the required CO_2 transportation infrastructure (see Chapter 15).

A number of retrofit capture projects are either in the feasibility-study stage or already planned to be operational in the 2010–2015 time scale, as shown in Table 4.17.

Table 4.13 Technical factors for capture-ready power plant design

Factor	Requirements
Space	Ensure space is available in the required location for additional or upgraded equipment and utilities (e.g., air separation unit for oxyfuel, water–gas shift reactors for precombustion gasification, absorption towers for postcombustion capture; installation or upgrade of FGD, CO_2 compression, possible make-up steam generation), as well as space required during construction and for maintenance access.
Plant capacity and flexibility	Consider the energy penalty for capture when sizing the plant, depending on required net capacity.
	Consider required load flexibility of both pre- and post-retrofit plant, and implications for plant design.
Utility capacity	Ensure spare utility capacity or expansion capability for post-retrofit operation (e.g., electrical, control and instrumentation systems, fire and cooling water capacities, waste treatment).
Capture process-specific factors	Oxyfuel-ready; materials and design impact of higher combustion temperature; ensure minimum air leakage into boilers; power requirement for air separation unit.
	Gasification precombustion capture-ready; integration of WGS reactor, H_2-firing capability of gas turbines.
	Postcombustion capture-ready; LP steam requirement for CO_2 stripping; impact on steam turbine operation of reduced LP steam rate.
Physical integration	Provision of tie-in points for new equipment and to existing utilities, process heat and cooling system
Heat integration	Consider both high- and low-grade heat requirements to ensure maximum efficiency of the post-retrofit plant (e.g., excess or expandable steam generation capacity).
Operation	Consider options to reduce the downtime incurred to install the retrofit and optimize plant operability post-retrofit.
CO_2 storage	Establish carbon storage options and requirements (e.g., access to new or existing CO_2 transport infrastructure, geological or other storage site).

4.7 Approaches to zero-emission power generation

In addition to the options described above, which essentially apply capture technologies to more or less conventional power generation systems, a number of alternative concepts have been proposed to achieve zero emission power (ZEP) generation using various novel components.

Table 4.14 Risks associated with capture-ready options

Capture option	Capture-ready risk	Considerations
Postcombustion	Low	Some viable technology options are already commercially deployed and requirements for these are well understood. Further developments may provide opportunities for easier retrofit at reduced costs, or for the use of new technologies.
Oxyfuel combustion	Medium	Oxyfuel combustion has reached the demonstration scale but is not yet commercially deployed, and requirements are therefore not yet fully understood.
Precombustion (IGCC plant)	Medium or high	Higher base cost of IGCC relative to conventional pulverized coal plant and major plant impact of capture readiness means that the choice of IGCC over PC is currently a major preinvestment.

Table 4.15 Capture-ready power projects under development

Project; operator and partners	Project description	Planned start-up
Karsto; Aker, Fluor, Mitsubishi, Norway	420 MW gas-fired plant, 1.2 Mt-CO_2 per year; postcombustion capture-ready	2012
Rotterdam; E.On, Netherlands	1070 MW coal-fired plant, 5.6 Mt-CO_2 per year; postcombustion capture-ready	2013
Kingsnorth; E.On, UK	300 MW coal fired plant, 3.0 Mt-CO_2 per year; postcombustion capture-ready	2014
Saline Joniche; SEI, Italy	1320 MW coal-fired plant, 3.9 Mt-CO_2 per year; postcombustion capture-ready	2015
Ferrybridge; Scottish & Southern Energy, UK	500 MW coal-fired plant, 1.7 Mt-CO_2 per year; postcombustion capture-ready	2015+
Tilbury; RWE, UK	1600 MW coal-fired plant, 9.6 Mt-CO_2 per year; postcombustion capture-ready	2016

4.7.1 AZEP concept: Norsk Hydro/Alstom

The Advanced Zero Emission Power Plant (AZEP) concept was originally proposed by Norsk Hydro in 2002 and is now under development by a consortium of companies also including Alstom Power, Siemens, ENI Tecnologie, and Borsig. The AZEP concept, illustrated in Figure 4.10, is a Brayton-cycle gas

Table 4.16 Output and efficiency impact of postcombustion capture retrofit

Plant type	Subcritical	Supercritical	Ultrasupercritical	Future
Plant parameters before postcombustion capture retrofit				
Net output (MW_e)	500	500	500	500
Efficiency	35%	40%	50%	55%
CO_2 emissions (t-C per MWh)	0.91	0.79	0.64	0.58
Postcombustion capture penalties				
Capture energy penalty (MW_e)	215	188	150	136
Net output derating	43%	38%	30%	27%
Plant parameters after postcombustion capture retrofit				
Net output (MW_e)	286	313	350	364
Efficiency	20%	25%	30%	40%
CO_2 emissions (t-C per MWh)	0.055	0.048	0.039	0.035

Table 4.17 Retrofit capture projects in feasibility or engineering

Project; operator and partners	Project description	Planned start-up
Boundary Dam; SaskPower, Canada	1 Mt-CO_2 per year, postcombustion retrofit. Vendor and technology to be decided.	2015
Powerton; Midwest Generation, USA	1.54 GW, feasibility study	
Coal Creek; Great River Energy, USA	1.1 GW, feasibility study	
Intermountain; Intermountain Power, USA	950 MW, feasibility study	
Lingan; Nova Scotia Power, Canada	320 MW, feasibility study	
Bay Shore Unit 1; FirstEnergy, USA	176 MW CFB, feasibility study	

turbine in which oxyfueled combustion of natural gas is achieved in a mixed conducting medium (MCM) membrane reactor. The underlying ion transport membrane technology is described in Chapter 8.

High efficiency is achieved using an HRSG plus steam turbine bottoming cycle, shown here on a single drive shaft with the gas turbine. Heat from the

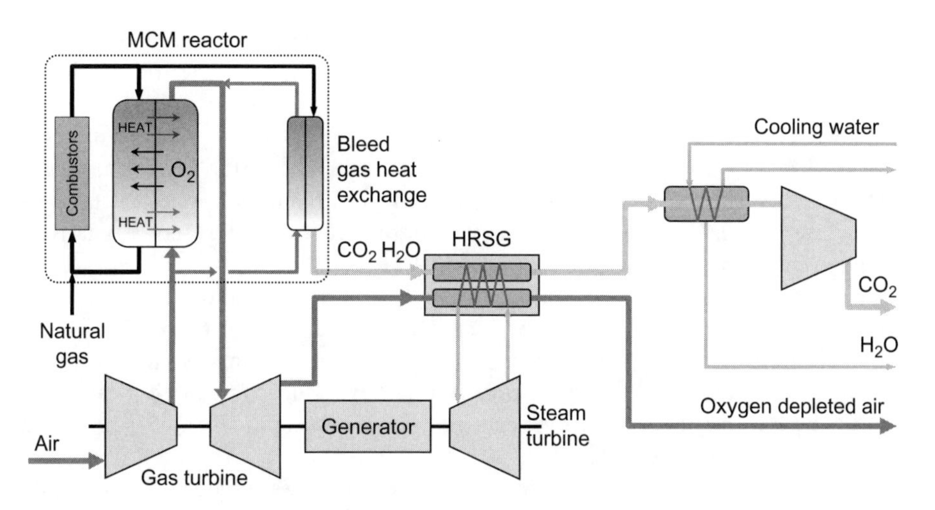

Figure 4.10 AZEP schematic flow scheme with dual-input HRSG

MCM reactor combustion product stream is also recovered, either by driving an auxiliary CO_2 and steam turbine or, as shown here, as a secondary heat input to the HRSG.

Results from fabrication and initial testing of the MCM modules have achieved targets for the project, validated model predictions, and confirmed the feasibility of the AZEP concept. The test module was constructed from extruded square-channel monoliths, achieving a contact area $>500\,m^2$ per m^3. Under AZEP process conditions the reactor is predicted to give an oxygen production rate of ~37 mol-O_2 per second per m^3 of module volume, equivalent to a gross power density of ~15 MW per net m^3 of MCM volume. Firm plans for a demonstration-scale plant have yet to be announced.

4.7.2 ZEC concept: Los Alamos National Laboratory

The zero emission coal (ZEC) concept was originally proposed by the Los Alamos National Laboratory and was further developed by the Zero Emission Coal Alliance (ZECA), later the Zeca Corporation.

The ZEC concept integrates a number of advanced technologies:

- Coal gasification and steam methane reforming for hydrogen production
- Chemical looping ($CaO \leftrightarrow CaCO_3$ cycle) for CO_2 removal from syngas
- Solid oxide fuel cells for electricity production from hydrogen

As shown in Figure 4.11, the gasification step is also novel in that it starts with hydrogasification (i.e., reduction) rather than oxidation of the carbon-based fuel using a hydrogen slip stream.

Since this reaction is exothermic ($\Delta H = -74.9$ kJ per mol), no external heat input is required for the gasification step, and injection of water can be used to control the gasifier reaction temperature. In the carbonation reactor, steam reforming of methane and carbonation of lime result in two reaction products: hydrogen and calcium carbonate.

Hydrogen is oxidized in the solid oxide fuel cell, at an operating temperature of 1050°C, to produce electricity and steam. The latter feeds the steam reforming reaction, while additional heat recovered from the fuel cell is used to regenerate the CaO absorbent in the calcination reactor. The major reactions involved in the process are shown in Table 4.18.

Laboratory-scale investigations of the hydrogasification and steam methane reforming reactions have been conducted under conditions representative of the ZEC concept, and have achieved moderate to high conversion of bituminous coal to hydrogen. The final stage of the ZEC concept is the sequestration of CO_2 by a mineral carbonation reaction with magnesium silicates (serpentine or olivine), as discussed in Chapter 10.

The theoretical maximum efficiency of this process exceeds 90%, while accounting for heat losses and non-ideal fuel cell performance, an efficiency in excess of 70% was considered achievable in practice. Compared to other

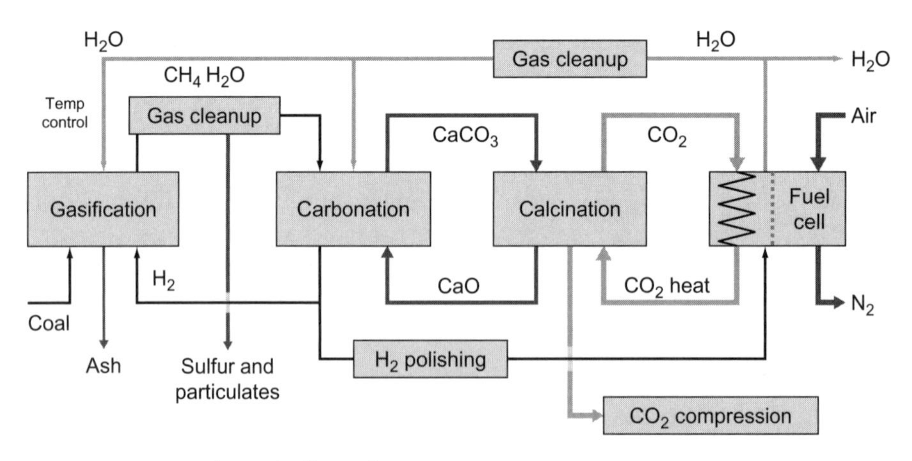

Figure 4.11 ZEC schematic flow scheme

Table 4.18 Major reactions in the ZEC process

Process	Reaction
Hydrogasification	$C + 2H_2 \rightarrow CH_4$
Steam methane reforming	$CH_4 + 2H_2O \rightarrow CO_2 + 4H_2$
Carrier carbonation	$CaO + CO_2 \rightarrow CaCO_3$
Fuel oxidation	$2H_2 + O_2 \rightarrow 2H_2O$
Carrier calcination	$CaCO_3 \rightarrow CaO + CO_2$

generation processes, this extremely high efficiency would result in a substantially reduced quantity, and therefore cost, of CO_2 to be disposed of per MWh generated.

Unfortunately, despite its apparent promise, Zeca Corp. seems to have disappeared without trace shortly after it was recognized by *Scientific American* as the "Business Leader in Environmental Science" for 2003. www.zeca.org is currently for sale!

4.7.3 FutureGen concept: FutureGen Alliance

The FutureGen project was launched by the U.S. Department of Energy in 2003 with the aim of constructing a demonstration-scale 275 MW_e IGCC power plant, including carbon capture and geological storage. The project was designed to support research, development, and demonstration of precombustion capture, use of hydrogen as a fuel in both gas-turbine and fuel-cell generation systems, as well as geological storage. The overall project concept is shown in Figure 4.12.

The proposed design of the FutureGen facility would enable testing and development of new technologies fully integrated with the demonstration-scale plant. Specific RD&D aims of the facility summarize the current (2010) research needs for IGCC with precombustion capture, and are shown in Table 4.19.

A site for the proposed plant in Mattoon, Illinois, USA was announced in December 2007 (the site-selection process employed by FutureGen is described in Section 11.4.3), and the project was initially planned to follow the timeline shown in Table 4.20.

U.S. Department of Energy funding was redirected in January 2008 away from the FutureGen project toward near-term CCS deployment at IGCC or other clean coal-powered plants. FutureGen and the people of Mattoon are continuing to lobby for reactivation of the project under the renewed clean-energy focus of the Obama presidency.

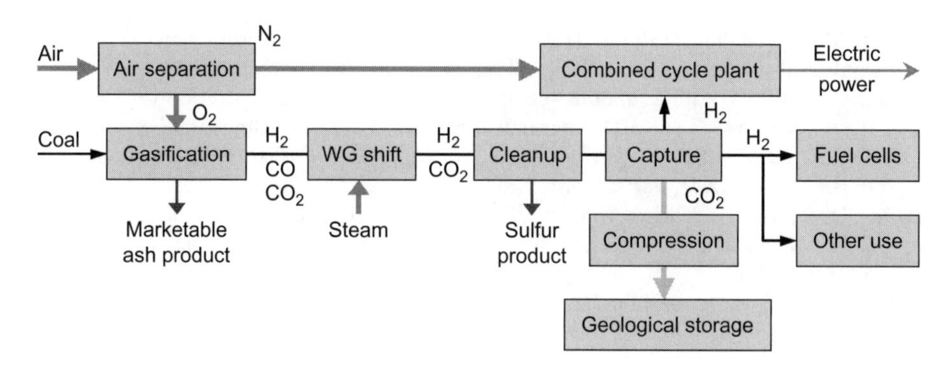

Figure 4.12 FutureGen CCS technology integration concept

Table 4.19 FutureGen technology development activities

System	Envisaged RD&D activities
Oxygen supply	Ion transport membrane production and operation (selectivity, durability, and operating flexibility)
Coal feed and gasifier	Evaluation of coal feed and gasifier performance and improvements, including coal type flexibility
Hydrogen treatment and separation	Testing ceramic, metal, and polymer membranes for hydrogen separation; evaluation of syngas clean-up for mercury and H_2S removal and testing advanced solvents and sorbents for CO_2 removal
Hydrogen-fueled gas turbine	Evaluate advanced hydrogen burners, including operability and impact on turbine performance
Advanced power options	Demonstrate test-bed scale integration of solid oxide fuel cell with hydrogen turbine cycle and carbon capture

Table 4.20 Planned FutureGen project schedule (at January 2008)

Period	Project Activity
2009–2012	Construction, commissioning, and start-up
2012–2015	Ramp-up and period of CO_2 storage
2015–2017	Postinjection site monitoring

4.8 References and Resources

4.8.1 Key References

The following key references, arranged in chronological order, provide a starting point for further study.

Ishida, M. and H. Jin. (1997). CO_2 recovery in a power plant with chemical looping combustion, *Energy Conversion and Management*, **38** (Supplement) (S187–S192).

IEA GHG. (2000). Leading options for the capture of CO_2 emissions at power stations, IEA Greenhouse Gas R&D Programme, Report PH3/14, Cheltenham, UK.

OECD/IEA. (2002). *Solutions for the 21st Century, Zero Emissions Technologies for Fossil Fuels*, OECD, Paris, France.

Bailey, D. W. and P. H. M. Feron. (2005). Post-combustion decarbonisation processes, *Revue de l'Institut Française du Pétrole, Oil & Gas Science and Technology*, **60** (461–474).

Buhre, B. J. P., *et al.* (2007). Oxy-fuel combustion technology for coal-fired power generation, *Progress in Energy and Combustion Science*, **31** (283–307).

FutureGen Alliance. (2007). *FutureGen Initial Concept Design Report*, Washington D.C. Available at www.futuregenalliance.org/publications.stm

4.8.2 Institutions and Organizations

The institutions that have been most active in the research, development, and commercialization of systems for carbon capture from power generation are listed below, with relevant web links.

Alstom Power (chemical looping technology): www.power.alstom.com

CASTOR (European Union-funded project for postcombustion capture development and field trial): www.co2castor.com

Chalmers University of Technology, Department of Energy and Environment (chemical looping combustion and reforming): www.chalmers.se/ee/EN/research/research-divisions/energy-technology/research/co2-capture

ENCAP (precombustion capture technology development project): www.encapco2.org

European Technology Platform for Zero Emission Fossil Power Plant (ETP ZEP): www.zero-emissionplatform.eu

FutureGen (U.S. public–private partnership aiming to build and operate the world's first coal-fueled, near-zero emissions power plant): www.futuregenalliance.org

HTC Purenergy (modular postcombustion capture system with capacity of 1000 t-CO_2 per day): www.htcenergy.com

IEA GHG Oxyfuel Combustion Network: www.co2captureandstorage.info/networks/oxyfuel.htm

Ohio State University Research Foundation (chemical looping technology): http://tlc.osu.edu/technologies/detail.cfm?TechID=231&CatID=7

Powerspan (ammonia-based postcombustion capture system development and demonstration): www.powerspan.com

Technische Universität Bergakademie Frieberg (COORIVA project coordination): http://tu-freiberg.de/fakult4/iec/index.en.html?int_fav=en

Vienna University of Technology (oxyfuel and chemical looping): http://www.oxyfuel.atwww.oxyfuel.at

5 Carbon capture from industrial processes

Although fossil-fueled power-generation plant account for the majority of large stationary sources emitting >0.1 Mt-CO_2 per year, such large sources are also a feature of several other industries and are therefore targets for the application of carbon capture technologies. The most important of these—cement production, iron and steel production, oil refining, and natural gas processing—are dealt with in this chapter.

5.1 Cement production

In 2008, worldwide cement production in excess of 2.5 Gt resulted in an estimated emission of ~2 Gt-CO_2 (at a global average CO_2 intensity of 0.8 t-CO_2 per t-cement), some 7% of total anthropogenic CO_2 emissions. These emissions are split roughly equally between CO_2 emitted from the calcination process (~52%) and emissions from the combustion of fuel to fire cement kilns (~48%).

Portland cement is a mixture of predominantly di- and tricalcium silicates ($2CaO \cdot SiO_2$, $3CaO \cdot SiO_2$), with smaller amounts of other compounds such as calcium sulfate ($CaSO_4$), magnesium, aluminum and iron oxides, and tricalcium aluminate ($3CaO \cdot Al_2O_3$). The first stage in the cement production process (Figure 5.1) is the sintering of a controlled mixture of raw materials in a kiln, typically a horizontal rotary kiln, at a temperature of ~1450°C. Calcium carbonate ($CaCO_3$) is the primary raw material, and may be in the form of crushed limestone, shells, or chalk. A variety of secondary raw materials are used as the source of silica and other minerals, including sand, shale, clay, blast furnace slag, and coal ash. The latter may be introduced directly by firing the cement kiln with pulverized coal or by importing fly or bottom ash from a coal-fired power plant.

The main reaction taking place in the cement kiln is the conversion or calcining of calcium carbonate to calcium oxide, a highly endothermic reaction requiring 3.5–6.0 GJ of energy per tonne of cement produced, depending on plant efficiency. Calcination and the other main chemical reactions proceed as follows:

$$CaCO_3 + \text{heat} \rightarrow CaO + CO_2 \tag{5.1}$$

Doi: 10.1016/B978-1-85617-636-1.00005-5

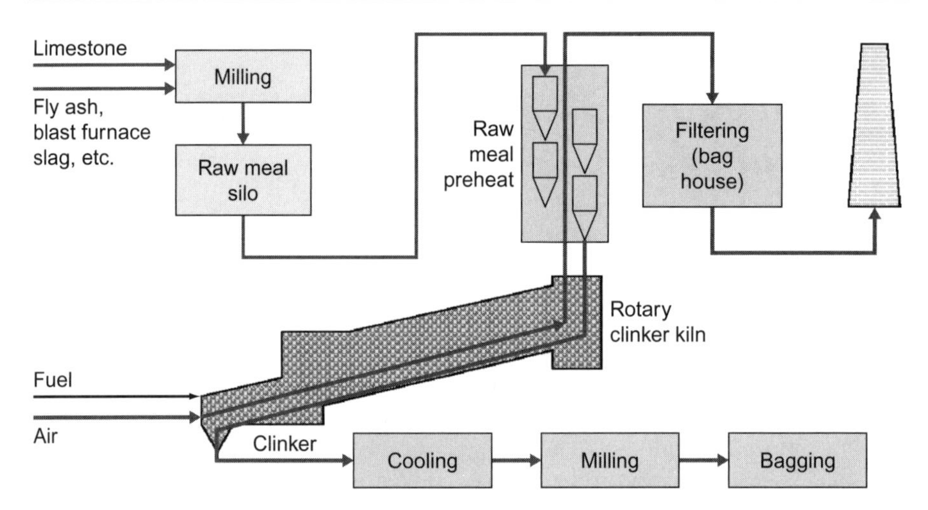

Figure 5.1 Cement production process

$$2CaO + SiO_2 \rightarrow 2CaO \bullet SiO_2 \quad \text{(Bellite)} \tag{5.2}$$

$$3CaO + Al_2O_3 \rightarrow 3CaO \bullet Al_2O_3 \tag{5.3}$$

$$4CaO + Al_2O_3 + Fe_2O_3 \rightarrow 4CaO \bullet Al_2O_3 \bullet Fe_2O_3 \tag{5.4}$$

$$CaO + 2CaO \bullet SiO_2 \rightarrow 3CaO \bullet SiO_2 \quad \text{(Alite)} \tag{5.5}$$

The CO_2 released in the initial calcining reaction in Equation 5.1, known as the process CO_2, combined with the CO_2 from combustion of the kiln fuel (coal, oil, or natural gas) yields a flue gas with a [CO_2] of typically 14–33%.

5.1.1 "Postcombustion" capture from cement plant

Capture of CO_2 from this process can be achieved using the technologies applicable to postcombustion capture from power-generation plants, including chemical and physical solvents (Section 6.2); sorbents, particularly high-temperature sorbents (Section 7.1); and membranes (Section 8.7). Compared to "conventional" postcombustion applications, the higher [CO_2] enables these technologies to achieve higher separation efficiencies in a cement plant application.

Table 5.1 Research and development issues for oxyfuel cement production

R&D	Description
[CO_2] impact on reaction kinetics	Assessing the impact of high [CO_2] on calcination and other clinker production reactions
Burner technology	Development of oxygen-drafted burners for clinker kilns, paralleling work on burners for oxyfuel steam boilers
Plant and process simulation	Process modeling, addressing the impact of oxyfuel on flue gas enthalpy and recirculation, heat transfer, energy and material flows

5.1.2 Oxygen enrichment and oxyfuel processes

The addition of oxygen to the air drafting the burners—so-called oxygen enrichment—has been practiced in cement kilns since the early 1960s and results in increased clinker production and reduced fuel consumption. Fuel consumption is reduced in part because nitrogen in the air does not need to be heated up to combustion temperature. Oxygen enrichment increases flame temperatures (to ~3500°C), which has an impact on kiln refractory design and materials, and also increases [CO_2] in the kiln flue gas stream, which, as noted previously, further aids postcombustion capture.

A full oxyfuel approach to cement production has also been proposed in which the kiln burners are drafted with oxygen rather than air, resulting in an offgas stream that would be 80% CO_2 and could be directly compressed for transportation and storage after steam condensation and possibly flue gas desulfurization (FGD). Technologies required for oxyfuel cement production are currently at the R&D stage, with projects addressing the key issues summarized in Table 5.1.

5.1.3 Cement production from carbon capture processes

Alongside carbon capture from conventional cement production, a number of venture capital companies are developing proprietary postcombustion capture processes that aim to produce cement or cement additives by the precipitation of calcium and magnesium carbonates from seawater reacted with power plant flue gas CO_2. This process is described in Chapter 14.

5.2 Steel production

Integrated coal-fueled steel mills accounted for ~60% of the 1.4 Gt of global steel produced in 2008, consuming 15–18 GJ per t-steel produced and emitting a global average of ~1.4 t-CO_2 per t-steel produced. In the first stage of the steel-making process, high-grade coal (anthracite) is used to fuel a blast furnace in which iron is extracted by reduction from the ore hematite (Fe_2O_3), using carbon monoxide as the reducing agent. The key reactions taking place in the furnace are as follows. Coal is combusted with oxygen to produce carbon dioxide and heat, while limestone, introduced as a fluxing agent to remove impurities from the iron, is calcined to produce CaO plus CO_2 via the same reaction as in cement production (Equation 5.1). The CO_2 product from these two reactions then reacts with more carbon, producing carbon monoxide:

$$CO_2 + C \rightarrow 2CO \tag{5.6}$$

Carbon monoxide reduces the hematite ore to produce molten iron (pig iron), which is collected at the base of the furnace:

$$Fe_2O_3 + 3CO \rightarrow 2Fe + 3CO_2 \tag{5.7}$$

The calcined limestone combines with impurities to form slag, primarily calcium silicate, which floats on top of the molten iron and can be removed:

$$CaO + SiO_2 \rightarrow CaSiO_3 \tag{5.8}$$

In the second stage of the steelmaking process, known as basic oxygen steelmaking (BOS), the carbon content of pig iron is reduced from a typical 4–5% to 0.1–1% in an oxygen-fired furnace. The excess carbon is oxidized to carbon monoxide, which can be recycled as a fuel gas or used as the reducing agent. At the same time other impurities, such as phosphorus and sulfur, are oxidized to form acidic oxides, neutralized by the addition of lime, and recovered as a slag that has a variety of recycling uses (see, for example, cement production, earlier, and mineral carbonation, in Section 10.3.1). Alloying elements such as chromium, manganese, nickel, and vanadium are also added at this stage to achieve the required steel composition and properties.

Blast furnace gases contain close to 30% CO_2, after full combustion of the CO fraction, while the overall flue gas stream from an integrated steel mill is ~15% CO_2. The same CO_2 capture options introduced above for power-generation plant can therefore also be applied to a steel mill:

- CO_2 capture from the overall flue gas stream
- Firing the blast furnace with oxygen rather than air, yielding a furnace offgas that is a pure CO and CO_2 mixture
- Capturing CO_2 in a precombustion step and using hydrogen instead of carbon monoxide as the reducing agent in Equation 5.7; that is:

$$Fe_2O_3 + 3H_2 \rightarrow 2Fe + 3H_2O \tag{5.9}$$

Recycling of scrap steel, using electric-powered arc or induction furnaces, accounts for ~35% of overall shipped steel volume worldwide, with production of 500 Mt-steel reported from this type of process in 2008. Recycled steel is significantly more energy-efficient than new-steel production, requiring only ~25% of the energy input per unit of steel shipped. A mini-mill typically consumes 4.0–6.5 GJ per tonne of steel produced, reducing CO_2 emissions by ~80% to ~0.3 t-CO_2 per t-steel. In this case the reduction of related CO_2 emissions reverts to the discussion of CCS in the power-generation sector.

5.3 Oil refining

Worldwide, CO_2 emissions from oil refineries account for ~5% of global anthropogenic emissions, and amounted to ~850 Mt-CO_2 emitted to the atmosphere in 2008. As noted in Table 2.3, the IPCC analysis of large point sources identified 638 refineries emitting an average of 1.25 Mt-CO_2 per year.

The crude oil feed to an oil refinery is a mixture of many hydrocarbon components from methane, the lightest with a molecular weight of 16, out to long-chain molecules with molecular weights in the hundreds. The refining process, shown schematically in Figure 5.2, starts by separating out up to 10 "fractions" of this mixture by a distillation process under atmospheric pressure. Crude oil is heated to 500–700°C and fed to the base of a distillation tower. As the vapor rises and cools, first the heavier and then progressively lighter components condense and are recovered as liquid fractions, with gases recovered

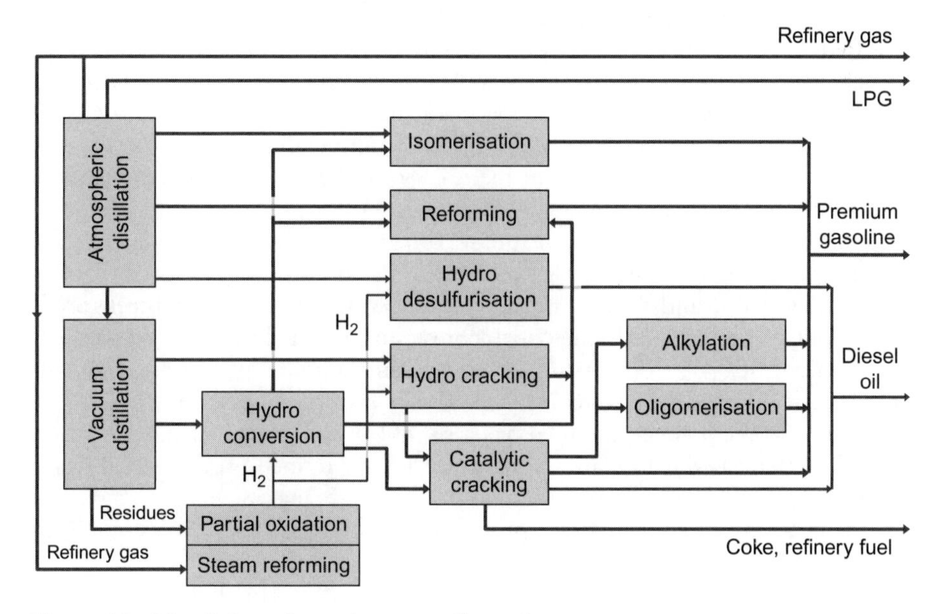

Figure 5.2 Oil refining schematic process flow scheme

from the top of the tower. The heavy residues recovered at the base of this initial distillation still contain significant lighter components, which are recovered in a further distillation under vacuum.

The second stage in the process, known as conversion, is the breaking down of larger molecules in the heavy fractions to meet the demand for lighter and higher-value products. This "cracking" process requires the presence of either catalysts (catalytic or cat-cracking), commonly zeolite, aluminum hydrosilicate, and bauxite; steam (steam cracking); or hydrogen (hydrocracking); and temperatures ranging from 400°C (catalytic) up to 850°C (steam). Further distillation is used to separate the products resulting from the cracking process.

Other important steps in the conversion process are:

- Catalytic reforming: a platinum or platinum–rhenium catalyst is used to promote the conversion of distillates in the 100- to 150-molecular weight range (light naphtha) into heavier aromatics for use in gasoline blending and petrochemicals. Hydrogen is a by-product of this reaction and is commonly used for hydrocracking.
- Alkylation: a catalyst such as hydrofluoric acid or sulfuric acid is used to convert low-molecular-weight compounds, such as propylene (C_3H_6) and butylene (C_4H_8), into high-octane hydrocarbons used in gasoline blending. Sulfuric acid used in alkylation may be a by-product from desulfurization (see following).

In upgrading, the final step in the refining process, undesirable compounds are removed and product characteristics are adjusted to comply with delivery specifications. Hydrodesulfurization (HDS), or hydrotreating, is an important upgrading step required to meet stringent environmental standards, for example in producing low-sulfur diesel to reduce SO_2 emissions. HDS is achieved by contacting the unfinished products with hydrogen at 370°C and a pressure of 6.0 MPa in the presence of a catalyst such as nickel molybdate ($NiMoO_4$). Sulfur atoms in the hydrocarbons bond with hydrogen to produce hydrogen sulfide (H_2S) and are then recovered as elemental sulfur or sulfuric acid.

An oil refinery is fueled by burning top-gas from the distillation process, supplemented as needed by additional fuel oil. Some 50% of consumed energy is used to generate process heat while the remaining 50% is used for power generation, for hydrogen production for hydrotreatment and hydrocracking, and for plant utilities. Refinery energy consumption and attendant CO_2 emissions are highly variable and depend strongly on the complexity of the refining processes employed, particularly the "deep conversion" capability required to process heavier crude oils. Typical self-consumption ranges from 6–8% by weight of the crude oil processed for conventional conversion processes to 11–13% by weight for deep conversion, which has a significantly higher hydrogen requirement. The trend toward greater demand for lighter refined products will result in upward pressure on self-consumption in the future, making energy efficiency, process integration, and carbon capture important if growth in emissions from this sector is to be avoided.

Options for capturing CO_2 in the refining processes include the integration of power generation and hydrogen production in an IGCC plant, which achieves precombustion capture of CO_2. Emissions from process heating can be captured either by oxyfueling or by postcombustion capture from the heater flue gases, or by also integrating process heat production into an IGCC-CHP plant.

5.4 Natural gas processing

Natural gas, produced either from gas fields or as associated gas with oil production, contains varying amounts of nonhydrocarbon gases, of which CO_2, N_2, and H_2S are the most common examples. Many large natural gas fields have CO_2 concentrations of up to 20%, while some fields with >50% CO_2 are also produced. Indonesia's Natuna field is an extreme example for commercial natural gas production and contains 71% CO_2.

The volume of CO_2 currently produced with natural gas has been estimated at 50 Mt-CO_2 per year and is therefore modest compared to global emissions. However, the removal of acid gases such as CO_2 and H_2S from natural gas—a process known as gas sweetening—is important because the technologies developed in this area have broader application for CO_2 capture in areas such as power generation. Natural gas consumption is also projected to grow more rapidly than other fossil fuels, in part due to its low carbon intensity as a power-plant fuel. Exploitation of more difficult gas reserves, often with higher CO_2 content, will also require capture and storage if increasing emissions from this sector are to be avoided.

CO_2 removal from natural gas is now a mature technology that has been applied on an industrial scale since the 1980s. The end-use specification, whether for pipeline transport into a supply grid or for liquefaction and onward transport as LNG, requires the CO_2 concentration to be reduced to 2–3 mol%. Various technologies have been applied, including physical and chemical solvents, membranes, and cryogenic separation, with the preferred technology in any specific application depending on the feedstock and required output compositions. The details of specific technologies and applications in natural gas processing are covered in Chapters 6–9.

5.5 References and Resources

5.5.1 Key References

The following key references, arranged in chronological order, provide a starting point for further study.

IEA GHG. (1999). *The Reduction of Greenhouse Gas Emissions from the Cement Industry*, Report PH3/7, IEA Greenhouse Gas R&D Programme, Cheltenham, UK.

Decroocq, D. (2003). Energy conservation and CO_2 emissions in the processing and use of oil and gas, *Revue de l'Institut Français du Pétrole, Oil & Gas Science and Technology*, **58** (331–342).

Gielen, D. J. (2003). CO_2 removal in the iron and steel industry, *Energy Conversion and Management*, **44** (1027–1037).

Wilkinson, M. B., *et al.* (2003). CO_2 capture from oil refinery process heaters through oxyfuel combustion, *Proceedings of the Sixth International Conference on Greenhouse Gas Control Technologies*, J. Gale and Y. Kaya (eds.), Elsevier, Oxford, UK.

Hoenig, V., H. Hoppe, and B. Emberger, (2007). *Carbon Capture Technology—Options and Potentials for the Cement Industry*, European Cement Research Academy, Report TR 044/2007. Available at www.ecra-online.org/downloads/Rep_CCS.pdf.

5.5.2 Institutions and Organizations

The institutions that have been active in the research, development and commercialization of systems for carbon capture from industrial processes are listed below, with relevant web links.

Technische Universität Hamburg-Harburg (oxyfuel cement production RD&D): www.tu-harburg.de/alt/iet/research.html

World Steel Association; CO_2 Breakthrough program (global steel industry approach to CO_2 emissions reduction): www.worldsteel.org/index.php?action = storypages& id = 226

StatoilHydro (capture from CHP plant and cracker unit at the Mongstad refinery): www.statoilhydro.com/en/NewsAndMedia/News/2009/Pages/11FebMongstad.aspx

6 Absorption capture systems

Processes for removing CO_2 from a gas stream, based on chemical or physical absorption, have been applied on an industrial scale for more than 50 years, driven by the requirement to treat natural gas to a sales gas specification with low content of acid gases, such as CO_2 and H_2S, or by the need to condition syngas from coal gasification as a feed for other chemical processes such as Fischer–Tropsch liquids production.

These processes have typically been applied to gas streams with a relatively high acid gas partial pressure and have provided an extensive body of experience from which to address the more challenging problem of CO_2 recovery at lower partial pressure, for example from combustion flue gases at typically 10–20 kPa CO_2 partial pressure or, in the extreme, the direct capture of CO_2 from air at a partial pressure of 38 Pa.

6.1 Chemical and physical fundamentals

Processes to capture CO_2 based on absorption are distinguished depending on whether the solvent either reacts chemically with the sorbate (CO_2) to form chemical compounds from which the CO_2 is subsequently recovered, or is chemically inert and absorbs the sorbate without a chemical reaction.

These distinct processes are termed chemical and physical absorption respectively.

6.1.1 Chemical absorption

Chemical absorption for CO_2 capture is based on the exothermic reaction of a sorbent with the CO_2 present in a gas stream, preferably at low temperature. The reaction is then reversed in a so-called stripping, or regeneration, process at higher temperature. Chemical absorption is particularly suitable for CO_2 capture at low partial pressure, with amine or carbonate solutions being the predominant solvents.

Doi:10.1016/B978-1-85617-636-1.00006-7.

Amine-based absorption

Amines are organic compounds derived from ammonia (NH_3) in which one or more of the hydrogen atoms are replaced by organic components or substituents. Depending on the number of substituents these compounds are termed primary, secondary, or tertiary amines. The simplest primary amine is methylamine (CH_3NH_2 or CH_5N), with one hydrogen replaced by a methane (CH_3) group. To simplify the nomenclature, the organic group is commonly replaced by R, giving R^1-NH_2, R^1R^2-NH, and $R^1R^2R^3$-N for primary, secondary, and tertiary amines respectively.

The most commonly used amine for CO_2 capture is ethanolamine (or monoethanolamine [MEA]), a primary amine with R = CH_2CH_2OH. In an aqueous solution, MEA acts as a weak base, which can neutralize an acidic molecule such as CO_2. In this reaction, a weakly bonded compound called a carbamate is formed (Figure 6.1):

$$2\text{R-NH}_2 + CO_2 \rightarrow \text{R-NH}_3^+ + \text{R-NHCOO}^- + \text{heat} \tag{6.1}$$

the heat of absorption for CO_2 in MEA being 2.0 MJ per kg-CO_2.

Two other reactions also occur that result in CO_2 dissolution, base-catalyzed hydration of CO_2:

$$\text{R-NH}_2 + CO_2 + H_2O \rightarrow \text{R-NH}_3^+ + HCO_3^- + \text{heat} \tag{6.2}$$

and the formation of carbonic acid:

$$CO_2 + H_2O \rightarrow H_2CO_3 \tag{6.3}$$

However, compared to the rate of the reaction in Equation 6.1 for MEA, the other two reactions make a minimal contribution to the overall CO_2 absorption rate.

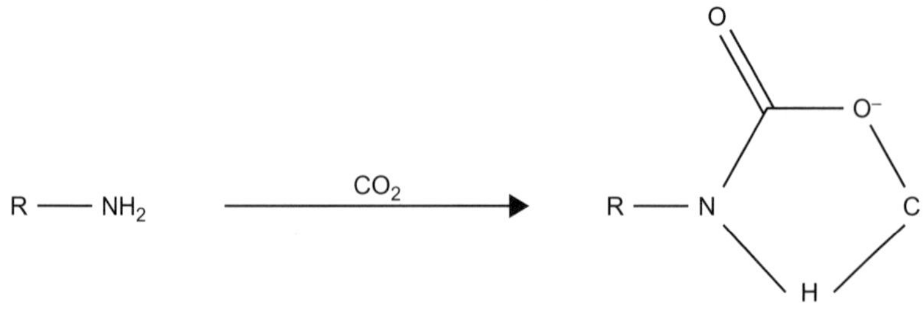

Figure 6.1 Carbamate ion structure

For a secondary amine, the equivalent carbamation reaction would be:

$$2R^1R^2\text{-}NH + CO_2 \rightarrow R^1R^2\text{-}NH_2^+ + R^1R^2\text{-}NCOO^- + \text{heat} \quad (6.4)$$

While for tertiary amines, the reaction is the base-catalyzed hydration of CO_2:

$$R^1R^2R^3\text{-}N + CO_2 + H_2O \rightarrow R^1R^2R^3\text{-}NH^+ + HCO_3^- \quad (6.5)$$

The carbamate ion nitrogen–carbon bond is easily broken down by the application of heat, leading to the reverse reaction in which the original solvent is regenerated. For example, for MEA regeneration:

$$R\text{-}NH_3^+ + R\text{-}NHCOO^- + \text{heat} \rightarrow 2R\text{-}NH_2 + CO_2 \quad (6.6)$$

Regeneration of the products of the reaction in Equation 6.2 similarly proceeds by the reverse reaction:

$$HCO_3^- + R\text{-}NH_3^+ + \text{less heat} \rightarrow R\text{-}NH_2 + CO_2 + H_2O \quad (6.7)$$

The application of these reactions to the removal of CO_2 from flue gases is described in Section 6.2.1.

The maximum CO_2 loading of the MEA solvent that can be theoretically achieved in the reaction in Equation 6.1 is 0.5 (i.e., one mole CO_2 in two moles of solvent), while the reaction in Equation 6.2 could achieve twice this loading (one mole of CO_2 per mole of solvent). This higher loading can be achieved in practice by steric hindrance of the carbamation reaction—that is, the use of a larger amine molecule in which the carbamation reaction is hindered because of the physical structure of the amine molecule. Although the energy required for regeneration is lower for these hindered amines, this is compensated by the slower overall reaction rates. Sterically hindered amine solvents are used in the KEPCO/MHI process summarized in Table 6.4 below.

When considering the three key requirements for effective solvents—fast reaction rate, low regeneration energy, and high loading capacity—the choice between primary, secondary, and tertiary amines is not straightforward. In general, the relative merits are as follows. For reaction rate, the most preferred would be primary, followed by secondary and tertiary. For regeneration energy and loading capacity, the most preferred would be tertiary, followed by secondary and primary.

Mixing primary, secondary, and tertiary amines in blended amine solvents is one approach to combine the advantages of the different amine types—for example, increasing the reaction rate of a tertiary amine such as methyldiethanolamine (MDEA) by adding a secondary amine activator to produce activated MDEA (aMDEA).

Amines are subject to degradation in the presence of oxygen, with primary amines being more susceptible than secondary or tertiary amines. Oxidative deamination results in the formation of a range of reaction products, including formic acid (HCOOH or CH_2O_2) and ammonia. SO_2 and NO_2 also react with amines to form heat-stable salts (sulfates and nitrates), which lead to a loss of solvent absorption capacity and additional requirements for solvent make-up and waste stream disposal.

Carbonate-based absorption

The dissolution of carbonate rocks by carbonic acid—formed from CO_2 and rainwater—is a process that removes CO_2 from the atmosphere as part of the geochemical carbon cycle (Chapter 1), and proceeds according to the reactions:

$$H_2O + CO_2 \rightarrow H_2CO_3 \tag{6.8}$$

$$CaCO_3 + H_2CO_3 \rightarrow Ca^{2+} + 2HCO_3^- \tag{6.9}$$

Acceleration of this natural process has been proposed as a method of removing CO_2 from power plant or cement plant flue gases, using a reactor in which the gas is flowed through a bed of crushed limestone, wetted by a continuous flow of water. Rather than regenerating the limestone sorbent, the concept is that the bicarbonate-rich effluent stream would be disposed of into the ocean. In view of the ocean storage aspect, this accelerated weathering of limestone (AWL) process is described further in Section 12.4.

Regenerable aqueous carbonate solutions have also been proposed for CO_2 capture by chemical absorption. A potassium carbonate (K_2CO_3) absorption system is based on the following two-stage reaction:

$$K_2CO_3 + H_2O \rightarrow KOH + KHCO_3 \tag{6.10}$$

$$KOH + CO_2 \rightarrow KHCO_3 \tag{6.11}$$

These reactions have the advantage of lower desorption energy requirements compared to amine-based systems (0.9–1.6 MJ per kg-CO_2 vs. 2.0 MJ per kg-CO_2 for MEA) balanced by the disadvantage of lower rates of reaction at the low CO_2 pressures typical for flue gases. However, the rate of absorption and solvent absorption capacity can be increased by adding piperazine to the reaction. Piperazine is a cyclohexane ring with two opposing carbon atoms replaced by amine functional groups (Figure 6.2).

The piperazine (PZ) enhanced reaction scheme can be simplified as follows, following the reactions in Equations 6.10 and 6.11:

$$KHCO_3 \rightarrow K^+ + HCO_3^- \tag{6.12}$$

$$PZ + HCO_3^- \leftrightarrow PZCOO^- + H_2O \tag{6.13}$$

$$PZCOO^- + HCO_3^- \leftrightarrow PZ(COO^-)_2 + H_2O \tag{6.14}$$

The carbamate reactions (Equations 6.13 and 6.14) dominate the CO_2 absorption process and, as for the straight K_2CO_3 system, require significantly lower heat input than MEA for the regeneration reaction.

Aqueous ammonia-based absorption

The reaction of ammonia and its derivatives with CO_2 also has the advantage of lower heat of reaction than the equivalent amine-based reactions, opening up the possibility of significant energy-efficiency improvement and cost reduction compared to an amine-based baseline system.

An aqueous solution of ammonium carbonate ($(NH_4)_2CO_3$) reacts with water to form ammonium bicarbonate (called the AC/ABC reaction):

$$(NH_4)_2CO_3 + CO_2 + H_2O \leftrightarrow 2NH_4HCO_3 \tag{6.15}$$

Ammonia-based scrubbing is also applied to flue gas SO_2 and NO_2 removal, as described in Chapter 3. The use of ammonia for CO_2 capture potentially

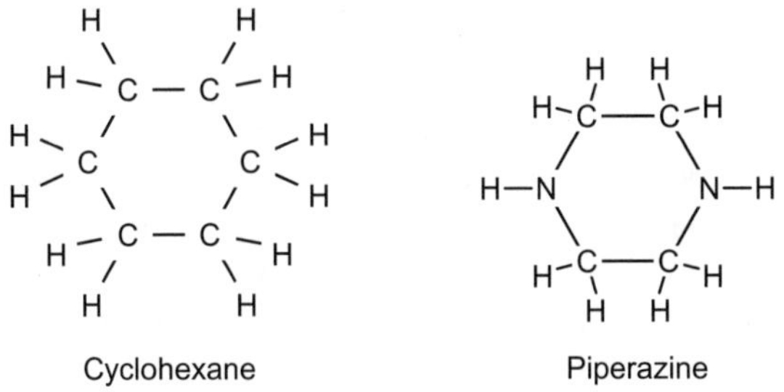

Cyclohexane

Piperazine

Figure 6.2 Cyclohexane (C_6H_{12}) and piperazine ($C_4H_{10}N_2$)

allows a single solvent system to be used for multicomponent flue gas emissions control.

Sodium hydroxide-based absorption

The principal reactions taking place in the absorption of CO_2 by sodium hydroxide are firstly carbonic acid formation from CO_2 and water, and secondly sodium bicarbonate or carbonate formation from sodium hydroxide (NaOH) and carbonic acid according to:

$$H_2CO_3 + NaOH \rightarrow NaHCO_3 + H_2O \quad (6.16)$$

$$NaHCO_3 + NaOH \rightarrow Na_2CO_3 + H_2O \quad (6.17)$$

The relative proportions of the carbonate and bicarbonate end products can be controlled by adjusting the pH in the absorption stage, for example by adding hydrochloric acid (HCl).

Regeneration of sodium hydroxide can be achieved by the addition of lime (CaO) to the carbonation products:

$$CaO + H_2O \rightarrow Ca(OH)_2 \quad (6.18)$$

$$Ca(OH)_2 + Na_2CO_3 \rightarrow 2NaOH + CaCO_3 \quad (6.19)$$

yielding a calcium carbonate slurry. This is in turn regenerated by heating to drive off the water, followed by calcination at ~900°C:

$$CaCO_3 + heat \rightarrow CaO + CO_2 \quad (6.20)$$

Depending on the way in which heat is provided to the calciner, the offgas may be either a pure CO_2 stream that can be compressed for storage, or a CO_2-enhanced flue gas stream (~20% CO_2), from which CO_2 would then have to be captured.

This process has the advantage of using chemicals that are inexpensive and abundantly available, but a significant disadvantage is the high energy requirement for $CaCO_3$ recovery in the calciner (15 GJ per t-C). An alternative solvent recovery route has been investigated using the solid-state reaction between sodium carbonate and sodium trititanate ($Na_2O{\cdot}3TiO_2$), the so-called titanate cycle. The solid carbonate and trititanate are heated to ~860°C, at which point the pentatitanate is formed and CO_2 is released:

$$5(Na_2O \cdot 3TiO_2)_{(s)} + 7Na_2CO_{3(s)} \rightarrow 3(4Na_2O \cdot 5TiO_2)_{(s)} + 7CO_{2(g)} \quad (6.21)$$

The enthalpy of reaction 6.21 is $\Delta H = 65\,kJ$ per mol-CO_2, ~35% of that required for calcination. The sodium pentatitanate is then cooled and leached with water to regenerate the caustic solvent:

$$3(4Na_2O \cdot 5TiO_2)_{(s)} + 7H_2O \rightarrow 14NaOH_{(aq)} + 5(Na_2O \cdot 3TiO_2)_{(s)} \quad (6.22)$$

with an enthalpy of reaction of $\Delta H = 15\,kJ$ per mol-C at 100°C. Overall, including heating and heat recovery, the titanate cycle has approximately half the energy requirement of the calcination route at ~130 kJ per mol-CO_2 versus ~250 k per mol-CO_2 for calcining (~2.9 GJ per t-CO_2 vs. ~5.7 GJ per t-CO_2).

An alternative to regeneration of the caustic solvent is to continuously produce sodium hydroxide by electrolysis of NaCl brine:

$$2NaCl + 2H_2O \rightarrow Na^+ + Cl^- + H^+ + OH^- \rightarrow 2NaOH + Cl_2 + H_2 \quad (6.23)$$

This has been proposed for a capture system that would produce solid sodium bicarbonate as a product stream (Section 6.3.4).

6.1.2 Physical absorption

Physical absorption processes use organic or inorganic physical solvents to absorb acid gas components rather than reacting with them chemically. CO_2 absorption by a solvent is determined by the vapor–liquid equilibrium of the mixture via Henry's law, which states that, at a given temperature, the amount of a gas dissolved in unit volume of a solvent is proportional to the partial pressure of the gas in equilibrium with the solvent.

Thus the solubility of CO_2 (K_{CO_2}) is expressed as:

$$K_{CO_2} = C_{CO_2}/P_{CO_2} = 1/K_{HCO_2} \quad (6.24)$$

where C_{CO_2} is the dissolved concentration of CO_2, P_{CO_2} is the CO_2 partial pressure, and K_{HCO_2} is the Henry's law constant for CO_2. Equation 6.24 shows that the concentration of a solute such as CO_2 in a solvent is proportional to its partial pressure above the solvent. In physical absorption processes for CO_2 capture, high solvent loading will therefore be achieved for feed gas streams at high pressure, with a high CO_2 content, or both.

Methanol (CH_3OH) is used as a physical solvent in CO_2 separation from natural gas in the Rectisol process (Section 6.2.2). The dependence of CO_2 solubility in methanol on temperature and partial pressure is illustrated in Figures 6.3 and 6.4. Solubility rises steeply at reducing temperatures below –40°C, enabling very effective separation of CO_2 at operating temperatures in the range from –60°C to –70°C.

Figures 6.3 and 6.4 also illustrate that from an absorption operating point at low temperature, high pressure, or both, a change in operating conditions to higher temperature or lower pressure, will result in release of the solute as a result of lower solubility. Regeneration of a physical solvent can thus be achieved by an increase in temperature (temperature swing) or a reduction in pressure (pressure swing).

Other physical solvents that have been applied to CO_2 capture include propylene carbonate ($C_4H_6O_3$) in the Fluor process, and dimethyl ethers of polyethylene glycol ($CH_3(CH_2CH_2O)_nCH_3$) in the Selexol® process (see Table 6.4). Unlike chemical sorbents, physical sorbents are chemically inert to the treated gas stream, avoiding the formation of heat-stable salts that can be problematic for chemical absorption processes.

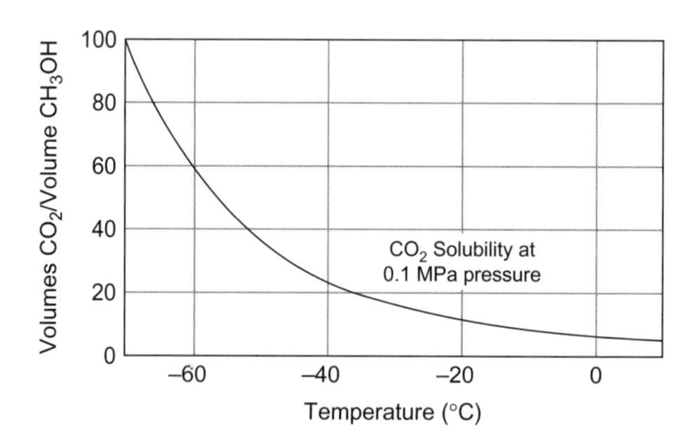

Figure 6.3 Temperature dependence of CO_2 solubility in methanol

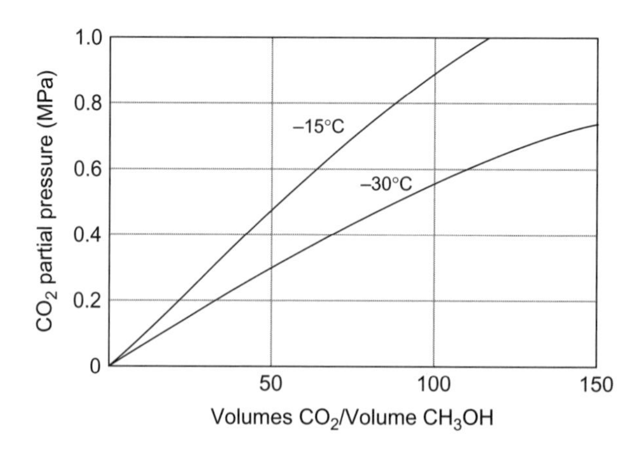

Figure 6.4 Partial-pressure dependence of CO_2 solubility in methanol

6.2 Absorption applications in postcombustion capture

The capture of CO_2 from postcombustion flue gas is complicated by the low partial pressure of CO_2 and the presence of various contaminants in the gas stream. The key requirements for sorption processes to achieve capital and energy efficiency are therefore:

- High loading of CO_2 per unit volume of sorbent, coupled with low sorbent cost
- Low heat of desorption to reduce the energy penalty for sorbent regeneration
- Sorbent tolerance to contaminants, to reduce degradation and byproduct formation

6.2.1 Chemical absorption applications

Amine-based chemical absorption

Chemical absorption processes using a variety of amine-based solvents have been deployed for postcombustion CO_2 capture, with single-train capacities approaching 500 t-CO_2 per day. Figure 6.5 illustrates the process flow scheme of a typical amine-based capture system.

Flue gas entering the process at close to atmospheric pressure is cooled to the required operating temperature in the region of 40–60°C. Cooling by direct water contact will also beneficially remove fine particulate matter from the gas stream. The lean solvent (low content of CO_2 reaction products) is brought into contact with cooled flue gas in a packed absorber tower (amine scrubber). Flue gas exiting the top of the absorber is water washed, to reduce the entrainment and carryover of solvent droplets and vapor, and is then vented to the atmosphere.

Rich solvent (high content of CO_2 reaction product) exits the base of the scrubber and is pumped to the top of the amine stripping tower. A heat

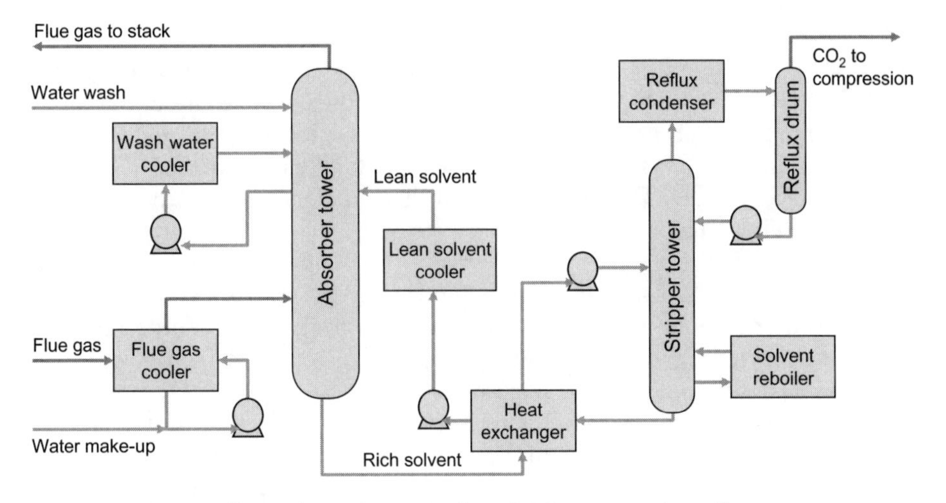

Figure 6.5 Process flow scheme for amine-based CO_2 capture from flue gas

exchanger heats the rich solvent, recovering heat from the regenerated solvent cycling back to the absorber. The stripping tower typically operates at 100–140°C and at marginally higher pressure than the absorber. The heat required to reverse the absorption reaction, releasing pure CO_2 and regenerating the lean solvent, is supplied by a reboiler, which would typically be integrated into the steam cycle of the host plant. Steam and released CO_2 exit the top of the stripping tower, where the steam is condensed from the CO_2 product stream, while the lean solvent from the base of the tower is cooled and cycled back to the absorber.

Due to the low tolerance of amine-based solvents to the presence of SO_2 and NO_2 in the flue gas, removal of these components to low levels is required ahead of the CO_2 capture process, using the techniques described in Chapter 3.

A number of amine-based processes have been commercialized, the key features of which are summarized in Table 6.4 later in the chapter.

Ammonia-based chemical absorption

A flue gas CO_2 capture process using a chilled slurry of dissolved and suspended ammonium carbonate and ammonium bicarbonate in ammonia is being developing by Alstom Power Systems and the Electric Power Research Institute (EPRI).

The system uses a typical absorber tower configuration, operating at near-freezing conditions (0–10°C), in which cooled flue gas flows upward in countercurrent to the absorbent slurry. The low operating temperature allows high CO_2 loading of the solvent slurry and reduces "ammonia slip"—the carryover of entrained ammonia droplets and suspended solids with the clean flue gas exiting the tower. Ammonia slip is further reduced by a cold-water wash of the cleaned flue gas, which consists mainly of nitrogen, excess oxygen, and a low residual concentration of CO_2.

The solvent slurry regenerator operates at temperatures >120°C and pressures >2 MPa. Ammonia slip on regeneration is also controlled by water washing, yielding a high-pressure CO_2 stream with low moisture and ammonia content. This high-pressure regeneration has the advantage of reducing the energy requirement for subsequent compression and delivery of the CO_2 product stream for storage. The flow scheme of the chilled ammonia process (CAP) is shown in Figure 6.6.

This process is the subject of an extensive development program, including a pilot test on a 5 MW power plant, capturing 15 kt-CO_2 per year. The pilot commenced operation in February 2008 and aims to demonstrate the capabilities of the technology on actual flue gas, gather operating data, and evaluate the process energy consumption (Figure 6.7).

Two demonstration-scale trials are also planned: a 20 MW test at AEP's Mountaineer Plant in West Virginia, capturing up to 110 kt-CO_2 per year, and a test at StatoilHydro's Mongstad refinery in Norway, capturing 100 Kt-CO_2 per year from the refinery's cracker unit flue gas or from a new natural gas

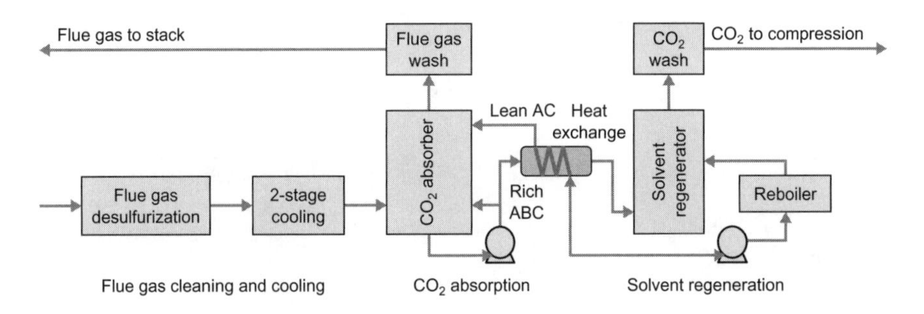

Figure 6.6 Chilled ammonia process flow scheme

Figure 6.7 3D graphic and picture of CAP pilot installation (Courtesy; Alstom Power Inc.)

combined-cycle (NGCC) power plant. The 12- to 18-month test is scheduled to start by 2010 and is expected to lead to construction of a unit that may eventually capture >1 Mt-CO_2 per year.

6.2.2 Physical absorption applications

Capture of CO_2 by physical absorption relies on the solubility of CO_2 in the solvent, which depends on the CO_2 partial pressure and temperature of the feed gas. Physical absorption has the advantage over chemical absorption in that the heat required for desorption is significantly lower; however, this also necessitates low operating temperatures at the absorption stage to achieve adequate solvent loading.

Figure 6.8 illustrates the relative solvent loading of chemical and physical solvents as a function of the partial pressure of the sorbate gas. As shown in

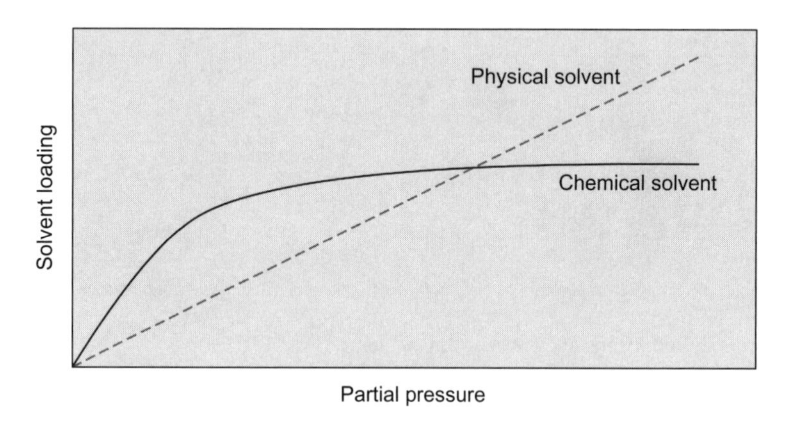

Figure 6.8 Relative solvent loading of chemical and physical solvents

Equation 6.24, the solvent loading capacity of physical solvents increases with the partial pressure of the sorbate. Although chemical solvents can achieve high loading at low partial pressure, physical solvents are favored at high pressure. Physical absorption is therefore typically applied to CO_2 separation at high pressure, such as CO_2 recovery from a produced natural gas stream, while chemical absorption is preferred for low-pressure applications, such as CO_2 capture from flue gases. Use of physical absorption for CO_2 capture from flue gas would require compression of large volumes of gas, the major component being nitrogen, which would subsequently be blown down for release to the atmosphere. The attendant energy penalty would make this process uneconomical.

Selexol® process

The Selexol process uses a liquid physical solvent to remove acid gas from synthetic or natural gas streams and has been in commercial application since the early 1970s, with well over 50 units currently operating worldwide. The proprietary Selexol solvent was originally developed by Allied-Signal (now Honeywell) in the 1950s and is a mixture of the dimethyl ethers of polyethylene glycol with the formula $[CH_3(CH_2CH_2O)_nCH_3]$, where n is 3–9.

Table 6.1 shows the relative solubilities of typical gas stream components in the Selexol solvent. Because of the high solubility of water, the partial pressure of water vapor in the feed gas stream to a Selexol plant must be kept low in order to avoid impairing the CO_2 loading capacity of the solvent.

Although the solubilities of propane (C_3H_8) and heavier hydrocarbon components exceed the solubility of CO_2, the actual removal rate of these components from a hydrocarbon gas feed stream will depend on the partial pressure, operating temperature, solvent to gas ratio in the absorber, and other operating parameters.

Figure 6.9 illustrates a typical Selexol process for CO_2 removal from a natural gas stream.

Table 6.1 Relative solubilities of sorbates in the Selexol solvent at 25°C

Component		**Relative solubility (CO_2 = 1)**
Hydrogen	H_2	0.013
Methane	C_1	0.066
Ethane	C_2	0.42
Carbon dioxide	CO_2	1.00
Propane	C_3	1.01
nButane	nC_4	2.37
nPentane	nC_5	5.46
nHexane	nC_6	11.0
nHeptane	nC_7	23.7
Water	H_2O	730

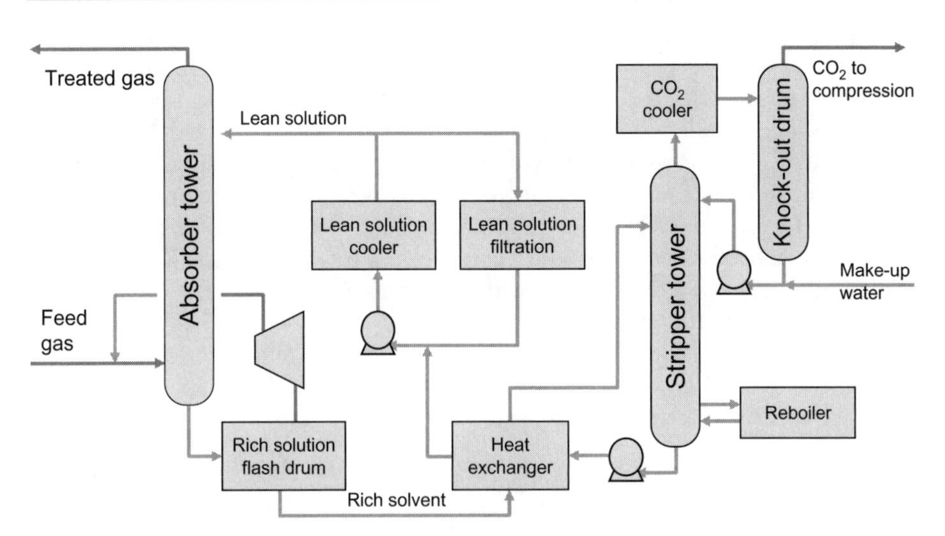

Figure 6.9 Selexol process scheme

The basic flow scheme is very similar to the amine scrubbing process described in the previous section, with cooled feed gas at 0–20°C contacting lean solvent in an absorber, and rich solvent being regenerated in a stripping tower, heated by reboiling of the lean solvent. Feed gas composition typically is 5–60% or more acid gas by volume, at feed pressures of 2–15 MPa and CO_2 partial pressure of 0.7–3 MPa. The process can be optimized for either trace or bulk removal of acid gases (H_2S, COS, and CO_2), while the high solubility of heavy hydrocarbon components means that the process can also be used for dew point control of hydrocarbon gas feed to pipeline or LNG feed specifications.

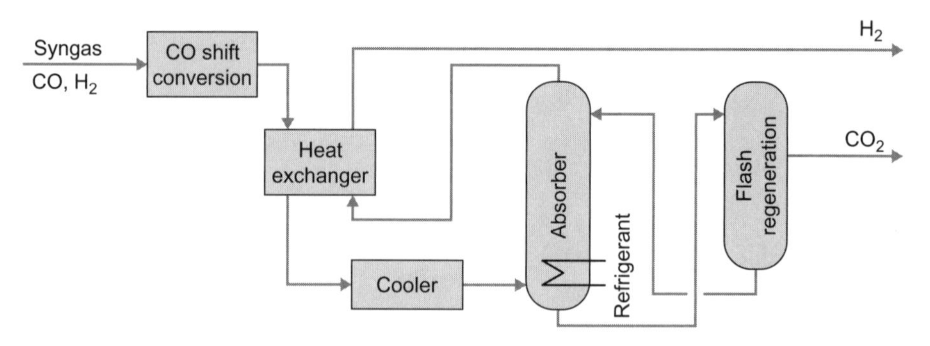

Figure 6.10 CO_2 removal from water–gas–shifted syngas

Rectisol process

The Rectisol process, using refrigerated methanol as a physical solvent, was initially developed by Lurgi GmbH (now part of Air Liquide SA) and found its first commercial application in 1955 in the purification of coal gasification syngas for ammonia and Fischer–Tropsch liquids production. It has become the dominant process applied worldwide for syngas purification and now purifies 75% of syngas produced globally.

The Rectisol process can be configured in many different ways, depending on the types of contaminants to be removed and the required product streams—for example, whether food-grade CO_2 is required. When applied to CO_2 capture from a syngas production process after water–gas shift conversion (Section 3.1.2), raw syngas, produced by partial oxidation of the input fuel, is cooled and enters the purification plant where, if necessary, trace components such as HCN and NH can be removed in a cold methanol prewash stage. H_2S is then removed in an absorber using CO_2-rich methanol. Regeneration of the H_2S-rich solvent takes place in two stages:

a) A medium-pressure flash recovers H_2 and CO, which are returned to the syngas stream, and
b) Hot regeneration using methanol vapor as a stripping gas recovers H_2S, which can be fed to a Claus plant for sulfur recovery

Following desulfurization, the raw syngas is water–gas–shifted (Figure 6.10), producing a stream with typically 33% CO_2 content. This is cooled and enters the CO_2 absorber, where CO_2, typically at a partial pressure >1 MPa, is removed by contact with refrigerated methanol at –10°C to –70°C. Two contacting stages are used, with flash-regenerated methanol in the lower stage to initially reduce CO_2 content to ~5%, followed by refrigerated hot-regenerated methanol in the upper stage, further reducing CO_2 content to ~3%.

Fluor process

The Fluor solvent process was developed in the late 1950s by Fluor and the El Paso Natural Gas Company and was the first physical absorption process

Table 6.2 Relative solubilities of sorbates in propylene carbonate at 25°C

Component		Relative solubility (CO_2 = 1)
Hydrogen	H_2	0.0078
Methane	C_1	0.038
Ethane	C_2	0.17
Carbon dioxide	CO_2	1.00
Propane	C_3	0.51
nButane	nC_4	1.75
nPentane	nC_5	5.0
nHexane	nC_6	13.5
nHeptane	nC_7	29.2
Water	H_2O	300

for CO_2 removal from natural gas. Propylene carbonate ($C_4H_6O_3$) was chosen as the solvent for this process in view of its high solubility of CO_2 relative to methane (Table 6.2), and also for its noncorrosivity, allowing plant fabrication from inexpensive non-alloy steel.

The process flow scheme for the Fluor process is shown in Figure 6.11. CO_2 is removed from the natural gas stream in a high-pressure contactor operating at below ambient temperature, and the solvent is regenerated by pressure swing in a series of flash vessels at successively lower pressures. Because of the reactivity of propylene carbonate with CO_2 and with water at temperatures >90°C, operating temperature is limited to a maximum of 65°C, while process performance can be improved by chilling the feed gas stream to about −18°C, to condense C5+ hydrocarbons and increase the CO_2 loading capacity of the solvent. This reduces the solvent circulation rate, leading to lower plant capital and operating costs, but, depending on the water content of the feed gas, may require glycol injection upstream of the absorber to prevent hydrate formation.

Offgas from the first flash stage is recycled to reduce methane loss. Energy input to the process is limited to gas recycle compression and solvent circulation pumps, with hydraulic turbines used to recover ~50% of the required recompression energy. Solvent loss is low due to the low vapor pressure of the solvent, eliminating any need for solvent recovery by water washing of the treated gas and CO_2 streams.

6.3 Absorption technology RD&D status

Although absorption processes for CO_2 capture have over half a century of industrial application, it remains an area of significant ongoing RD&D effort, applied both to the continuous improvement of existing processes and also to

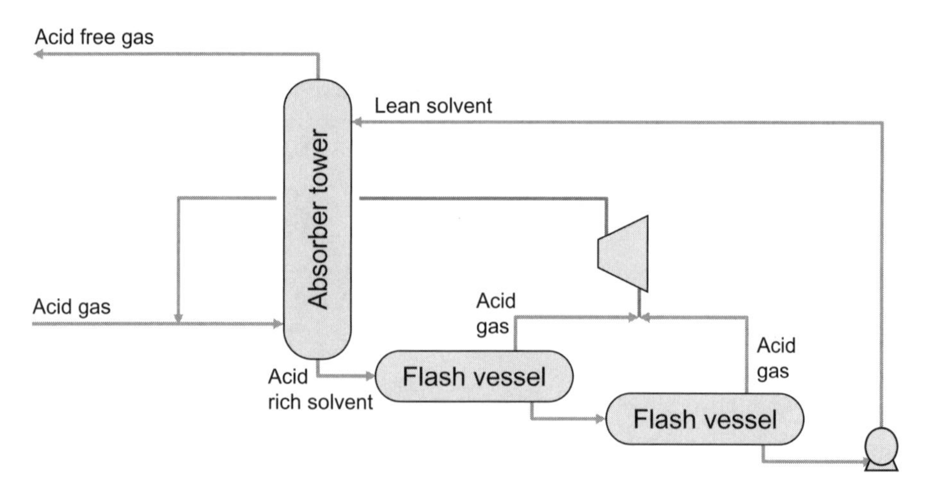

Figure 6.11 Fluor solvent process scheme for CO_2 capture from natural gas

the development of novel systems aiming for a step-change improvement in the cost-effectiveness of absorption as a capture technology.

6.3.1 Improved amine-based systems

All of the major suppliers of amine-based systems, summarized in Table 6.3, have ongoing development programs aimed at improving the effectiveness and market competitiveness of their proprietary processes.

Areas of research and development to improve the performance of these systems include:

- Improving solvent formulations to reduce energy requirement and solvent degradation
- Increasing contact area and reducing pressure drop by modified tower packing
- Reducing energy requirements by increasing heat integration with the host plant
- Producing higher amine concentration with reduced corrosion by using new additives
- Improving processes for solvent regeneration

6.3.2 Dry sorbent-based process R&D

As part of the NETL sequestration technology development program, RTI International is developing a process that uses a dry, regenerable, sodium carbonate-based sorbent that captures CO_2 in a carbonation reactor at ~30°C and in the presence of water to form sodium bicarbonate. The sorbent is regenerated by heating (thermal swing) to ~120°C to produce a gas stream containing CO_2 and water, the latter being removed by condensation to produce a pure CO_2 stream for storage. Rapid reaction and release of the heat of absorption results in heating of the dry sorbent, which must be uniformly controlled within the carbonation reactor to prevent localized decarbonation.

Table 6.3 Amine-based CO_2 capture systems in commercial use and under development

Process	Description	Application limits/Operational conditions	Further development focus areas
Kerr-McGee/ABB Lummus Crest Technology process	MEA solvent, using low amine concentration (15–20%) to limit corrosion. Single train capacity up to 400 t-CO_2 per day.	+ SO_2 tolerant to 100 ppm − High solvent flow rate and regeneration heat requirement −Lower oxygen tolerance leading to solvent degradation	Higher solvent concentration Process improvements to reduce oxygen degradation of solvent
Fluor Daniel Econamine FG^{SM} process	Inhibited aqueous MEA solvent (30% MEA by weight). Single train capacity up to 320 t-CO_2 per day.	+ Improved oxygen tolerance with low corrosivity − Higher solvent consumption (1.5–2.0 kg per t-CO_2) Low SO_2 tolerance (<10 ppm)	Improved solvent formulation to increase reaction rate and reduce absorber tower height Higher solvent capacity leading to lower solvent flow rate and reduced regeneration energy cost Process improvements to reduce energy costs
Kansai/Mitsubishi KEPCO/MHI process	Sterically hindered amine solvents (KS1, KS2, KS3). Single train capacity up to 200 t-CO_2 per day.	+ Low regeneration heat requirement + Low solvent flow rate, degradation, and losses − Higher solvent cost compared to MEA − Low SO_2 tolerance (<10 ppm)	Further development of hindered amines with lower heat of desorption Reducing other process heat requirements Reducing solvent losses
Praxair process	Blended amine solvent with additional flash deoxygenation or oxygen scavenging stage ahead of the solvent stripping tower	+ Improved oxygen tolerance without the use of inhibitors	Blended amine solvents (e.g., 10–20 wt.% MEA +20–40 wt.% MDEA), allowing greater capacity and reaction rates without incurring operational problems due to corrosion

Table 6.4 Absorption technologies and RD&D summary

Absorption technology	Advantages	Disadvantages	RD&D focus areas
Physical absorption			
Fluor process (propylene carbonate)	Well-proven technology with 50+ years of commercial application High selectivity for CO_2 relative to methane Noncorrosive solvent Lower hydrocarbon losses and recycle gas compression requirement compared to Selexol	Low H_2S tolerance in feed gas stream Feed gas must be dehydrated due to high water solubility Irreversible reaction with CO_2 and water at 90°C precludes temperature swing regeneration	
Rectisol process (chilled methanol)	High selectivity for H_2S and CO_2 Able to remove many contaminants in a single process Low solvent cost Long-proven commercial application in gasification projects	High refrigeration energy cost	
Selexol process (dimethyl ethers of polyethylene glycol [DMEPG])	Well-proven technology with 30+ years of commercial application Chemically inert solvent not subject to degradation Nonaqueous solvent allows low-cost carbon steel plant construction	Feed gas must be dehydrated due to high water solubility Required high partial pressure of CO_2	

Chemical absorption			
Amine systems (e.g., MEA)	Well-proven technology with extensive commercial application, leading to low capital costs Range of solvents have been developed	Low amine concentrations to resist corrosion limit loading capacity Low tolerance to SO_2, NO_x, and O_2 in flue gas	Identification of new amine-based solvents with lower heat of adsorption Improvements in energy integration with host plant
	Technical and economic benchmark for other solvent systems	High energy requirement for absorbent regeneration	Additives to improve system performance (e.g., to enable higher amine concentration)
			Engineering improvements to absorption and stripping towers to reduce energy requirements
Carbonate systems	Higher solvent loading capacity Lower regeneration energy requirement Low oxygen solubility, reducing corrosion	High cost of additives such as piperazine	Energy saving through improved tower packing and multistage stripping
			Catalysts or additives to improve CO_2 absorption rate and reduce stripping heat requirement
Dry sorbent systems	Potentially lower cost and heat requirement than amine systems	Mechanical complexity due to solids handling	Sorbent particle preparation and structure to maximize surface area and reduce particle attrition during fluidization and transport between reactors

(*Continued*)

Table 6.4 Continued

Absorption technology	Advantages	Disadvantages	RD&D focus areas
		Sorbent heating due to heat of formation requires accurate and uniform reactor temperature control	
Aqueous ammonia	Low regeneration energy requirement	High solvent volatility requires reduced operating temperature	Increasing solvent loading
	Lack of solvent degradation during the absorption and regeneration cycle	High solvent loss during regeneration at high temperature	Reducing solvent losses
	Low solvent cost		
	High-pressure regeneration		
	Tolerant to O_2 and other contaminants		
	Possibility of producing ammonium sulfate and ammonium nitrate fertilizer byproducts		
Chilled ammonia	Low regeneration energy requirement Lack of solvent degradation during the absorption and regeneration cycle	Near-freezing operating conditions required to limit solvent losses Potential for fouling due to ammonium bicarbonate solids	Demonstration scale plant in operation Reducing solvent emissions

	Low solvent cost		
Task-specific and reversible ionic liquids	Very low volatility and solvent loss Very low heat of adsorption (10–15% of MEA) High thermal stability Dual mode (physical plus chemical solvents) with high loading capacity	High production cost of ionic liquids High viscosity, reducing absorption rate and increasing parasitic energy load for fluid transport	Synthesis and performance testing of new salts Increased absorption capacity and rate Use of TSILs in membrane contactors
Sodium hydroxide	Low cost and abundance of required chemicals All process steps are currently proven technologies, applied in other industries	High energy requirement in calcining intermediate regeneration product ($CaCO_3$) High water and solvent loss in spray tower contactor	Alternative methods of solvent recovery (titanate cycle) Laboratory-scale spray tower optimization and commercial-scale concept development Process simulation and optimization Titanate cycle solvent regeneration
Oligomeric solvents	Potential for high solvent loading with very low regeneration energy requirement		Identification, modeling, synthesis, and testing of candidate oligomers

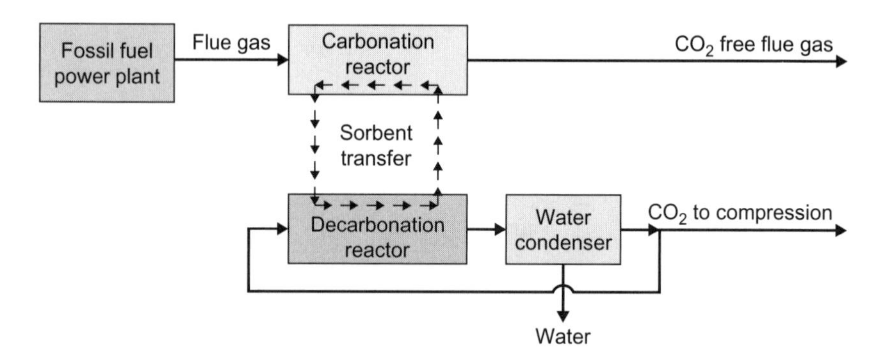

Figure 6.12 Dry carbonate sorbent-based capture process

The dry sorbent process is illustrated in Figure 6.12. The sorbent is conveyed respectively to and from the decarbonation reactor by steam-heated and water-cooled screw conveyors.

A small-scale laboratory demonstration prototype of the process has been constructed that is capable of treating up to $15\,m^3$ per hour of flue gas and circulating sorbent at up to 100 kg per hour. The demonstration module has completed a fully integrated test under actual flue gas conditions at the U.S. Environmental Protection Agency's Multi-Pollutant Control Research Facility.

Preliminary economic analyses indicate that a commercial-scale application of the dry carbonate process could be significantly less costly and less energy-intensive than conventional amine-based systems. The process is being specifically developed to enable retrofitting to existing coal-fired power plants and, with process modifications, is expected to be applicable to CO_2 capture from other sources such as natural gas plants, oil refineries, and cement plants.

6.3.3 Ionic liquid solvents

One of the disadvantages of liquid solvents for gas separation is the loss to the gas stream of solvent, and water in the case of aqueous solutions, due to the vapor pressure of the solvent solution. Room-temperature ionic liquids (RTILs), formed by the ambient-temperature melting of certain organic salts, have negligible vapor pressure as a result of the strong Coulombic attraction between the constituent ions, and could therefore enable gas capture without loss of solvent to the gas stream.

Ionic liquids (ILs) typically comprise a large organic cation and a smaller inorganic anion, and it is this asymmetry between the cation and anion sizes that reduces the strength of the binding force and results in a low melting point.

Task-specific ionic liquids

By preparing ionic liquids that include ions with specific functional groups, so-called task-specific ionic liquids (TSILs) can be designed for specific separation applications. Amine functional groups have been successfully used; capturing CO_2 as a carbamate ion in chemical absorption reactions equivalent to Equation 6.1. Without the specific functional groups, ionic liquids can also be applied to CO_2 capture as physical sorbents.

A current RD&D focus of work on TSILs is the engineering of new low-cost liquids with absorption capacity to match traditional amine systems. TSIL molecular structures are readily amenable to modification, so that structure-dependent physical and thermodynamic properties can be readily optimized. The potential to use hollow fiber membrane contactors (HFMC; Chapter 8) as an alternative to conventional column absorption is also under investigation, since this technology promises lower energy consumption and ease of operation.

Reversible ionic liquids

Reversible (also known as switchable or smart) ionic liquids are a class of IL solvents in which a key property, for example CO_2 solubility, undergoes a step change in response to an external stimulus, such as heat, light, or pH. This abrupt property change allows the regeneration method to be engineered into the structure of the solvent, resulting in substantially lower regeneration energy requirements due to the low enthalpy of solution. These novel solvents also offer higher CO_2 loading capacity when compared with more traditional solvents, since they can operate as "dual-mode" solvents, absorbing CO_2 by both chemical and physical absorption.

Ongoing research focuses on design of reversible IL solvents using molecular modeling to optimize the solvent properties for CO_2 capture and regeneration, followed by synthesis, characterization, and testing of the chemical and physical properties of the new solvents.

6.3.4 Oligomeric solvents

Oligomeric solvents are a further class of novel solvents that are under investigation and have the potential for high solvent loading and low regeneration energy requirements. Oligomers are short polymers, consisting of a limited number (typically 10–100) of monomer units. One compound that has been proposed is the oligomer of carbonic acid $(H_2CO_3)_n$, which, although unstable at ambient temperatures, becomes stable at roughly –80°C.

Initial investigation of these novel compounds is typically based on molecular modeling, using quantum mechanical calculations and semiempirical statistical mechanical models to evaluate the suitability of various oligomers for CO_2 capture. This initial modeling improves the cost-effectiveness of subsequent synthetic and laboratory screening experiments, which are used to evaluate potentially promising solvents.

6.3.5 Sodium hydroxide-based systems

Capture of CO_2 using chemical absorption by sodium hydroxide is under research and development both for flue gas and direct air capture applications.

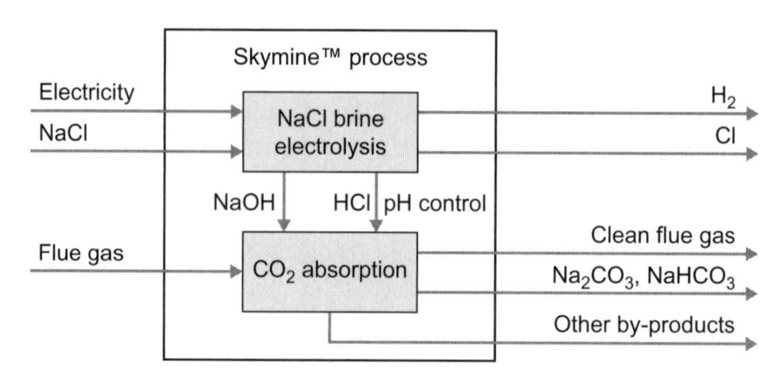

Figure 6.13 Skymine™ process schematic

Flue gas CO_2 capture using sodium hydroxide

A flue gas capture process, applying carbonation reactions in Equations 8.16 and 8.17 to cooled flue gas at 30°C, is under development by Skyonic Inc. in the Skymine™ process. Few technical details of the process have been released, but the concept shown in Figure 6.13 indicates that hydrogen and chlorine, together with carbon as sodium bicarbonate, in solid form or as a concentrated solution, are the process products.

The process also generates other intermediate components; sodium hydroxide is generated in the process by the electrolysis of acidified NaCl brine, and HCl is generated from hydrogen and chlorine products of the brine electrolysis and used to acidify the brine ahead of electrolysis as well as to control pH in the absorption columns. As noted above, pH in the absorption column determines the ratio of carbonate to bicarbonate reactions products.

Direct air CO_2 capture using sodium hydroxide

Unlike the capture of CO_2 from flue gas or from natural gas processing, which deals with CO_2 at partial pressures from ~15 kPa to >10 MPa, the direct capture of CO_2 from ambient air requires a capture technology that can operate efficiently at a CO_2 partial pressure of only 38 Pa. Nevertheless, direct air capture is thermodynamically feasible at an energy cost that could in principle be in the region of 10% of the energy content of fossil fuels.

The theoretical minimum energy requirement is determined by the enthalpy of mixing of CO_2 in ambient air and is given by:

$$\Delta H_{mix} = kT \ln(p/p_{CO_2}) \tag{6.25}$$

where k is the Boltzmann constant (8.314 J per mol-°K) and T is the ambient temperature (°K = °C + 273.15). For typical ambient conditions (25°C = 298°K), kT = 2.48 kJ per mol, and for CO_2 in air at a partial pressure of 38 Pa, ΔH_{mix} = 19.5 kJ per mol, or 1.62 GJ per t-C. Including the energy required to

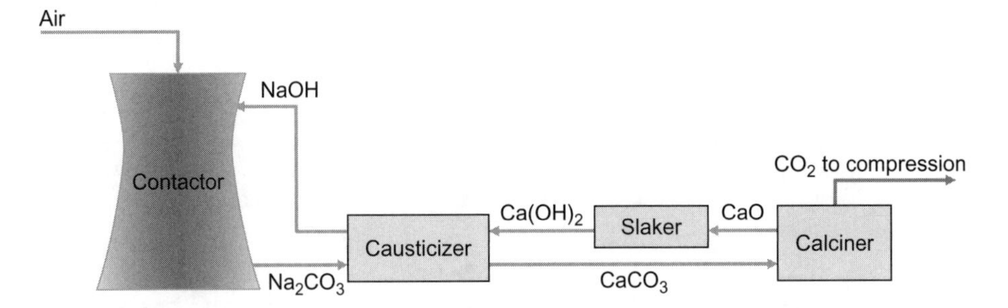

Figure 6.14 Process concept for air capture using sodium hydroxide

compress the captured CO_2 to ~10 MPa for geological storage (at 50% energy efficiency) yields an overall theoretical minimum energy requirement of ~4 GJ per t-C for capture from ambient air.

Figure 6.14 illustrates an overall process for air capture using NaOH as a solvent. Air is drawn through a contactor where CO_2 is captured by the solvent. Physically the contactor could be a packed tower configuration, in which the solvent drips down through packing material designed to maximize contact area, or an empty tower, similar to the familiar evaporative cooling tower of a power plant, with a solvent spray dropping in a co-current (downward) air flow. A 50% capture efficiency is estimated to be achievable with a 1.5 m high packed bed or a 100 m high spray tower.

Figure 6.15 shows a 6 m high packed tower under development at the Energy and Environmental Systems Group at the University of Calgary. The tower achieves a CO_2 reduction of ~200 ppm with a solvent flow rate of 5 l per second and an air flow speed of 1.25 m per second, at an energy cost of 290 MJ per t-CO_2, and has a capture capacity of 15 t-CO_2 per year per m^2 of absorption cross section. An interesting result achieved in these experiments is that the capture performance of the tower falls off only gradually if the solvent supply is interrupted. Intermittent supply of solvent to the tower saves on energy required to pump the solvent, and 85% of the capture performance can be achieved with a 5% solvent pump duty cycle. Evaporative water loss is a potentially significant issue for this type of contactor; a water loss of 20 mol-H_2O per mol-CO_2 (6.4 t-H_2O per t-CO_2) was determined for a previous version of a contacting tower (open spray tower rather than packed tower).

In the causticizer, NaOH is recovered by the addition of slaked lime ($Ca(OH)_2$) produced by the hydration of lime (CaO) according to the reaction in Equation 6.18. The calcium carbonate mud produced in the causticizer is heated in the calciner, driving off water as steam and yielding lime and CO_2. Calcining is used on an industrial scale in the production of cement, and a conventional high-efficiency calciner comprises a fluidized bed fired by the combustion of natural gas. The CO_2 content of the exhaust gas stream is enhanced as a result of the offgas from the calcining reaction, and CO_2 capture from the exhaust gas stream can be achieved by any absorption process.

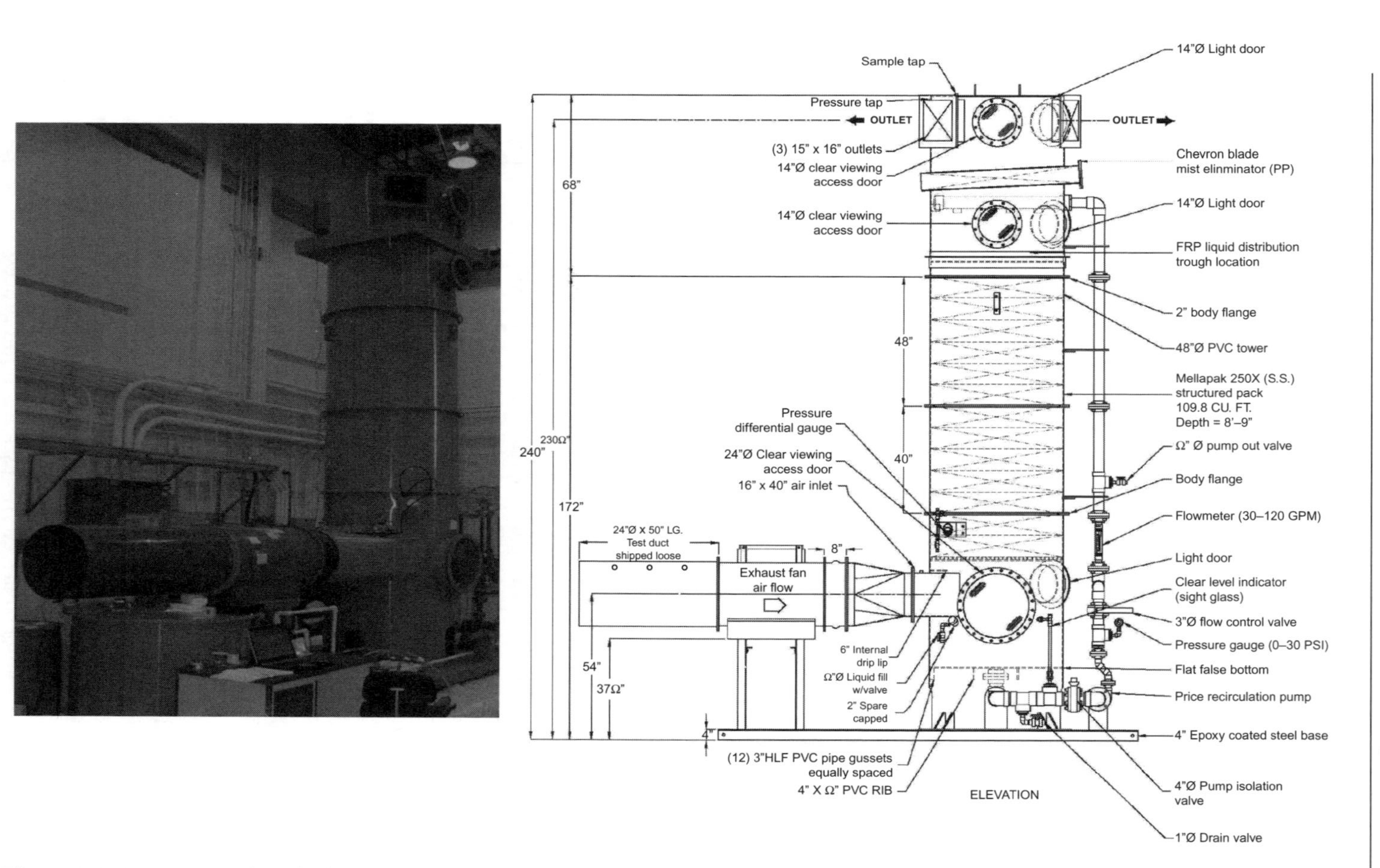

Figure 6.15 Experimental packed tower for air capture (Courtesy; University of Calgary (photograph) and Advanced Air Technologies Inc. (line drawing))

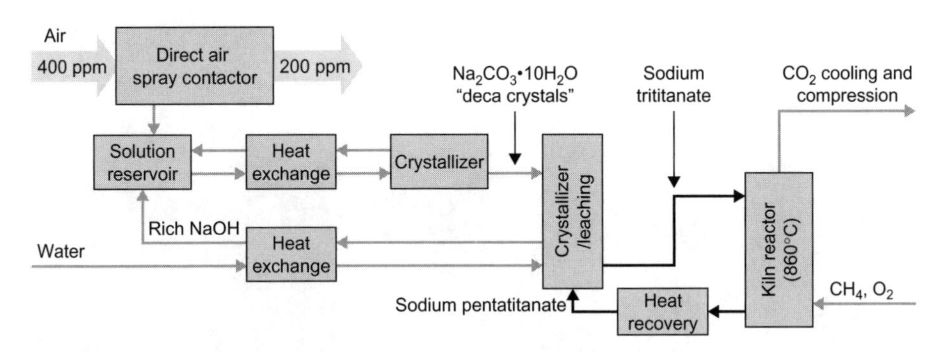

Figure 6.16 Direct air capture process using sodium titanate cycle

The energy efficiency of the process would be enhanced by integrating heat recovery from steam produced in the calciner, as well as recovering low-level heat from the exothermic slaking reaction ($\Delta H = -82$ kJ per mol).

All the elements of such an air capture system are available today as proven technology, and are currently applied on an industrial scale in the cement and paper industries. However, the high energy cost for calcining plus the capital cost of the contacting system, which must be added to the cost of a flue gas capture system, means that air capture using such a system cannot be cost-competitive for CCS compared to capture from large sources of emissions.

Figure 6.16 illustrates an alternative process scheme that uses the titanate cycle (Section 6.1.1) rather than calcination to regenerate the caustic solvent. Following the spray tower contactor, a first crystallization stage precipitates Na_2CO_3•$10H_2O$ ("deca-crystals") by cooling the rich solvent to ~10°C. Lean solvent is recycled to the spray tower reservoir, while the deca-crystals are conveyed to a second crystallizer/leaching unit.

In this unit, operating at 100°C, Na_2CO_3 is recrystallized in an anhydrous form as a result of the higher temperature and, together with sodium trititanate, is conveyed to a oxycombustion gas-fired reactor, where the trititanate is converted to sodium pentatitanate at 860°C (reaction in Equation 6.21) and CO_2 is released. By using oxycombustion in this reactor, the exhaust gas comprises CO_2 plus steam; cooling to 25°C condenses the steam to produce make-up water, while the pure CO_2 is compressed for transportation and storage. After cooling, the pentatitanate is leached in the crystallizer and leaching unit, regenerating NaOH and sodium trititanate and completing the titanate cycle. The NaOH is recycled to the spray tower reservoir.

Ongoing research on this process is focusing on improving the kinetics of the crystallization and titanate cycle reactions and the design of a contactor for a 1 Mt-CO_2 per year air capture unit. A contacting system of this capacity could take the form of a segmented vertical slab, as illustrated in Figure 6.17. Prevailing winds, assisted by a fan wall, would sustain a horizontal air flow of 2–5 m per second (7–18 km per hour) while the solvent falls though a packed wall with vertically oriented packing plates, optimized to maximize liquid

Figure 6.17 Conceptual design of intermittently wetted cross-flow air capture contactor (Courtesy; Eric Au, University of Calgary)

hold-up under intermittent flow conditions. Solvent supply would be switched sequentially between slab sections, allowing intermittent wetting of individual sections with continuous pump operation. For a capture capacity of 20 t-CO per m^2 per year, a 10 m high contactor slab of 5 km length would be required for a 1 Mt-CO_2 per year air capture unit.

The cost per ton of CO_2 captured achievable for such a system is unlikely to be competitive with CO_2 capture from power plants. However, it does provide an additional option for reducing atmospheric CO_2 inventory that may become relevant in some stabilization scenarios where negative future emissions are necessary.

6.4 References and Resources

6.4.1 Key References

The following key references, arranged in chronological order, provide a starting point for further study.

Barchas, R. and R. Davis. (1992). The Kerr-McGee/ABB Lummus Crest Technology for the recovery of CO_2 from stack gases, *Energy Conversion and Management*, **33** (333–340).

Sander, M. T. and C. L. Mariz. (1992). The Fluor Daniel Econamine FG™ process: Past experience and present day focus, *Energy Conversion and Management*, **33** (341–348).

Rao, A. B. and E. S. Rubin. (2002). A technical, economic and environmental assessment of amine-based CO_2 capture technology for power plant greenhouse gas control, *Environmental Science and Technology*, **36** (4467–4475).

Mimura, T., et al. (2003). Recent developments in flue gas CO_2 recovery technology, *Greenhouse Gas Control Technologies*, Proceedings of the 6th International Conference on Greenhouse Gas Control Technologies, Elsevier Science Ltd., Oxford, UK.

Keith, D. W., M. Ha-Duong, and J. K. Stolaroff. (2006). Climate strategy with CO_2 capture from the air, *Climate Change*, **74** (17–45).

Peeters, A. N. M., A. P. P. Faaij, and W. C. Turkenburg. (2007). Techno-economic analysis of natural gas combined cycles with post-combustion CO_2 absorption, including a detailed evaluation of the development potential, *International Journal of Greenhouse Gas Control*, **1** (396–417).

6.4.2 Institutions and Organizations

Institutions that have been active in the research, development, and commercialization of absorption systems are listed below.

Alstom Power Systems (chilled ammonia process): www.alstom.com/home/activities/power_generation

Cansolve Technologies Inc. (amine-based capture system): www.cansolv.com/en/co2 capture.ch2

Energy and Environmental Systems (EES) Group at the Institute for Sustainable Energy, Environment and Economy (ISEEE), University of Calgary (researching air capture technology and policy implications): www.ucalgary.ca/ees/ and www.ucalgary.ca/~keith

GE Global Research (novel solvent R&D): www.ge.com/research

Global Research Technologies (GRT; research and development of air capture technology): www.grtaircapture.com

Powerspan (ECO$_2$® chilled ammonia capture process): www.powerspan.com

7 Adsorption capture systems

In contrast to absorption, in which the absorbed component (the sorbate) enters into the bulk of the solvent and forms a solution, adsorbed atoms or molecules (known as adparticles) remain on the surface of the sorbent. Similar to absorption, however, the bonding of the adsorbate to the surface may be through either a chemical bond (chemical adsorption or chemisorption) or a weaker physical attractive force (physical adsorption or physisorption).

Gas separation or purification based on adsorption has a history of industrial application as long as that for absorption-based technologies, having been initially driven primarily by air purification applications. Adsorption processes using solid sorbents have a number of potential advantages when compared to absorption into liquid sorbents, including a very wide range of operating temperatures, a lack of liquid waste streams, and in many cases solid wastes that are environmentally benign and pose few problems for disposal.

A rapid expansion in the application of adsorption as a gas separation technology has occurred in the past decades, due in part to the development of new sorbent materials, and a diverse range of sorbents is now available or under development for CO_2 separation. These sorbents can be combined with a broad spectrum of process options, including steady-state and cyclic processes, yielding a very fertile field for performance optimization and innovation.

7.1 Physical and chemical fundamentals

The most important characteristic of a sorbent is the quantity of sorbate that a given quantity of sorbent can hold, at the relevant operating temperatures and pressures. For applications such as CCS that require separation of one specific component from a mixed feed stream, the selectivity of the sorbent for that component is equally important, and is determined by the relative adsorptive rates and capacities of the sorbent for the various components in the feed.

7.1.1 Physical adsorption thermodynamics

The process of physical adsorption is described by the Langmuir adsorption equation, or Langmuir isotherm, which can be derived from a simple model of the equilibrium between adsorption and desorption. For an adparticle of mass

Doi:10.1016/B978-1-85617-636-1.00007-9.

m (kg), at pressure P (Pa), the kinetic theory of gases gives the frequency of collisions, ν (s^{-1}), of the particle with the containing surface as:

$$\nu = P/(2\pi mkT)^{1/2} \tag{7.1}$$

where k is the Boltzmann constant (J per °K) and T is the absolute temperature (°K).

If θ is the fraction of adsorption sites that are occupied, and α is the probability of a particle sticking to the surface during any single collision, then the rate of adsorption (A) per unit surface area will be:

$$A = \alpha \nu (1 - \theta) \tag{7.2}$$

The rate of desorption, B, will be given by:

$$B = \beta \theta \, e^{-Q/RT} \tag{7.3}$$

where β is the rate constant for desorption, Q is the activation energy for desorption, which will be equal to the heat of adsorption ($-\Delta H_{abs}$), and R is the gas constant.

At equilibrium, the rates of adsorption and desorption are equal, giving:

$$\alpha \, \nu (1 - \theta) = \beta \, \theta \, e^{-Q/RT} \tag{7.4}$$

which can be solved to give the equilibrium coverage θ_{eq}:

$$\theta_{eq} = KP/(1 + KP) \tag{7.5}$$

where

$$K = \alpha \, e^{Q/RT} / \beta (2\pi mkT)^{1/2} \tag{7.6}$$

For an *n*-component system, the equilibrium surface coverage of component i is given by:

$$\theta_{eq,i} = K_i P_i / (1 + \sum\nolimits_{j=1}^{n} K_j \, P_j) \tag{7.7}$$

Adsorption from a multicomponent feed is thus a competitive process, since the equilibrium surface coverage of one component will be reduced as the summation in the denominator of Equation 7.7 is increased by the adsorptivity of other components.

The general form of the Langmuir isotherm is illustrated in Figure 7.1 for two different temperatures. The temperature dependence of the Langmuir constant is $K \sim e^{Q/RT}\, T^{-1/2}$. Since physical adsorption is always exothermic (i.e., ΔH_{abs} is negative), Q in Equation 7.3 is always positive and, as a result, K decreases rapidly with increasing temperature, as shown by the two adsorption isotherms in the figure.

The temperature and pressure dependence of θ_{eq} are exploited as the basis of the two main adsorption-based gas separation processes: temperature swing adsorption (TSA) and pressure swing adsorption (PSA). In their simplest form, these processes rely on the preferential adsorption of one gas component from a mixed feed stream at a certain operating pressure and temperature (P_H, T_L), followed by desorption of the adsorbed component at either a reduced pressure or increased temperature (P_L, T_H) (or both in the case of a pressure and temperature swing [PTSA]). The differential adsorptive capacity of the sorbent between P_H and P_L, or T_L and T_H, is called the working capacity (Δ) of the sorbent, expressed as mole per mole or mole per kilogram of sorbent (Figure 7.2).

For a given mass of sorbent (m_{abs}) of working capacity Δ, applied in an adsorption cycle of total duration t_c, the rate of production of the sorbate-rich product under ideal conditions is:

$$q = m_{abs}\,\Delta / t_c \tag{7.8}$$

For a given process throughput rate, the sorbent mass and related plant size and cost are therefore minimized by shortening the duration of the cycle. Rapid cycling is also desirable since release of the heat of adsorption results in heating of the sorbent bed during adsorption. If the bed is allowed to cool before pressure swing desorption occurs, the increase in adsorptive capacity of the sorbent as the temperature falls will reduce the working capacity of the sorbent bed.

In general, TSA cycles require relatively longer heating and cooling times, leading to large sorbent beds for a given throughput, while PSA cycles can be

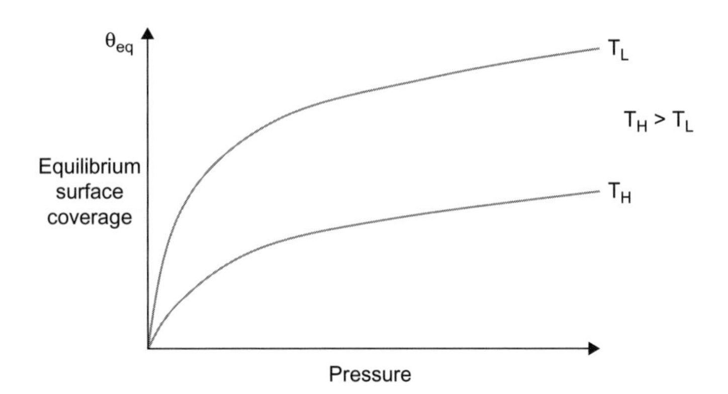

Figure 7.1 Langmuir adsorption isotherms versus temperature and pressure

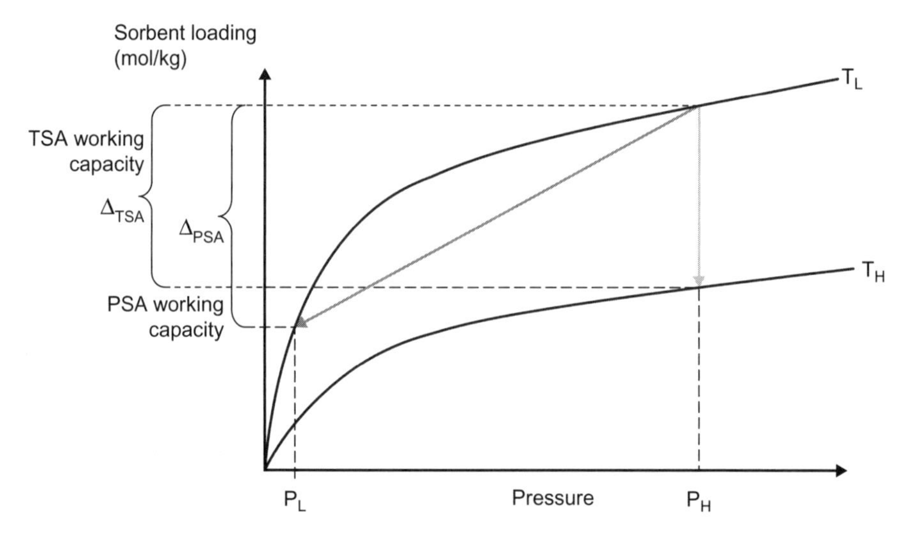

Figure 7.2 Sorbent working capacity for temperature swing and pressure swing adsorption

faster and employ smaller beds, since pressurization and depressurization can be achieved more rapidly.

7.1.2 Chemical adsorption

Metal oxide sorbents

Several of the absorbents described in Chapter 6 have also been applied in the solid state as chemical adsorbents, the most common examples being the carbonation of metal oxides, such as calcium, sodium, and potassium oxide, as well as the further carbonation to produce bicarbonates:

$$CaO_{(s)} + CO_{2(g)} \leftrightarrow CaCO_{3(s)} \tag{7.9}$$

$$Na_2O_{(s)} + CO_{2(g)} \leftrightarrow Na_2CO_{3(s)} \tag{7.10}$$

The high temperatures required for the decarbonation step make these sorbents particularly suited to flue gas or syngas processes, typically taking place at 400–600°C.

Sodium carbonate sorbent

At lower temperatures, the reversible reaction of sodium carbonate (Na_2CO_3) to form sodium bicarbonate (Na_2HCO_3) or Wegscheider's salt ($Na_2CO_3{\bullet}3NaHCO_3$)

has been investigated for CO_2 capture from postcombustion flue gases. Carbonation and regeneration occur as a result of the following reactions:

$$Na_2CO_3 + CO_2 + H_2O \leftrightarrow 2NaHCO_3 \quad (7.11)$$

$$Na_2CO_3 + 0.6\,CO_2 + 0.6\,H_2O \leftrightarrow 0.4\,(Na_2CO_3 \bullet 3NaHCO_3) \quad (7.12)$$

The carbonation reactions occur in the temperature range 60–70°C and are exothermic, with a heat of absorption of approximately $\Delta H_{abs} = -136$ kJ per mol-CO_2 (3.1 MJ per kg-CO_2). Regeneration of the sorbent is then achieved in the reverse reactions in the temperature range 120–200°C.

Solid amine sorbents

Amines and related organic absorbents have also been applied as adsorbents, by attaching the functional groups onto the surface of high-specific-area solids such as meso- or micro-porous silica, alumina, or polymer-based materials. Solid amine sorbents, using polyethyleneimine (PEI) bonded to acrylic-based polymer beads, have been used since the early 1990s for CO_2 removal in space shuttle life-support systems, with regeneration by vacuum swing desorption.

7.1.3 Physical sorbents

The desired properties of physical sorbents for CO_2 separation are summarized in Table 7.1.

Figure 7.3 shows adsorption–desorption isotherms at 25°C for three low-temperature CO_2 sorbents (activated carbon, natural zeolite, and zeolite-13x)

Table 7.1 Desired properties for CO_2 physical sorbents

Property	**Description**
Preferential CO_2 absorption	High selectivity for CO_2 relative to other CCS relevant gases (N_2, CO, CH_4) is essential for an efficient adsorption/desorption cycle.
Low heat of adsorption	Low heat of adsorption ensures that the drop of pressure or increase in temperature required in the desorption stage does not result in a high energy penalty.
High sorbent working capacity	Increased working capacity reduces the volume of sorbent required for a given throughput, reducing sorbent bed and related equipment size, capital costs, and energy requirements.
Low adsorption hysteresis	Hysteresis of the adsorption isotherm results in increased pressure drop or temperature increase to achieve the same sorbent working capacity, increasing energy penalty.
Steep adsorption isotherm	A sorbent with a steep adsorption isotherm delivers a given working capacity with the lowest pressure or temperature swing, and therefore the lowest energy penalty.

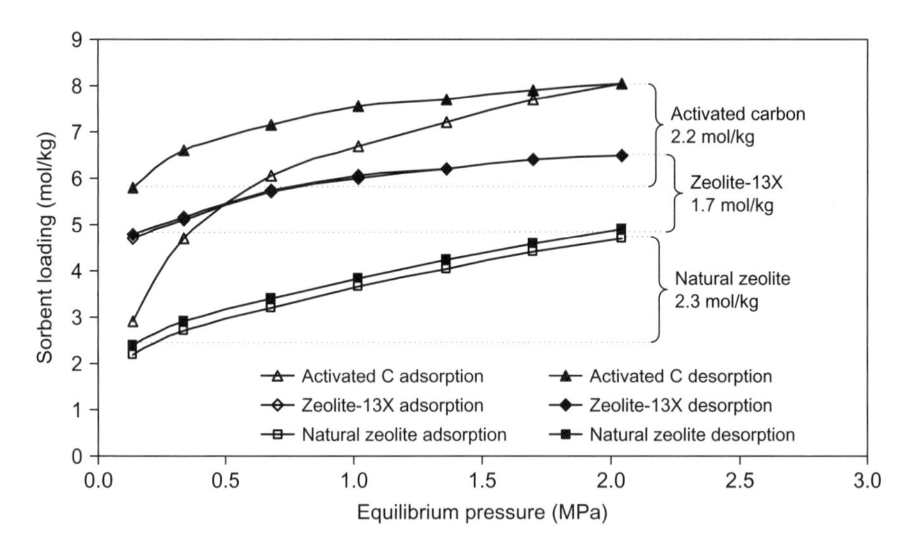

Figure 7.3 Adsorption–desorption isotherms of activated carbon, zeolite-13x, and natural zeolite at 25°C

in the pressure range 0.2–2.0 MPa. The three sets of curves illustrate the desired properties listed in Table 7.1.

Activated carbon has the highest CO_2 adsorption across most of the pressure range, reaching 8 mole per kg at 2 MPa, as well as the steepest adsorption isotherm. However, the desorption isotherm shows a high degree of hysteresis, reducing the working capacity over the pressure range shown to ~2.2 mole per kg. Zeolite-13x shows an intermediate adsorption capacity, with very low hysteresis, but the shallow gradient of the isotherm results in a lower working capacity of 1.7 mole per kg. Natural zeolite has the lowest adsorption capacity, little more than half that of activated carbon at 2 MPa, but the steeper isotherm and relatively low hysteresis result in a working capacity over this pressure range of 2.3 mole per kg, marginally higher than activated carbon.

Due to the weaker binding forces and resulting lower heat of adsorption, physical sorbents tend to find application at low temperatures, while chemical sorbents, with higher heats of adsorption, are able to sustain high capacity at higher temperatures. The division between physical and chemical sorbents is therefore also generally a division between low and high operating temperatures.

A wide variety of physical sorbents can be used for low-temperature CO_2 capture, including those summarized in Table 7.2. Others, such as zeolites and carbon molecular sieves, are described in Chapter 8.

These low-temperature sorbents are typically applied at temperatures from ambient up to ~100°C, in a few cases up to a few hundred degrees, since their capacity drops to very low levels at higher temperatures.

A range of sorbents has also been developed with high capacities at the operating temperatures typical of hot flue gas applications (400–600°C). Table 7.3

Table 7.2 Low-temperature CO_2 sorbents

Sorbent	Description
Activated alumina	Non-reactive synthetic amorphous sorbents with low heats of adsorption produced from aluminum trihydrate. The sorbent can be a beaded material or powder, and surface areas can be controlled during manufacturing to tailor the sorbent to various applications.
Activated carbon	Carbonaceous crystalline sorbents are available with a wide range of properties, depending on the raw material and activation method. Activated carbons are used in a wide range of purification processes, from purification of drinking water to various petrochemical applications.
Silica gel	Silica gels are amorphous sorbents made from silicon dioxide (SiO_2) with high porosity and internal surface areas onto which certain materials are selectively adsorbed. They have stable chemical properties and a range of microporous structures that can be controlled using different manufacturing processes.
Ion-exchange resins	Ionic polymers typically prepared as macroporous beads, in which either the anion or the cation is bound to the resin structure while the counter-ion is free. Ions with the same charge as the free counter-ion can be exchanged from a solute—for example, CO_2^- into an anion exchange resin. Able to work in humid conditions.
Surface-modified porous media	Silica or carbon-based meso-, micro-, or nanoporous media with high specific areas, modified by the incorporation of functional groups such as amines and related organic compounds.

Table 7.3 High-temperature CO_2 sorbents

Sorbent	Characteristics
Metal oxides (e.g., CaO)	High-temperature (600–1000°C) chemical sorbents, transformed to carbonates and bicarbonates and regenerated by calcining at temperatures of ~900°C
Hydrotalcites	Naturally occurring anionic clays that become CO_2 sorbents in the temperature range ~200–400°C, displaying superior stability than CaO but lower capacity and slower reaction kinetics
Lithium zirconate (Li_2ZrO_3)	Mid- to high-temperature (600–750°C) chemical sorbent. Capacity can be significantly increased by adding potassium carbonate (K_2CO_3)

summarizes a number of these high-temperature sorbents, while further novel materials under development are described in Section 7.3.3.

Hydrotalcites (HTCs) are a class of high-temperature chemical sorbents that have been widely investigated for application in sorption-enhanced reactions (Section 7.3.2). Also known as layered double hydroxides or Feitknecht compounds, hydrotalcites are a family of naturally occurring anionic clays with the general formula:

$$M^{2+}{}_{1-x}\ M^{3+}{}_{x}(OH)_2(A^{n-})_{x/n} \bullet mH_2O$$

Here M^{2+} and M^{3+} are divalent and trivalent metal ions, most commonly magnesium and aluminum respectively, A^{n-} is an anion, and the value of x is 0.20–0.33. The metal ions are held in an octahedrally coordinated hydroxide layer, similar to the mineral brucite, while the anions and water of hydration occur in an anionic interlayer. The net positive charge of the metal ion layer is compensated by the net negative charge of the anion–water interlayer.

The mineral hydrotalcite is the most common of the class of HTCs and has the formula:

$$Mg_{0.75}Al_{0.25}(OH)_2(CO_3)_{0.125} \bullet 0.5H_2O$$

The thermal decomposition (calcining) of HTC is a complex three-stage process, starting in the temperature range from ambient up to ~300°C, with the dehydration of water molecules held in the anion–water interlayer. At temperatures up to 700°C, this is followed by dehydroxilation, in which OH groups plus oxygen from the metal ion layer are released as H_2O, while two oxygenes from the carbonate ion are included into the metal ion layer. This leads to the start of decarbonation and the release of CO_2. Up to about 500°C, the so-called hydrotalcite memory effect ensures that the layered structure is restored on rehydration, either in a carbonate solution or by steam in the presence of CO_2. Above 700°C, decarbonation of the interlayer is completed, resulting in the final decomposition products MgO and $MgAl_2O_4$ in the case of mineral hydrotalcite.

The calcining process is accompanied by a rapid increase in the specific surface area as the progressive loss of the anion–water causes the clay layers to separate. The resulting high density of strong basic sites on the layer surfaces make HTCs effective as high-temperature CO_2 sorbents. The CO_2 loading capacity of HTC can be significantly increased by impregnating (or promoting) the mineral with 10–20 wt% of potassium carbonate (K_2CO_3); however, loading capacity and carbonation and decarbonation rates remain inferior to CaO, and it is only in the area of sorbent stability that HTCs exceed the performance of the simpler metal oxides.

7.1.4 Adsorption process modes

A variety of different physical configurations can be used to achieve the sorption–desorption process, the main distinction being between fixed and moving sorbent beds.

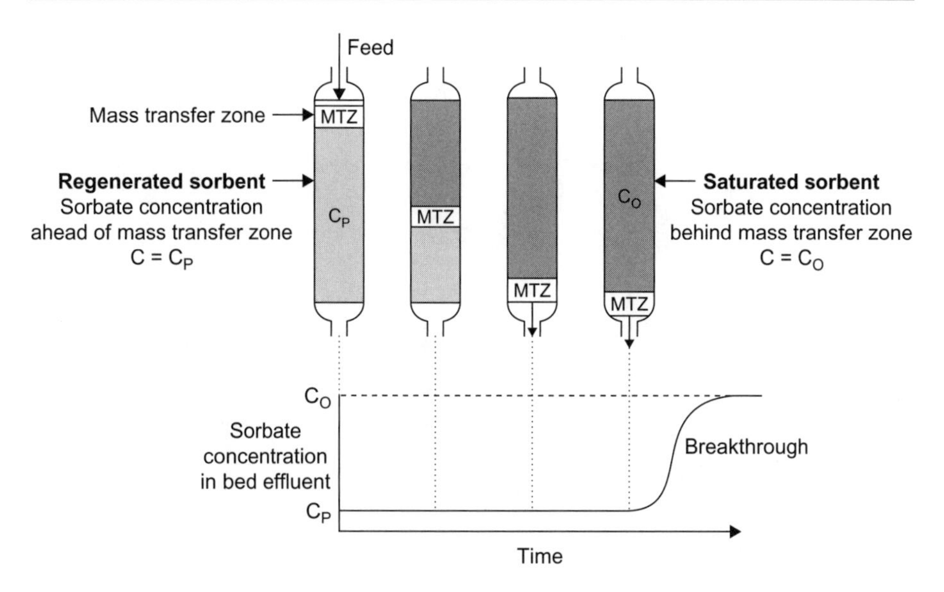

Figure 7.4 Movement of the mass transfer zone through a fixed sorbent bed

In fixed-bed systems the sorbent is held in a vertical cylinder with feed gas entering at one end and light product, from which the sorbate has been removed, produced at the other. When a bed is first put into feed mode, transfer of the sorbate from the feed gas to the sorbent starts at the feed end of the bed. Figure 7.4 illustrates the progressive movement of this mass transfer zone through a fixed bed.

The figure shows that throughout the adsorption period, only a small part of the sorbent inventory is actively adsorbing, while the majority of the sorbent is either fully loaded, and therefore waiting for regeneration, or not yet contacted by the sorbate. The release of the heat of adsorption within the localized mass transfer zone also means that heat transfer requires careful attention in fixed-bed design. This is needed in order to avoid overheating of the sorbent, leading to a reduction in carrying capacity, or to a waste of the liberated heat, which would increase the energy penalty for regeneration.

Moving-bed processes overcome these difficulties, since each parcel of sorbent can be more promptly exported from the adsorption bed for regeneration once it has been fully loaded, while mass transfer between sorption and desorption zones provides an extra degree of freedom for overall heat management within the process. The moving-bed concept was initially patented in the 1920s as a method of separating components from coal gasification syngas, and was renamed hypersorption when renewed interest in the late 1940s led to the construction of a number of commercial moving-bed units. As illustrated in Figure 7.5, the moving-bed or hypersorption process is functionally equivalent to a distillation tower, and one of the feature of the process is that the flexible operating conditions—temperature, pressure, and sorbent circulation rate—can be tuned to perform sharp separations.

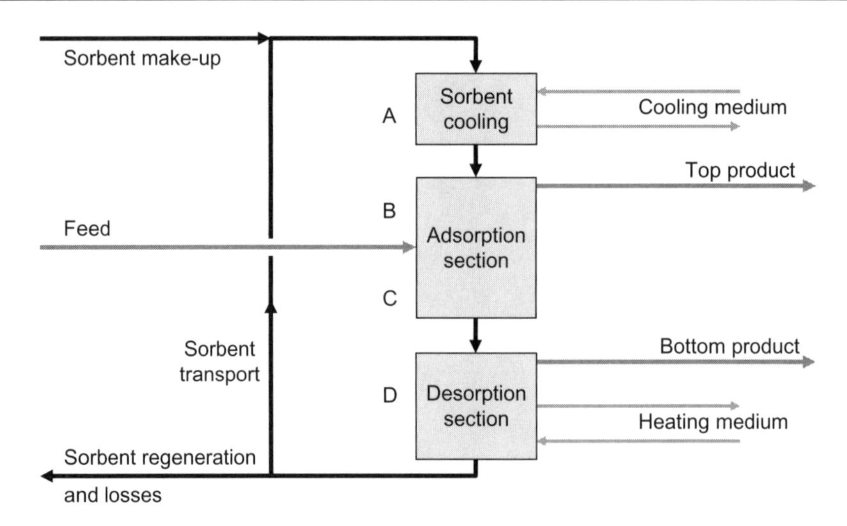

Figure 7.5 Schematic of hypersorption gas separation unit

Feed is introduced into the middle of the column and contacts sorbent, which is moving down under gravity. More-adsorbing components in the feed move down with the sorbent, while less-adsorbing components move up and exit the top of the column. The decreasing temperature toward the top of the column, controlled by cooling the circulating regenerated sorbent, increases the adsorptive capacity of the sorbent and improves separation.

Heat is provided to the bottom of the column by a heated reflux of the heavy product, equivalent to a distillation column reboiler. The increasing temperature toward the bottom of the column progressively releases the sorbate, with the potential for recovery of intermediate as well as heavy product streams. Reflux of product side streams back into the hypersorption columns can be used to achieve increased purity of these products, while multistage hypersorption, using two or more such columns in series, can also be used for separating multiple feed components.

Fluidized beds are the most common form of moving bed adsorber, and have been developed to a high state of refinement due to their widespread application in many industrial processes, including power generation combustion and gasification, as described in Section 3.1. Figure 7.6 illustrates a state-of-the-art fluidized bed absorber.

Fluidized beds also introduce additional technical requirements, including the need for the sorbent particle to resist attrition, and for equipment to circulate sorbent and to remove attrition products.

An intermediate step between fixed and moving beds is a simulated moving bed, in which sorbent in a fixed bed is cycled through the flows, temperatures, and pressures as experienced by a parcel of sorbent circulating in a moving-bed process. This is illustrated in Figure 7.7, where the moving bed system of Figure 7.5 is broken into four process steps. The fixed bed here undergoes the

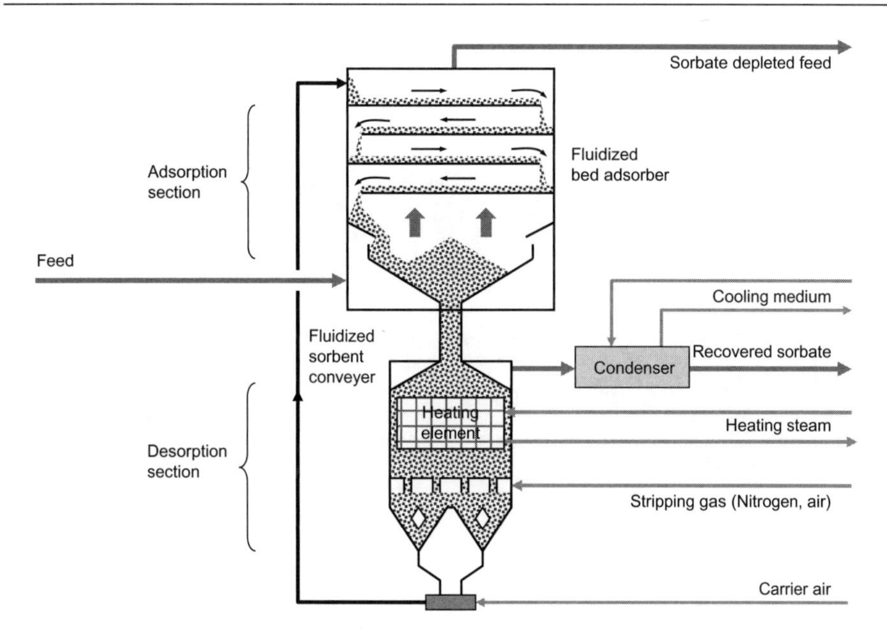

Figure 7.6 Fluidized adsorption bed configuration

same adsorption–regeneration process experienced by a parcel of sorbent falling through the hypersorption column:

Zone	Process step
A	Sorbent cooling by chilled light product reflux
B	Feed and light product recovery
C	Heavy product recovery
D	Further heavy product recovery by temperature swing

This scheme is conceptually identical to the cyclic schemes described in Section 7.2.2, where each individual fully loaded bed is taken through a sequence of regeneration steps and then rejoins the cascade of adsorbing beds.

A moving bed can also be simulated using a long fixed bed by sequentially moving the inlet and outlet feed and regeneration points along the bed. This technique is applied to small-scale fractionation for chemical analysis and is known as moving-port chromatography.

7.2 Adsorption process applications

7.2.1 Temperature swing adsorption

The use of CaO as a regenerable sorbent for CO_2 is an example of a temperature-swing process that was first proposed in the 19th century. In the temperature

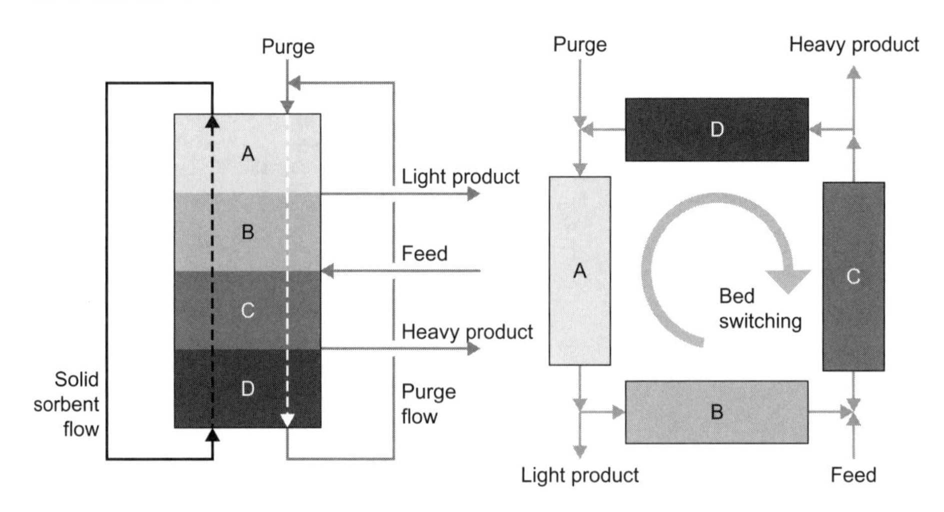

Figure 7.7 Simulated moving bed using cascaded fixed beds

range 600–800°C the carbonation reaction proceeds rapidly and high sorbent capacity can be achieved, while the familiar calcining reaction, releasing CO_2, is favored at >900°C. A pilot-scale postcombustion separation plant has been demonstrated, using interconnected fluidized bed carbonation and calcining reactors, as described in Section 7.3.5.

Natural calcium-based sorbents, commonly derived from limestone, have the disadvantage that sorbent capacity degrades with repeated regeneration, with a typical loss of 50% of capacity after 5 cycles and 80% after 20 to 40 cycles. This is largely due to the loss of micropore volume (pore diameter < 200 nm) and surface area as a result of physical changes in the sorbent, such as the sintering of micrograins and the collapse of the high-surface-area micropore structure. Sorbents derived from calcined dolomite ($CaCO_2 \bullet MgCO_3$) or huntite ($CaCO_3 \bullet 3MgCO_3$) exhibit improved durability under repeated calcining. This is attributed to the presence of MgO in the calcined product, which remains inert in the carbonation reaction at high temperature.

Electric swing adsorption

One disadvantage of desorption by temperature swing is the relatively long time needed to heat the sorbent bed during the desorption step and then cool the bed for the next adsorption step. As noted earlier, extended cycle times lead to larger adsorption beds for a given throughput rate and to higher capital costs.

Electric swing adsorption (ESA), or electrothermal desorption, is a variant of temperature swing adsorption in which the sorbent is heated by Joule heating—that is, by passing a current through the resistive sorbent material. This allows very fast heating of the sorbent bed, overcoming the long desorption period typically inherent in TSA cycles.

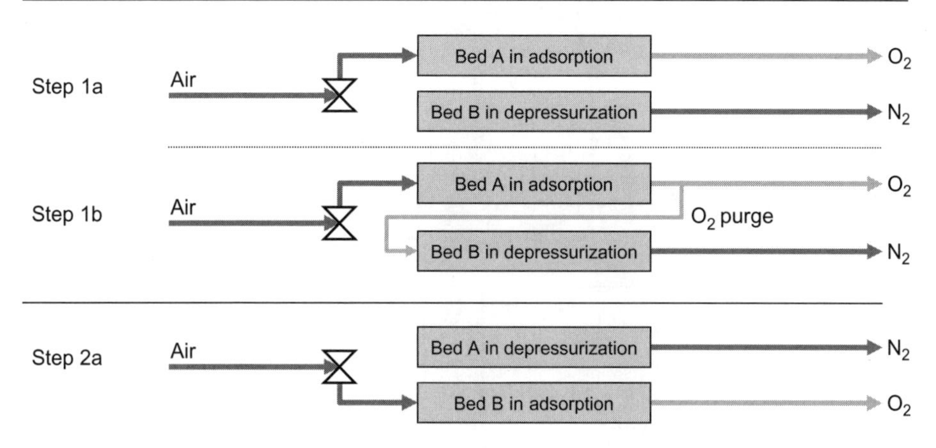

Figure 7.8 Skarstrom PSA cycle

When integrated into a larger process, TSA can usually benefit from the cheap availability of low-grade waste heat to provide energy for heating the sorption bed, whereas ESA requires high-grade electrical energy. This will increase both the operating costs and the carbon penalty of the process, unless the electric power is delivered from a zero-emissions source.

7.2.2 *Pressure swing adsorption processes*

The pressure swing adsorption concept was developed by the Esso Research and Engineering Company and published in a 1960 patent that disclosed a dual-bed, two-step cycle known as the Skarstrom cycle, illustrated in Figure 7.8.

Feed air is pretreated by compression, drying, and filtering and then flows into one of two parallel adsorption beds (A). Here nitrogen is adsorbed and oxygen passes through the bed and is recovered as the first product stream. While the first bed is operating in this mode, the sorbent material in the second bed (B) is being regenerated by releasing the pressure, recovering nitrogen as the second product stream. After depressurization, nitrogen recovery is increased by purging bed B with a fraction of the bed A product stream, the cycle ending before the purge stream breaks through the bed.

In the second part of the cycle, the functions of the two beds are reversed, with repressurization and adsorption taking place in bed B and regeneration (N_2 release and purge) in bed A. The process can be represented by the scheme shown in Figure 7.9, in which the upper schematic shows (from left to right) the successive cycle steps for bed A, while the table shows how these are phased over the two beds. This format will be useful in describing more complex PSA cycles in the following sections.

The nomenclature of this schematic, including terms that will appear in the later discussion, is summarized in Table 7.4.

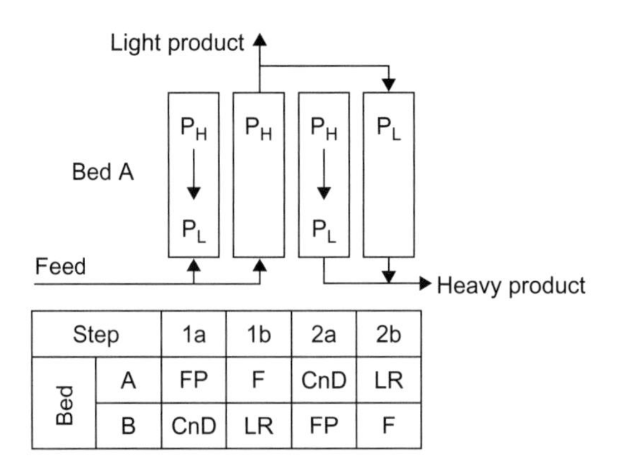

Figure 7.9 Schematic and cycle sequence chart of the two-bed Skarstrom PSA cycle

Table 7.4 PSA process schematic nomenclature

Symbol	**Description**
CnD	Counter-current depressurization
CoD	Co-current depressurization
$EQ_{1,2,3}$	Pressure equalization step 1, 2, 3
F	Feed
FP	Feed pressurization
HP	Heavy product (sorbate-rich)
HR	Heavy product reflux
LP	Light product (sorbate-lean)
LPP	Light product pressurization
LR	Light product reflux
P	Purge
P_H	High operating pressure
P_I	Intermediate pressure
$P_{1,2,3}$	Intermediate pressures 1, 2, 3
P_L	Low operating pressure

Although the simple Skarstrom cycle has not been applied to commercial air separation, due to low product recovery and high energy requirements, it is widely used for air drying using synthetic zeolites or activated alumina to adsorb water vapor.

A disadvantage of this simple cycle is that the light product reflux step causes a dilution of the heavy component relative to the initial feed, and as a result it is difficult to achieve high purity for the heavy product. Several improvements have been made to the Skarstrom cycle, including initial pressurization using light product, co-current depressurization after the initial feed step, one or more pressure-equalization steps, and a heavy product reflux step, as described in Table 7.5. While the addition of reflux steps increases product purity and recovery, either the flow rate or duration of the reflux step must be controlled to avoid breakthrough of the reflux into the other product stream, reducing throughput of the overall system.

To illustrate these improvements, Figure 7.10 shows the Skarstrom cycle with light product pressurization (LPP) and co-current depressurization. Further improvements to these adsorption cycles are being developed to optimize the process for CO_2 capture from flue gases, and are described in Section 7.3.1.

Pressure swing adsorption has been used since the 1980s for the purification of hydrogen produced by steam methane reforming (SMR) followed by the water–gas shift (WGS) reaction. This process is commonly used in oil refining to generate hydrogen for hydrotreating heavier crude oils.

The SMR and WGS product stream is typically delivered at a pressure of ~2 MPa and temperature of 850–950°C. CO_2 removal from the hydrogen product stream using pressure swing adsorption can recover 90% of the hydrogen in the feed stream with a purity of >99.99%. Systems can simultaneously deliver H_2 and CO_2 at high purity and typically operate with a 3- to 10-minute cycle time. Continuous development has resulted in commercial systems that deliver high reliability at low capital and operating costs.

This separation can also be achieved using a chemical looping sorption approach as described in Section 7.3.5.

Table 7.5 Skarstrom cycle improvement steps

Cycle step	Description
Light product pressurization	Light product purity can be improved by using a stream of light product rather than raw feed to bring the sorption bed up to high pressure before the main feed step. This necessitates a shift from two-bed to four-bed configuration for continuous operation, since with LPP the feed is active in only one of the four steps.
Co-current depressurization	Co-current depressurization reduces the mass of feed plus LP reflux in the pore space of the sorption bed, and allows the heavy product to start desorption while the light product is being produced, resulting in a higher-purity heavy product.
Pressure equalization	Interconnection of two beds to achieve pressure equalization improves energy efficiency by transferring momentum from a bed that is in the pressure-reducing stage after feed to another bed in the pressure-increasing stage leading up to the feed step.
Heavy reflux	Addition of a heavy product reflux to the co-current depressurization step increases heavy product purity in the later counter-current depressurization and light product reflux steps, but requires recompression of the heavy product up to the full feed pressure.

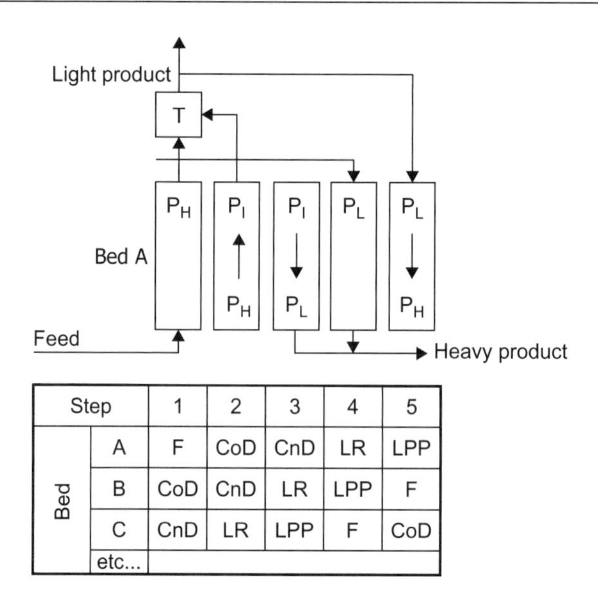

Step		1	2	3	4	5
Bed	A	F	CoD	CnD	LR	LPP
	B	CoD	CnD	LR	LPP	F
	C	CnD	LR	LPP	F	CoD
	etc...					

Figure 7.10 Five-step Skarstrom PSA cycle with LPP and CoD

Vacuum swing adsorption

If the feed gas to a separation process is at or near ambient pressure, the energy penalty associated with compression to achieve an elevated feed pressure typical of a PSA cycle can be minimized or avoided by operating the cycle with P_L below ambient pressure. Vacuum swing adsorption (VSA) is such a variant of PSA in which a partial vacuum is applied to the sorbent bed during the desorption stage. The product purity and separation efficiency of a PSA system may also be increased by applying a vacuum to the bed under regeneration, resulting in a hybrid vacuum and pressure swing adsorption (VPSA) cycle.

A VSA cycle incorporating a depressurization step to an intermediate pressure followed by evacuation at a pressure below ambient was patented in 1957 by French engineers P. Guerin de Montgareuil and D. Domine. This cycle can be implemented with a wide range of operating schemes and bed numbers, and a schematic of two-bed version is shown in Figure 7.11.

The Guerin–Domine cycle performs significantly better in air separation than the Skarstrom cycle and has been commercially applied for midsized air separation units (up to ~150 t-O_2 per day) since the 1970s.

7.3 Adsorption technology RD&D status

7.3.1 Advanced PSA cycles

Optimization of the PSA cycle configuration is an active area of research, with the aim of achieving high purity of both light and heavy product streams. Early

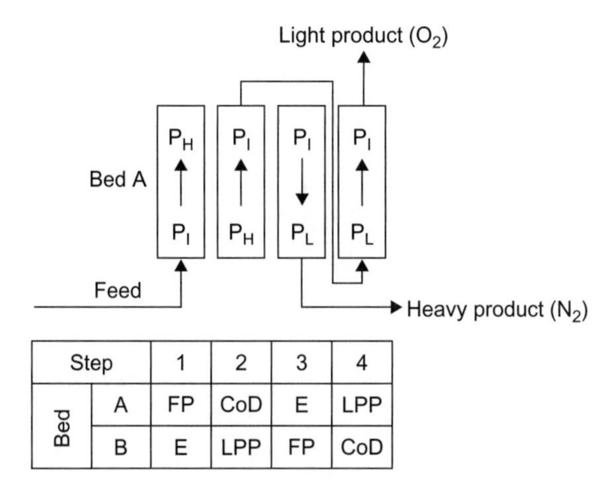

Step		1	2	3	4
Bed	A	FP	CoD	E	LPP
	B	E	LPP	FP	CoD

Figure 7.11 Two-bed air separation unit using the Guerin–Domine cycle

improvements described earlier included the introduction of refluxes of either light (LR) or heavy product (HR) to improve the purity of one or other product streams. The combination of LR and HR into a single dual-reflux cycle has been shown in process simulations to outperform other cycle configurations in maximizing recovery of a high-purity CO_2 heavy product stream from high-temperature flue gases.

The dual-reflux cycle is shown schematically in Figure 7.12. The heavy reflux can be delivered either from the counter-current depressurization step, as shown here, or from the subsequent light reflux step, the former having the advantage of requiring the reflux product to be compressed only from P_I to P_H rather than from P_L to P_H.

With the use of dual refluxes, the PSA cycle begins to mimic the operation of a distillation column. Optimization of the dual-reflux cycle is continuing, building on this analogy to develop even more flexible cycles.

7.3.2 Sorption-enhanced reactions

High-temperature adsorption has potential applications in precombustion CO_2 capture and is the subject of a number of R&D projects. The production of hydrogen through SMR and the WGS reaction (Equations 7.13 and 7.14) is an important example.

$$\text{SMR} \quad CH_4 + H_2O \leftrightarrow CO + 3H_2 \tag{7.13}$$

$$\text{WGS} \quad CO + H_2O \leftrightarrow CO_2 + H_2 \tag{7.14}$$

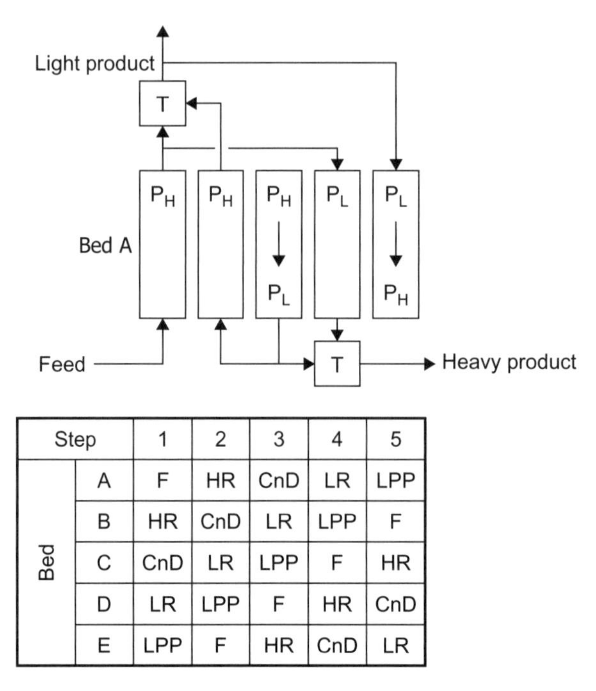

Step		1	2	3	4	5
Bed	A	F	HR	CnD	LR	LPP
	B	HR	CnD	LR	LPP	F
	C	CnD	LR	LPP	F	HR
	D	LR	LPP	F	HR	CnD
	E	LPP	F	HR	CnD	LR

Figure 7.12 Five-bed dual-reflux PSA cycle with heavy reflux from countercurrent depressurization

$$CO_2 \text{ removal} \qquad CO_2 + CaO \leftrightarrow CaCO_3 \tag{7.15}$$

$$\text{Overall reaction} \qquad CH_4 + 2H_2O + CaO \rightarrow CaCO_3 + 2H_2 \tag{7.16}$$

Reversible chemical reactions, such as the WGS reaction, can reach an equilibrium state before the full consumption of the reactants as a result of the buildup of reaction products and the increasing rate of the reverse reaction. Removal of one of the reaction products from the gas phase in the reaction zone, for example by adsorption, enables the reaction to reach completion.

As described in Section 7.2, pressure swing adsorption has been commercially applied to purification of the hydrogen product stream from these reactions for many years. Current research aims to move the adsorption application further upstream, into the high temperatures and pressures of the SMR and WGS reactors.

The conditions in SMR and WGS reactors put severe requirements on sorbents, including:

- High CO_2 loading capacity and sufficiently fast adsorption and desorption kinetics
- Mechanical and chemical stability for long periods, at temperatures up to 900°C and pressures up to 4.0 MPa

Table 7.6 SEWGS feed and product stream compositions

Component	Steam-reformed feed gas	SEWGS product gas
H_2	57%	89%
H_2O	16%	8%
CO	16%	0.5%
CO_2	10%	2%
CH_4	0.5%	0.5%

- Tolerance to high stream partial pressures, with p_{steam}/p_{CO2} typically >20 in steam reforming

Sorption-enhanced water–gas shift

In conventional SMR, typically 70–80% of the methane feed is converted to hydrogen in a reactor at a temperature of 700–950°C and a pressure of 1.5–4.0 MPa, yielding a product stream with the composition shown in Table 7.6. A sorption-enhanced WGS (SEWGS) reactor, using a WGS catalyst and high-temperature CO_2 sorbent, can remove 90% of the carbon from the fuel gas product stream. If higher-purity hydrogen is required, for example for zero-emission combustion, an additional PSA stage can be added downstream of the sorption-enhanced reactor.

A six-bed laboratory-scale SEWGS reactor has been developed as part of the European Union-funded CACHET project. The prototype uses a pressure swing cycle with three pressure-equalization steps, as shown schematically in Figure 7.13.

The pressure-equalization steps improve the efficiency of the cycle by transferring energy between depressurizing and repressurizing beds, while the sequencing of the steps maximizes the purity of the two product streams. However, the low delivery pressure of the CO_2 plus steam product stream (~0.2 MPa) and the requirement for high-pressure CO_2 (~3.0 MPa) for the heavy reflux, step are challenges for the energy efficiency of the overall process.

The reactor in these trials uses a potassium carbonate (K_2CO_3)-promoted hydrotalcite sorbent and an iron–chromium catalyst, both deposited on a structured support. Further RD&D work is focused on:

- Verifying the stability of the sorbent and catalyst
- Optimizing flow rates of the feed, heavy reflux, and purge steps
- Minimizing CO_2 slip into the fuel gas product stream
- Maximizing CO_2 purity and delivery pressure
- Minimizing catalyst requirement to reduce reactor vessel sizes and costs

Sorption enhanced steam reforming

As an alternative to enhancing the WGS reaction, the overall SMR plus WGS reaction can also be achieved in a single sorbent-enhanced reaction step (Equation 7.16). The removal of CO_2 using CaO shifts the overall reaction to

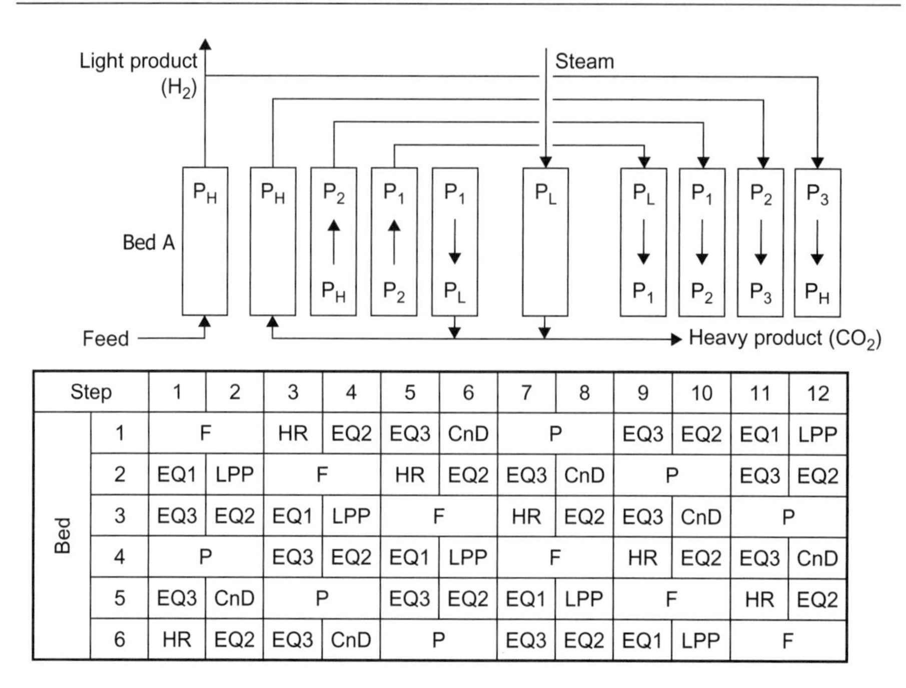

Step		1	2	3	4	5	6	7	8	9	10	11	12
Bed	1	F		HR	EQ2	EQ3	CnD	P		EQ3	EQ2	EQ1	LPP
	2	EQ1	LPP	F		HR	EQ2	EQ3	CnD	P		EQ3	EQ2
	3	EQ3	EQ2	EQ1	LPP	F		HR	EQ2	EQ3	CnD	P	
	4	P		EQ3	EQ2	EQ1	LPP	F		HR	EQ2	EQ3	CnD
	5	EQ3	CnD	P		EQ3	EQ2	EQ1	LPP	F		HR	EQ2
	6	HR	EQ2	EQ3	CnD	P		EQ3	EQ2	EQ1	LPP	F	

Figure 7.13 SEWGS cycle schematic

the right, and the presence of a catalyst allows the reaction to proceed without the need for a separate WGS stage, enabling almost complete conversion at significantly lower temperatures (400–600°C). The lower operating temperature can result in significant cost benefits compared to conventional reforming, since expensive alloy steels are not required and the overall size of heat exchange equipment can be reduced.

Figure 7.14 shows a process schematic for a sorption-enhanced reactor (SER)-based hydrogen production system, using a simple dual-bed reactor. A methane plus steam mixture is fed into reactor bed 1, containing a high-temperature CO_2 sorbent and a SMR catalyst. The sorption-enhanced reaction results in a hydrogen plus steam product stream, while CO_2 is adsorbed within the reactor. The reaction step continues until the H_2 effluent purity drops to a preset value. Concurrently, reactor 2 is fed with steam as a purge gas to remove CO_2, with desorption enhanced either by a pressure or temperature swing, or both. Water is condensed from the desorbed stream, yielding a CO_2 product stream suitable for compression and storage.

Laboratory-scale proof-of-concept experiments have confirmed the high H_2 and very low CO and CO_2 concentrations in the light product stream compared to conventional SMR. Further development is ongoing, focusing on:

- Identifying sorbent–catalyst combinations with improved properties
- Natural sorbent selection criteria and pretreatment for low attrition
- Fluidized bed operating conditions to reduce sorbent attrition

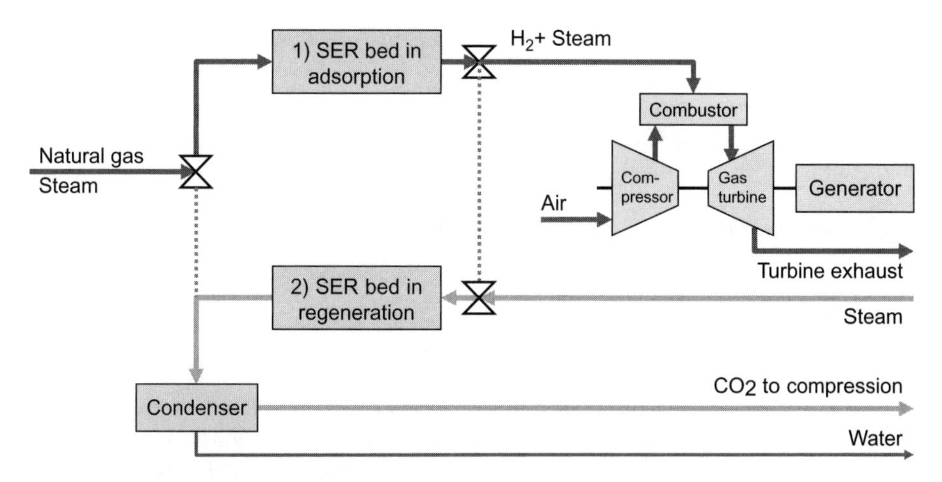

Figure 7.14 Hydrogen production using sorption-enhanced steam reforming

7.3.3 Novel sorbent materials

Natural calcium-based minerals, such as limestone and dolomite, have a number of advantages as high-temperature CO_2 sorbents, such as fast carbonation and calcination reaction kinetics, natural abundance, and low cost. However, they suffer from a gradual loss of sorption capacity with repeated cycling, and degradation in the presence of sulfur due to the formation of non-regenerable calcium sulfate.

In applications where a catalyst is added, such as in SMR, any mismatch between catalyst and sorbent lifetimes is problematic, requiring either the premature disposal of still-active catalyst or the separation of mixed solid catalyst and sorbent. Considerable research effort has been aimed at developing more reliable high-temperature sorbents, based on lithium and sodium, as well as calcium. Among the candidate sorbents that have been tested are:

- Lithium zirconate (Li_2ZrO_3)
- Lithium orthosilicate (Li_4SiO_4)
- Sodium zirconate (Na_2ZrO_3)
- Calcium aluminate ($CaAl_2O_4$)

These sorbents can be carbonated in SMR reactions, such as:

$$Li_4SiO_4 + CH_4 + 2H_2O \rightarrow Li_2SiO_3 + Li_2CO_3 + 4H_2 \tag{7.17}$$

and regenerated through the analogous calcining reaction:

$$Li_2SiO_3 + Li_2CO_3 + \text{heat} \rightarrow Li_4SiO_4 + CO_2 \tag{7.18}$$

The zirconate sorbents are found to give high hydrogen yield in SMR experiments but suffer from slow reaction kinetics (lithium zirconate) and catalyst poisoning (sodium zirconate). Further investigation of carbon-based sorbents also continues, investigating synthetic CaO sorbents and treatment methods to address the performance issues of natural carbon-based sorbents. One such area is the chemical modification of high-surface-area carbon-based adsorbents by the introduction of CaO and CaO • MgO, resulting in significantly enhanced high-temperature sorption capacity for CO_2.

Surface-modified meso- and nanoporous silicas

This technology incorporates mono- and diethanol amines onto the pore surface of porous silica, creating a chemical adsorbent that has a number of advantages over the analogous aqueous chemical absorption.

The rate and efficiency of adsorption is increased as a result of the high surface area of the silica, which allows easy access of CO_2 to the functional amine groups and results in high sorbent loading. Oxidative degradation and loss of amine by evaporation are minimized as a result of the surface binding, while the operational problems typical of aqueous amine systems, notably corrosion, foaming, and contaminated wastewater, are eliminated in the dry gas–solid system.

7.3.4 High-frequency pressure cycling

In pressure swing adsorption, the amount of sorbent required to achieve a given sorbate-throughput rate is directly related to the frequency of the pressure cycle (Equation 7.8), with a high frequency resulting in a higher throughput rate. Raising the cycling frequency therefore has the potential to reduce the capital cost of PSA systems by reducing sorbate volume and the related physical system size. Since mechanical switching of valves can result in early mechanical failure, alternative methods of rapidly cycling the system pressure are under investigation. Possible approaches include the use of reciprocating (piston) or pressure wave-based systems.

7.3.5 Chemical looping

CaO looping

CaO looping is being investigated for postcombustion CO_2 capture, using the carbonation and calcining of CaO (Equation 7.15) in interconnected fluidized-bed chemical adsorption and regeneration reactors. Figure 7.15 illustrates the physical configuration, in which hot flue gases enter the base of the carbonation reactor (A), along with a feed of hot CaO. The fine sorbent particles rise in the column along with the flue gas stream and are separated in a cyclone separator (C1).

Depleted flue gases are released while the carbonated sorbent ($CaCO_3$) is conveyed to the base of the regenerating bed (B). Calcination at 900°C results in the release of CO_2, which is separated from regenerated sorbent (CaO) by a second cyclone separator (C2). The loop is completed by conveying the regenerated sorbent to the base of the carbonation reactor. An equivalent CaO looping concept can also be used to remove CO_2 from the $H_2 + CO_2$ syngas product stream from the WGS reaction, commonly achieved using fixed-bed systems.

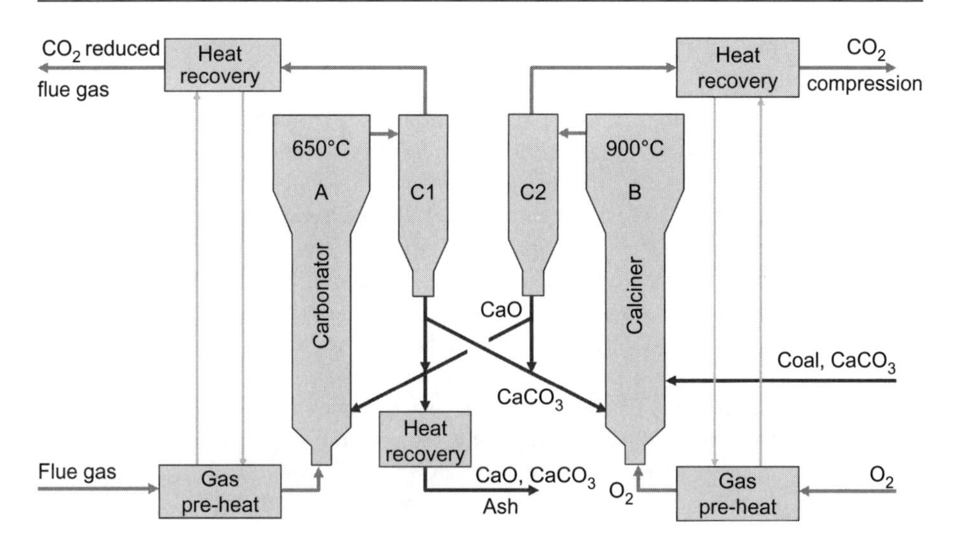

Figure 7.15 Flue gas capture using CaO looping

Ongoing research into synthetic calcium-based sorbents for chemical looping aims to sustain sorbent capacity through repeated regeneration cycles. Mixed sorbents such as $CaO(75\%)/Ca_{12}Al_{14}O_{33}(25\%)$ or $CaO(90\%)/Al_2O_3(10\%)$ have been shown to sustain higher carrying capacity over repeated cycles than pure CaO.

The sorption capacity depends on the surface properties of the sorbent particles, notably the specific area (m^2 per kilogram) of the sorbent, which can be varied by synthesizing the CaO sorbent from a range of precursors. Synthesis by drying and calcining calcium acetate-based solutions has been shown to yield high-capacity sorbent material with a fluffy morphology. Selectivity for CO_2 over N_2, O_2, and H_2O can also be tailored by doping the precursor with alkali metals, doping with cesium (Cs) yielding the highest selectivity compared with sodium, potassium, or rubidium dopants.

Alternatives to CaO, such as lithium silicate (Li_4SiO_4), have also been investigated for high-temperature chemical looping reactors, with the aim of identifying sorbents with lower regeneration temperatures. Lithium silicate can be regenerated at a temperature of ~800°C, significantly lower than the temperature required for calcining $CaCO_3$.

Hybrid combustion–gasification by chemical looping

The versatility of chemical looping is well illustrated by the proposed combination of two calcium-based chemical loops into a hybrid combustion–gasification process. The first loop uses $CaSO_4$ as an oxygen carrier to gasify coal in a partial oxidization step, completing the loop by oxidation of CaS in air:

$$4C + CaSO_4 \rightarrow 4CO + 4CaS \tag{7.19}$$

$$CaS + 2O_2 \rightarrow CaSO_4 \tag{7.20}$$

In the presence of steam, the partial oxidation product is water-shifted to produce H_2 and CO_2, and a second CaO loop is then used to remove CO_2 from the syngas stream.

A schematic of this hybrid system, which is currently at the pilot-testing stage, is shown in Figure 7.16. Three interconnected fluidized-bed reactors are shown: a reducer, calciner, and oxidizer. The reducer is fed with pulverized coal or other carbonaceous fuel, steam, and the chemical sorbents $CaSO_4$ and CaO. The fuel gasification, WGS, and CaO carbonation reactions take place in this reactor, producing a product gas stream of pure hydrogen. The solid reaction products are carried out of the top of the reactor with the gas product stream and separated in the first of three cyclone separators. The solids are then conveyed to the second reactor, where the $CaCO_3$ is calcined and CO_2 released. Heat for the calcination reaction is provided by a heat exchange loop from the combustion reactor, using bauxite as the heat transfer medium (not shown in the figure). Solid products are then conveyed to the final reactor, where CaS is oxidized in a heated air stream.

Attrition products from the $CaSO_4$ and CaO particles, as well as other inert combustion products, are removed from circulation, while make-up carbonate is added to the calciner as required. Reflux of the solid reaction products from the reducing and oxidation reactors can be used for additional process control.

The two chemical loops have been proven in laboratory-scale tests, and a small-scale pilot facility integrating both loops has been developed by Alstom Power.

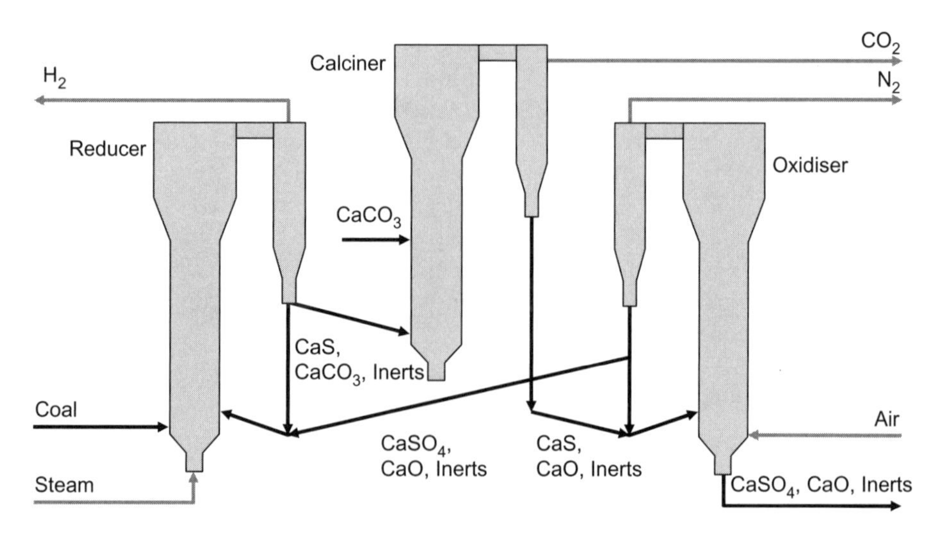

Figure 7.16 Hybrid combustion–gasification chemical looping process

7.4 References and Resources

7.4.1 Key References

The following key references, arranged in chronological order, provide a starting point for further study.

Yang, R. T. (1987). *Gas Separation by Adsorption Processes*, Butterworth Publishers, London, UK.

Ruthven, D. M., S. Farooq, and K. S. Knaebel. (1994). *Pressure Swing Adsorption*, John Wiley & Sons, Inc., New York.

Ishibashi, M., *et al.* (1996). Technology for removing carbon dioxide from power plant flue gas by the physical adsorption method, *Energy Conversion and Management*, **37** (929–993).

Reynolds, S. P., A. D. Ebner, and J. A. Ritter. (2005). New pressure swing adsorption cycles for carbon dioxide separation, *Adsorption*, **11** (531–536).

Hufton, J. R., *et al.* (2005). Development of a process for CO_2 capture from gas turbines using a sorption enhanced water gas shift reactor system, *Proceedings of the 7th International Conference on Greenhouse Gas Control Technologies*, Elsevier, Oxford, UK.

Chaffee, A. L., *et al.* (2007). CO_2 capture by adsorption: Materials and process development, *International Journal of Greenhouse Gas Control*, **1** (11–18).

Grande, C. A. and A. E. Rodrigues. (2007). Electric swing adsorption for CO_2 removal from flue gases, *International Journal of Greenhouse Gas Control*, **2** (194–202).

H.-W. Häring (ed.). (2007). *Industrial Gases Processing*, Wiley-VCH Verlag GmbH, Weinheim, Germany.

7.4.2 Institutions and Organizations

Institutions that have been active in the research, development, and commercialization of adsorption systems are listed here.

Alstom Power (development of chemical looping combustion systems): www.power.alstom.com/home/about_us/clean_power_today

CACHET (an integrated research project, funded by the European Union, aiming to develop technologies to reduce power station GHG emissions by 90%): www.cachetco2.eu

ECN (Energy Research Centre of the Netherlands; novel sorbents and sorption-enhanced reactions): www.ecn.nl

IFE (Institute for Energy Technology; high-temperature sorbents for sorption-enhanced reactions): www.ife.no

RTI (Research Triangle International; novel sorbents): www.rti.org/page.cfm/Climate_Change

7.4.3 Web Resources

PSA Plants (business-to-business marketplace for the PSA process): www.psaplants.com

Adsorption.org (knowledge sharing site for adsorption): www.adsorption.org

8 Membrane separation systems

The first application of membranes for gas separation was in the 1980s, when techniques were developed for hydrogen separation, for example from hydrotreaters in refineries, for oxygen/nitrogen separation, and for the separation of CO_2 from natural gas, the latter either to process natural gas to sales quality or for enhanced oil recovery (EOR) applications. A number of early membrane gas separation applications are summarized in Table 8.1.

The technological breakthrough that brought membranes into large-scale industrial use was the development of composite polymer membranes, in which a thin selective layer is bonded to a thicker, non-selective and inexpensive layer that provides mechanical support.

Membranes have a number of potential applications in carbon capture, from hydrogen separation in precombustion capture systems to CO_2 separation from postcombustion flue gases. A wide variety of different membrane techniques are potentially applicable, using proven technologies that are already deployed on an industrial scale as well as novel technologies, based on advanced materials, that are under research and development.

8.1 Physical and chemical fundamentals

Conceptually, a membrane acts as a filter, separating one specific component (the permeate) from a mixture of gases in a feed stream. This "filtration" process can involve a number of different physical and chemical processes, depending on the membrane design and materials, and these processes are discussed in this section.

The two common membrane separation flow schemes are illustrated schematically in Figure 8.1. A feed gas stream enters the module and the permeate exits on the downstream side, having passed through the membrane. The feed stream minus permeate is termed the retentate, and may either exit the module having been depleted of the permeate (as in the cross flow example [a]) or be continuously replenished (as in the dead-end flow case [b]).

The key characteristic of membranes that determines the mechanism for transporting permeate molecules is porosity, and membranes are classified as either porous or nonporous. For porous membranes, the pore size determines the size of particles (molecules at the smallest scale) that can pass through the membrane, as shown in Table 8.2.

Doi:10.1016/B978-1-85617-636-1.00008-0

Table 8.1 Early membrane separation applications and technologies

Application	Company	Technology
Hydrogen recovery from ammonia production	Monsanto	Prism® hollow-fiber polymeric membranes
Nitrogen separation from air	Dow	Generon® hollow-fiber polymeric membranes
CO_2 separation from methane and ethane	Cynara/NATCO	Acetyl activated nonporous cellulose acetate membrane
CO_2 separation from methane and ethane	UOP-Honeywell	Separex™ spiral-wound and hollow-fiber cellulose acetate membranes

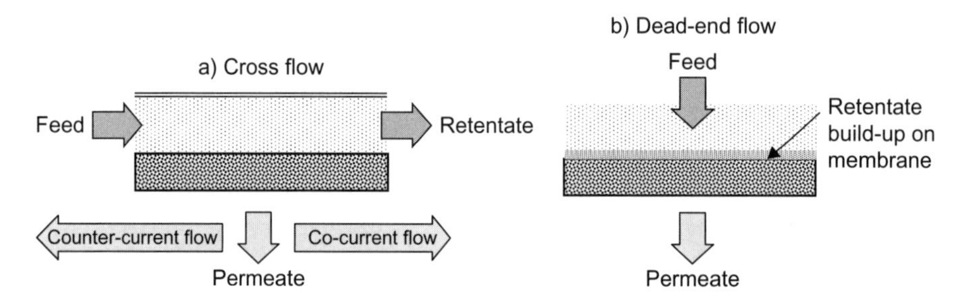

Figure 8.1 Cross flow (a) and dead-end flow (b) membrane configurations

Table 8.2 Porous membrane classification

Pore classification	Pore size range	Filtration classification
Macroporous	>50 nm (500 Å)	Microfiltration
Mesoporous	2–50 nm (20–500 Å)	Ultrafiltration
Microporous	1–2 nm (10–20 Å)	Nanofiltration
Nanoporous	<1 nm (10 Å)	Molecular sieving

At the macropore scale, gas flow is dominated by viscous forces, giving way to diffusion as the pore size reduces below the mean free path of gas molecules. Below 10 Å, as the pore size approaches molecular dimensions, flow is increasingly dominated by surface interaction on the pore walls, and is termed molecular sieving. Once the pore size drops to the 1 Å scale, a solution–diffusion transport mechanism occurs rather than molecular sieving. These transport mechanisms are described in the following sections.

8.1.1 Porous membrane transport processes

The physical nature of the gas transport process occurring in porous membranes depends on the size of pores in the membrane and on the temperature and pressure of the feed gas. The different transport regimes are characterized by the Knudsen number N_{Kn}, which is equal to the ratio of the mean free path of gas molecules for component i (λ_i) to the average pore diameter in the membrane (d_0):

$$N_{Kn} = \lambda_i/d_0 \tag{8.1}$$

Viscous Capillary Flow

If the membrane pore diameter is very large compared to the mean free path of the gas molecules ($N_{Kn} < 0.001$), the gas flux is governed by Poiseuille's equation for viscous flow through capillaries:

$$J_i = r^2 \Delta P P_{av}/(8 \mu RTL) \tag{8.2}$$

where J_i is the flux of gas component i, ΔP is the differential pressure across the membrane, P_{av} is the average pressure within a capillary, r is the representative pore radius, μ is the gas viscosity, R is the gas constant, T is the temperature (°K), and L is the length of the capillary (which, ignoring pore tortuosity, equals the membrane thickness). The flux can also be expressed in terms of the porosity (ε) of the membrane as:

$$J_i = r^2 \Delta P \varepsilon/(8 \mu L) \tag{8.3}$$

A mixture of two gases in capillary flow will be characterized by a single average pressure and a single viscosity that will depend on the ratio of gases in the mixture. The ratio of fluxes of the two components (J_i/J_k) will depend on their partial pressure in the mixture and the transport will exhibit no selectivity between the components, since selectivity requires a transport mechanism that depends on characteristics of the individual components in the mixture.

Knudsen Diffusion

If the mean free path is very much larger than the membrane pore radius ($N_{Kn} > 100$), gas molecules will collide more frequently with the pore walls than they will with other gas molecules. This situation is termed free-molecule flow, or Knudsen flow (Figure 8.2), and the flux rate is determined by the Knudsen equation:

$$J_i = D_i \Delta P/RTL = 2r \Delta P(8RT/\pi M_i)^{1/2}/3RTL \tag{8.4}$$

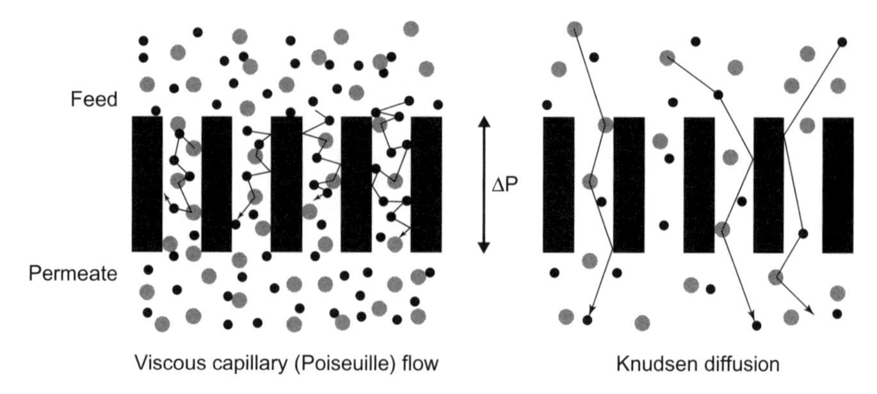

Figure 8.2 Viscous capillary flow and Knudsen diffusion

where $D_i = 2/3\ r\ (8RT/\pi M_i)^{1/2}$ is the Knudsen diffusion coefficient for component i and M_i is the molecular weight of the gas. Equation 8.4 shows that if two gas species are being transported through a membrane under Knudsen flow, the flux ratio (J_i/J_k) varies inversely with the ratio of the molecular weights:

$$(J_i/J_k) = (M_k/M_i)^{1/2} \tag{8.5}$$

In the transition region between Knudsen flow and Poiseuille flow ($0.001 < N_{Kn} < 100$) correction factors are required to account for the varying effect of molecular and boundary collisions. Equations 8.4 and 8.5 apply to pure gases moving in straight cylindrical pores, and a correction factor is also needed to account for pore size distributions and for tortuous flow paths through actual membrane pore networks. Equation 8.4 can be corrected with a porosity–tortuosity factor ε/T by replacing D_i for a single pore by D_i for a porous medium:

$$D_{i\ \text{Porous medium}} = D_{i\ \text{Single pore}} \cdot \varepsilon/\tau \tag{8.6}$$

The molecular weight dependence in Knudsen flow introduces the possibility of selective flow of gases through the membrane. However, as shown in Table 8.3, the selectivity that can be achieved for common CCS gas separation problems is limited due to the relatively small molecular weight differences. Meso- and macroporous membranes exhibiting Knudsen flow are therefore not attractive for direct gas separation applications.

However, these membranes are proposed for use in gas–liquid membrane contactors that are being developed as an alternative to amine absorption towers for CO_2 recovery from flue gas. In this application flue gas flows through the bores of a hollow-fiber bundle membrane module (Section 8.2.1), with a CO_2 absorbent liquid, such as an amine or amino acid salt solution, flowing

Table 8.3 Porous membrane selectivity under Knudsen flow

Gas components	Molecular weight ratio	Knudsen flux ratio
CO_2/H_2	21.83	1/4.67
CO_2/CH_4	2.75	1/1.66
CO_2/N_2	1.57	1/1.25
O_2/N_2	1.14	1/1.07

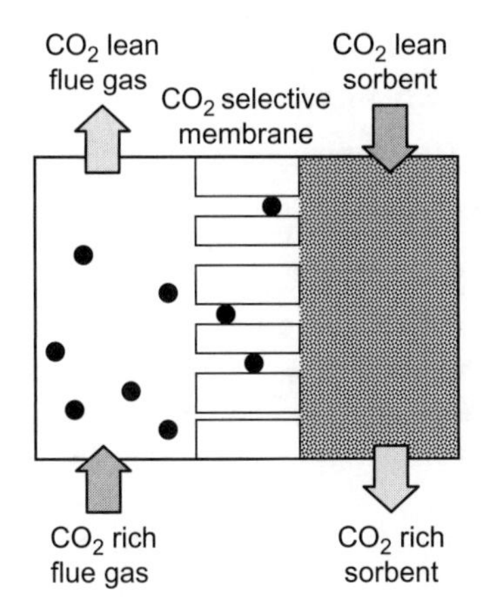

Figure 8.3 Gas–liquid membrane contactor for CO_2 removal from flue gas

in counter-current through the shell side of the module. Figure 8.3 illustrates this membrane gas absorption (MGA) process. In this application the membrane itself offers no selectivity, having a nominal pore diameter typically in the 0.2-μm range. However it serves to provide an interface between the gas and the sorbent liquid that offers very high surface-to-volume ratio for mass exchange between the gas and sorbent. This application is discussed further in Section 8.7.1.

An important assumption underlying the calculation of flux in Knudsen flow is that the interaction between permeate molecules and the pore surface is simply one of momentum exchange in collisions. Adsorption and desorption are assumed to be in equilibrium and surface diffusion effects are ignored. As the membrane pore diameter approaches the dimensions of the permeate molecules, these assumptions are no longer valid and the interactions between permeate molecules and the pore surfaces become significant factors in determining flux.

Surface diffusion and capillary condensation

In Knudsen flow of a gas mixture, a component with a higher molecular weight and with higher polarity and polarizability will tend to be preferentially adsorbed onto the pore surface (Figure 8.4). This will increase the permeation

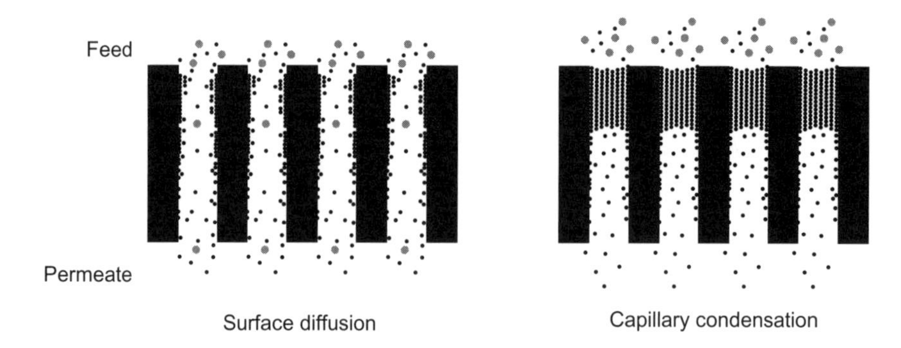

Figure 8.4 Surface diffusion and capillary condensation effects

rate of the adsorbed component by adding a surface diffusion flux to the Knudsen diffusion flux. Selectivity can also be increased, since the adsorbed species can substantially reduce the pore space available for diffusion of other components, particularly if the pore size is in the range two to three times the kinetic diameter of the adsorbed species. Surface effects can be promoted by modifying pore surfaces, for example by the addition of functional groups that promote adsorption.

If adsorption continues beyond a monolayer, capillary condensation can take place (Figure 8.4). Ultimately the pores may become blocked with the condensate and diffusion of other gas components will be prevented, substantially increasing selectivity.

Molecular sieving

Molecular sieving is the process that takes over from Knudsen flow when the pore diameter approaches the diameter of the molecules attempting to pass through a porous membrane. Molecular sieves are materials with pore sizes of molecular dimensions within which, ignoring surface effects, selectivity is a function of molecular size, with smaller molecular species having higher diffusion rates.

In molecular sieve studies, permeate molecules are commonly known as "guest" molecules. At this scale, neither the pore walls nor the guest molecules can be considered rigid; thermal vibration of the lattice structure of the membrane material and of the guest molecule will give both a degree of flexibility (Figure 8.5). The effective size of a hydrogen, nitrogen, or CO_2 molecule is determined by a combination of the bond length between the constituent atoms plus the Van der Waal's radius, which characterizes the range of the electron cloud surrounding each atomic nuclei (Table 8.4). The size of a guest molecule is characterized by its minimum diameter when in an equilibrium state and is known as the minimum kinetic diameter.

Some important examples of molecular sieve materials, zeolite, activated carbon (carbon molecular sieves), and microporous silica, are compared in Table 8.5.

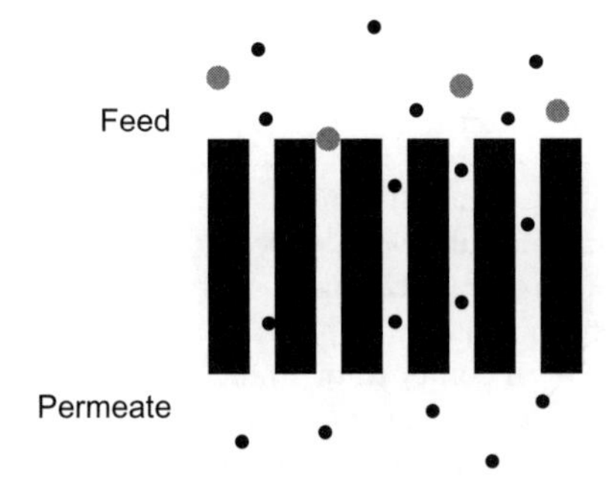

Figure 8.5 Molecular sieving

Table 8.4 Common molecular dimensions

Element	Bond length (Å = 10^{-10} m)	Van der Waal's radius (Å)	Kinetic diameter (d_m Å)
Hydrogen	1.97	1.20	2.89
Nitrogen	1.04	1.55	3.64
Oxygen	1.21	1.52	3.46
CO_2	1.42 (single bond)	1.70 (carbon)	3.35
Methane	1.09 (single bond)	2.00	3.80
H_2O	0.96 (single bond)	3.12	2.65

Transport of gas molecules through the tortuous pore network of a molecular sieve takes place through a combination of diffusion within the pore space and adsorption and surface diffusion on the pore surfaces. Adsorbed gas atoms or molecules (called adatoms or adparticles) diffuse on the pore surfaces (surface diffusion) down the concentration gradient between the upstream (feed) and downstream (permeate) sides of the membrane. These surface diffusion effects will be important only at lower temperatures since adsorption will no longer occur once the thermal energy of the gas molecule exceeds the heat of adsorption.

For a binary gas mixture, diffusivity within the pores will be dependent on the molecular sizes. If the adsorption strengths of the two species are similar, the membrane will be selective for the smaller molecule. However, if molecular sizes are similar, selectivity will be driven by adsorption strength, since the component with the higher adsorption strength will dominate the available adsorption sites and the surface diffusion contribution to permeate flux.

Among gases important for CCS (CO_2, N_2, CH_4, H_2) CO_2 has the highest heat of adsorption onto zeolites and other minerals with polar surfaces, and membranes made of these materials are therefore selective for CO_2 at low temperatures. If adjacent adsorption sites are separated by an energy barrier, as is the case for many adsorbates on zeolite, an activation energy is required to

Table 8.5 Molecular sieve materials compared

Material	Description	Example applications
Zeolite	Microporous aluminosilicate minerals with well-defined pore sizes in the range 3–13 Å, determined by the specific chemical composition and crystal structure. Selectivity can be manipulated over a wide range by modifying the structure and chemistry of synthetic zeolites	Hydrogen separation for 3 Å and CO_2/N_2 or CO_2/CH_4 separation for 4 Å sieves
Activated carbon	Activated carbon is an amorphous, disordered network of graphitic carbon platelets with interstitial spaces due to the presence of cross-linking chains of carbon atoms and foreign atoms between the layers. Carbon molecular sieves (CMS) use activated carbon with a molecular scale pore network	H_2/CO_2 separation from water–gas shift reaction products in gasification process. Oxygen separation from air using CMS and pressure-swing adsorption
Composite silica–alumina	Microporous silica (pore diameter ~3 Å) deposited on a mesoporous alumina support	H_2/CO_2 separation from water–gas shift reaction products in gasification process

enable an adparticle to hop from one adsorption site to another. This variant of surface diffusion is known as activated diffusion. Another example of activated diffusion is the passage of a guest molecule through a pore aperture that is the same size as or slightly smaller than the diameter of the diffusing molecule.

Multicomponent, multiphase transport through a molecular sieve, including surface diffusion, is described mathematically using the Maxwell-Stefan formulation, which accounts for both membrane–guest and $guest_1$–$guest_2$ interactions. An example of $guest_1$–$guest_2$ interactions is the competitive adsorption effect. This occurs when the available adsorption sites become saturated by the most strongly adsorbing component, blocking the transport of the other component either due to the lack of adsorption sites or because the adsorbed layers reduce the free pore size below the kinetic diameter of the second component.

When simplified for a single component, the Maxwell-Stefan expression for surface diffusivity is also known as the Darken equation, and gives the flux due to surface diffusion in the form:

$$J_S(T) \sim \rho\, q_{sat}\, D_S^0(0)\,(1-\theta)^{-1}\, \exp(-E_{D,S}/RT)\, d\theta/dx \qquad (8.7)$$

Table 8.6 Zeolite ion-exchange membranes: CO_2–N_2 molecular sieving performance

Material	CO_2–N_2 selectivity	CO_2 Permeance (mol m^{-2} s^{-1} Pa^{-1})
Li(20%)Y	10	7×10^{-7}
LiY	4	2×10^{-6}
K(62%)Y	39	5×10^{-7}
KY	30	1.4×10^{-6}

where J_S is the flux due to surface diffusion at temperature T, ρ is the density of the membrane material, q_{sat} is the saturated molar volume of gas adsorbed by the membrane, $D_S{}^0(0)$ is the limiting surface diffusivity, θ is the fraction of available adsorption sites that are occupied, $E_{D,S}$ is the surface diffusion activation energy, R is the gas constant, and x is the linear dimension parallel to the flow through the membrane.

The dependence of permeance and selectivity of a molecular sieve on the surface interactions between the guest molecule and the pore surfaces opens up the opportunity to achieve increased selectivity and permeance by "nano-engineering" the pore surfaces to increase the surface diffusion rate of the desired permeate. Synthetic zeolites have been widely investigated in this respect since the ionic properties of the surface can be controlled by varying the composition of the zeolite. Some examples of the CO_2/N_2 selectivity and CO_2 permeance for these ion-exchange zeolite molecular sieve membranes in the temperature range 35–40°C are shown in Table 8.6, and illustrate the generally inverse relationship between permeance and selectivity for a particular material type.

The molecular sieve with pore diameters of a few Angstroms is the finest form of microporous membrane. As pore sizes reduce further, and finally disappear, the transport process that takes over in a nonporous membrane—solution–diffusion—is a far simpler process than the complex, surface mediated adsorption–surface diffusion occurring in the finest-scale porous membranes.

8.1.2 *Solution–diffusion transport process*

The process of gas separation using nonporous membranes occurs in a two-step process in which the permeate first dissolves into the membrane and then diffuses through it (Figure 8.6). Hydrogen separation using a dense metal membrane is one such example, with adsorption of hydrogen on the feed side of the metal membrane surface followed by dissociation of hydrogen molecules and diffusion of protons in the membrane lattice. Hydrogen molecules are then regenerated on the permeate side of the membrane and desorbed into the permeate stream.

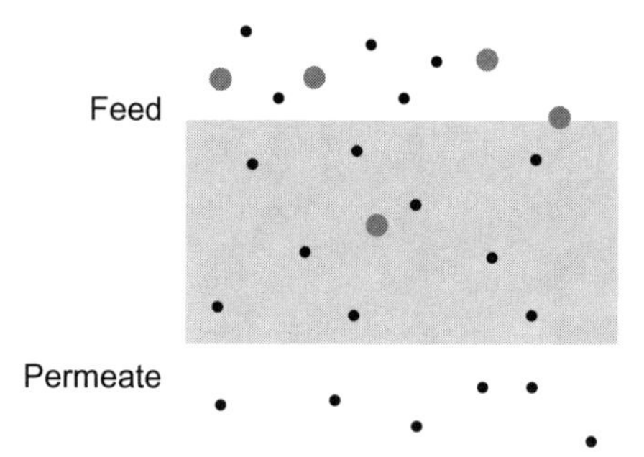

Figure 8.6 Solution–diffusion transport process

The effectiveness of the solution–diffusion process is determined by how well the permeate (e.g., CO_2 or hydrogen) is absorbed into the membrane and then how fast it diffuses through it.

The solubility of the permeate in the membrane is controlled by Henry's law, which states that, at a given temperature, the amount of a gas dissolved in unit volume of a medium is proportional to the partial pressure of the gas in equilibrium with the medium. This is expressed for permeate component i as:

$$C_i = P_i/K_{Hi} = P_i K_i \tag{8.8}$$

where C_i is the dissolved concentration, P_i is the partial pressure, K_{Hi} is the Henry's law constant, and K_i is the solubility of component i.

Considering CO_2 separation, the gas solubility (K_{CO_2}) is expressed as:

$$K_{CO_2} = C_{CO_2}/P_{CO_2} = 1/K_{HCO_2} \tag{8.9}$$

Diffusion through a nonporous membrane is governed by Fick's first law of diffusion, which states that:

$$J_{CO_2} = D_{CO_2} \Delta C_{CO_2}/L \tag{8.10}$$

where J_{CO2} is the flux of CO_2 through unit area of the membrane, D_{CO_2} is the diffusion coefficient or diffusivity of CO_2 in the membrane, and ΔC_{CO_2} is the difference in CO_2 concentration in the membrane between the upstream and downstream surfaces. Replacing ΔC_{CO_2} using Equation 8.9 gives:

$$J_{CO_2} = K_{CO_2} D_{CO_2} \Delta P_{CO_2}/L \tag{8.11}$$

where ΔP_{CO2} is the difference in CO_2 partial pressure across the membrane. The product of solubility and diffusivity is the permeability of the membrane to the molecular component under consideration.

Other gases in the feed stream will also dissolve in and diffuse through the membrane, depending on their specific solubility and diffusivity. The ratio of the flux rates for two components i and j is known as the selectivity of the membrane:

$$\alpha_{i/j} = J_i/J_k \tag{8.12}$$

Polymeric membranes are an important class of nonporous membranes operating through the solution–diffusion mechanism, and typically consist of a thin dense selective layer supported by a less dense, porous, and therefore nonselective supporting layer (Section 8.2.1). The thinness of the selective layer is dictated by the need to maximize permeance (and therefore flux through the membrane) and to minimize cost (the selective layer being high cost due to the customized materials and preparation requirements).

The performance of polymeric membranes can be enhanced by increasing the solubility and diffusion rate of the desired permeate. Solubility is controlled by changing the chemical composition of the polymeric material, while diffusion rate is controlled by changing the physical packing of polymer within the membrane—a characteristic that is strongly dependent on the method of membrane fabrication. Very high selectivities can be achieved by these means but can be sustained only at relatively low temperatures, typically up to ~100°C.

The physical arrangement of polymers within the membrane depends on whether the operating temperature of the membrane is above or below the glass transition temperature of the membrane material (T_g, see Glossary). Polymeric membranes operating above T_g are termed rubbery, and are flexible and generally of higher permeability and lower selectivity. Although hybrid (polymer + inorganic) membranes have been reported operating at up to 400°C, the rapid drop in selectivity with increasing temperature above T_g imposes a limit on the applicability of many polymeric membranes.

Below T_g, polymeric membranes are glassy, hard, and less permeable, but also have microscopic voids between the polymer chains as a result of imperfect packing. This excess free volume has an important effect on the performance of glassy membranes as a result of increased solubility due to adsorption of the permeate onto the void surfaces. This adsorption is described by Langmuir's adsorption equation:

$$C_{Li} = C_{max.i}\, a_i\, P_i/(1 + a_i\, P_i) \tag{8.13}$$

where C_{Li} is the adsorbed concentration on the microvoid surfaces, $C_{max.i}$ is the maximum adsorbed concentration, a_i is the Langmuir adsorption constant, and

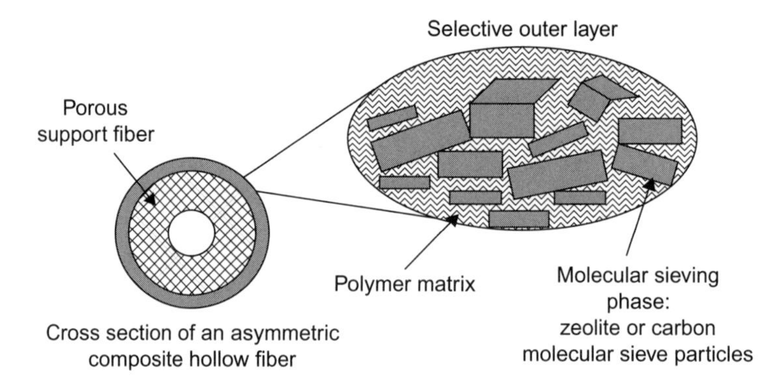

Figure 8.7 Mixed matrix hollow-fiber membrane structure

P_i is the partial pressure of component i. The total bulk concentration, from Equations 8.8 and 8.13, is therefore:

$$C_i = P_i K_i + C_{max.i}\, a_i\, P_i/(1 + a_i\, P_i) \qquad (8.14)$$

Glassy and rubbery materials have also been combined in composite (copolymer) polymeric membranes, which aim to exploit the selectivity of glassy polymers and the permeability of rubbery polymers to optimize overall membrane performance. Sections 8.5 to 8.7 describe a number of polymeric membrane applications and developments.

Mixed matrix membranes are an active area of research that aims to combine the best features of polymeric and molecular sieve materials. These are heterogeneous membranes with dispersed particles of molecular sieve material embedded in a continuous non-porous polymeric phase, which may be either glassy or rubbery (Figure 8.7). The key fabrication challenge is to ensure that the sieving particles are well bonded with the polymer to avoid microcavities within the membrane, which can result in a loss of selectivity.

Nonporous dense ceramic metal oxide membranes based on solution–diffusion have also been investigated for CO_2 separation at high temperatures. Oxides such as Li_2ZrO_3 are particularly suitable, having high CO_2 adsorption capacity (0.1 kg/kg at 600°C and $P_{CO2} = 70$ kPa) and negligible absorption of other gases such as hydrogen.

8.1.3 Facilitated transport membranes

The solution–diffusion process described above is an example of a passive membrane transport mechanism, in which the permeate travels by diffusion down a concentration gradient. The application of membranes based on this

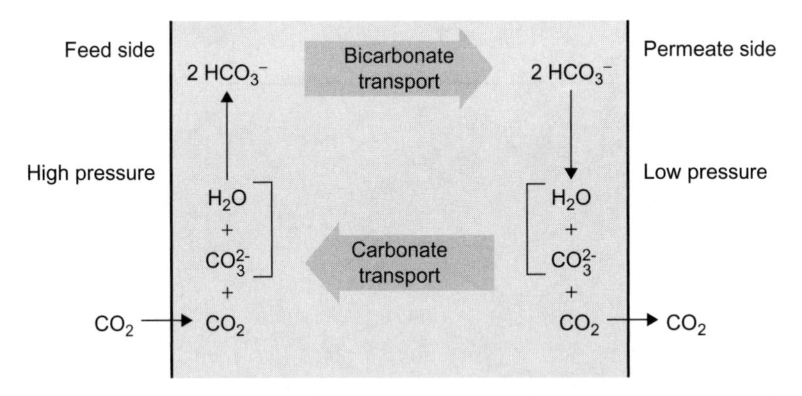

Figure 8.8 Facilitated transport of CO_2 in a supported liquid membrane with carbonate carrier

process is often constrained by low permeate flux rates due to a combination of low solubility and/or low diffusivity. In contrast, facilitated, or carrier-assisted, transport is an active transport mechanism that increases flux rate by transporting the permeate across the membrane attached to a carrier. Many biological processes rely on facilitated transport across cell walls, and in principle facilitated transport can enable the permeate to move against a concentration gradient (from a region of low concentration to one of high concentration).

In facilitated transport, the permeate reacts with a component present in the membrane (the carrier) to form complexes or reversible chemical reaction products at the feed side of the membrane. These complexes then diffuse across the membrane driven by a concentration gradient in the complex rather than a gradient in the permeate. At the downstream surface of the membrane the complexing process or reaction is reversed, liberating the transported component into the permeate stream.

Carriers can be either fixed within the membrane structure or mobile. Figure 8.8 illustrates the facilitated transport of CO_2 across a cellulose acetate membrane containing an aqueous carbonate solution. At the feed side of the membrane, bicarbonate anions are formed by the dissolution of CO_2 in water (Equation 8.15), which is energetically aided by the presence of the mobile carbonate anion carrier:

$$CO_2 + H_2O + CO_3^{2-} \leftrightarrow 2HCO_3^- + \text{heat} \quad (8.15)$$

At the permeate side of the membrane the reverse reaction occurs and CO_2 is liberated.

An example of a fixed carrier facilitated transport membrane for CO_2 separation is shown in Figure 8.9. Here CO_2 diffuses into the membrane and reacts with water and fixed amine groups, attached to the polymer backbone, to form bicarbonate ions:

$$CO_2 + H_2O + R\text{-}NH_2 \leftrightarrow HCO_3^- + R\text{-}NH_3^+ \quad (8.16)$$

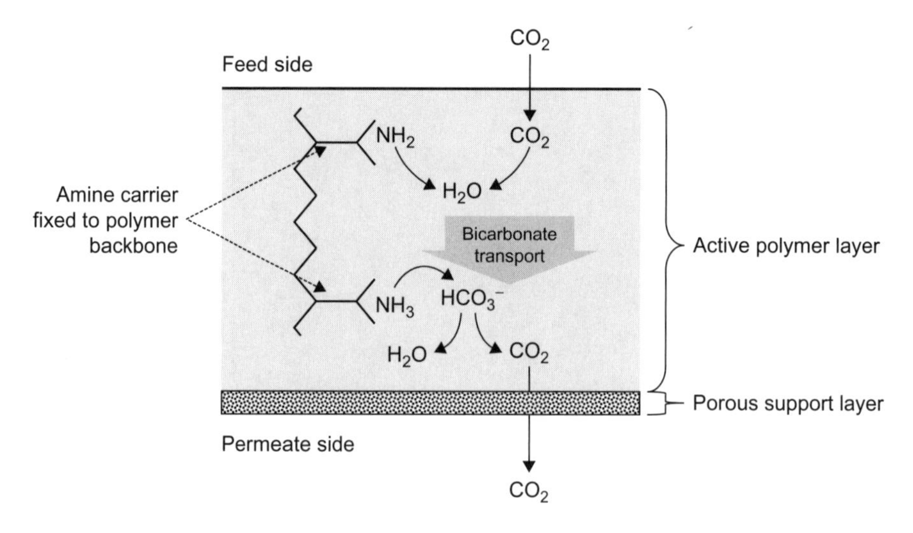

Figure 8.9 Amine facilitated transport membrane for CO_2 separation

Bicarbonate ions diffuse across the membrane between carrier sites, while protons are transported along the polymer backbone, with CO_2 being released and diffusing out of the membrane at the permeate side.

Secondary amines have also been investigated as carriers, the advantage being the lower binding energy and therefore faster desorption compared to primary amines.

The reaction of a secondary amine (R_1R_2NH) with dissolved CO_2 is described by a two-stage process:

$$CO_2 + R_1R_2NH \rightarrow R_1R_2NH^+CO_2^- \tag{8.17}$$

The $R_1R_2NH^+CO_2^-$ reaction product is known as a zwitterion, a polar molecule that is electrically neutral but carries formal charges on the amine and oxygen atom. In the second step, the zwitterion donates its hydrogen atom (proton) to a second amine, forming a carbamate ion that transports CO_2 across the membrane:

$$R_1R_2NH^+CO_2^- + R_1R_2NH \rightarrow R_1R_2NCO_2^- + R_1R_2NH_2^+ \tag{8.18}$$

The reverse reactions take place on desorption at the permeate side of the membrane.

Several examples of facilitated transport are summarized in Table 8.7. The permeability and selectivity values shown in the table illustrate the range of experimental results that have been achieved by varying the carrier chemistry for a given polymer type. The "barrer" is a commonly used unit for membrane permeability and is equal to $3 \cdot 10^{-16}$ mol m^{-1} s^{-1} Pa^{-1}.

Table 8.7 Experimental facilitated transport membrane performance

Membrane type: Facilitator	Experimental conditions	CO_2 permeability (barrer)	CO_2/X selectivity
Cellulose acetate: Pure water	$CO_2 + O_2$, 20°C, 4 kPa	400	22 (CO_2/O_2)
Cellulose acetate: 2 N $KHCO_3$ + 0.5 N $NaAsO_2$	$CO_2 + O_2$, 20°C, 4 kPa	2000	600–800 (CO_2/O_2)
Vinyl alcohol–acrylate copolymer: 2 N K_2CO_3	90%N_2, 10% CO_2, 20°C, 98 kPa	6100	317 (CO_2/N_2)
Vinyl alcohol–acrylate copolymer: 2 N K_2CO_3 + 0.5 N EDTA	90%N_2, 10% CO_2, 20°C, 98 kPa	2400	1417 (CO_2/N_2)
Polyvinylalcohol: Glycine + polyethylenimine	40/20/40 $H_2/CO_2/N_2$, 20°C, 200 kPa	194	28 (CO_2/H_2)
Polyvinylalcohol (cross-linked formaldehyde): Dimethylglycine + polyethylenimine + KOH	40/20/40 $H_2/CO_2/N_2$, 20°C, 200 kPa	338	1782 (CO_2/H_2)

8.1.4 Ion transport membranes

Ion transport membranes, also known as ion exchange membranes or mixed conducting membranes (MCM), are a class of facilitated transport membranes in which selective transport of either cations or anions is achieved as a result of the ionic structure of the membrane. Figure 8.10 shows an example of an ion exchange membrane in which oxygen ions are transported through a mixed metal oxide lattice by jumping between oxygen vacancies in the lattice structure. The vacancies are in effect fixed carrier sites facilitating oxygen transport. Electrons liberated at the permeate side as the ions recombine into oxygen molecules are conducted back through the matrix to maintain electrical neutrality. A high electrical conductivity is therefore a desired characteristic for such a membrane.

Minerals such as perovskite (see Glossary) and synthetic perovskite-like ceramics are suitable for this type of membrane, and can be engineered for selectivity and permeability by changing the stoichiometric ratios of the metal components and oxygen vacancies. These materials have the advantage of very high temperature operation (in the range 700–900°C) and are under development for use in syngas generation (Section 8.4) and oxygen separation for oxy-combustion (Section 8.5). In the latter application the oxygen separator may be either a low-temperature unit or a high-temperature reactor closely integrated with the combustion and heat cycle.

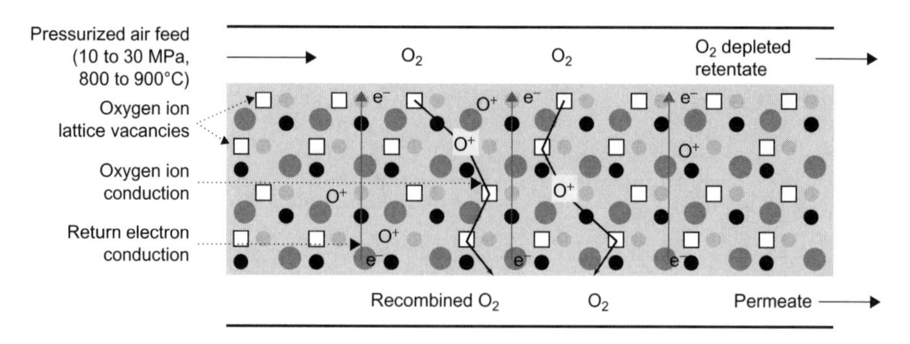

Figure 8.10 Schematic oxygen ion transport in a metal oxide lattice

Dual ion exchange membranes have also been investigated, with the return of the carrier to the feed side of the membrane being a second ion exchange process rather than the conduction of an electron. A molten carbonate dual ion exchange membrane has been proposed to enable CO_2 separation from high-temperature and high-pressure streams in power generation applications such as syngas production or IGCC (Section 8.6).

8.2 Membrane configuration and preparation and module construction

With the exception of natural gas processing, the use of membrane technology for CCS gas separation applications is still largely confined to the laboratory or pilot scale. Thus, as well as the physical and chemical investigation and optimization of membrane materials, the eventual application of membranes for CCS on an industrial scale will also require the further development of technologies for fabricating reliable and economical membrane modules.

The key challenges here are:

1. Reducing the thickness of the membrane selective layer to increase flux and reduce cost, while preserving selectivity
2. The large-scale production of defect-free membranes with the desired physical and chemical properties
3. Automation of membrane and module fabrication processes to cost-effectively package the membrane into reliable modules, achieving maximum membrane surface area within a given module size

Methods for industrial-scale preparation of membranes and fabrication of modules have been developed since the 1960s, driven by applications in reverse osmosis for water purification and desalination. Many of these techniques are directly applicable to gas separation problems (for example, hollow-fiber modules are applied in reverse osmosis as well as in natural gas sweetening),

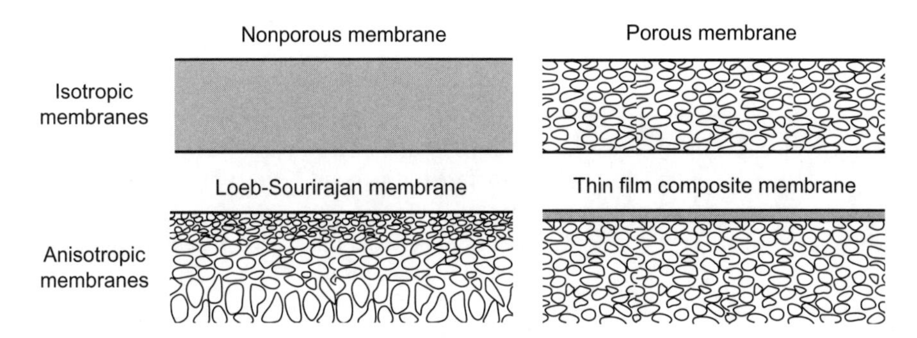

Figure 8.11 Physical structure of membranes

but novel membrane materials and applications in CCS will also require new developments in this area.

In this section membrane preparation and module fabrication methods are described, with reference to specific applications and the current areas of technology development.

8.2.1 Membrane types

Membrane preparation methods and suitable module configurations are strongly tied to the required structure of the membrane material. Several types of membrane structure can be identified, as shown in Figure 8.11, the key features of these structures being summarized in Table 8.8.

Most nonliquid membranes, whether polymeric, ceramic, or metallic, can be prepared as either symmetric or asymmetric structures. However, optimization for high flux rate generally drives toward asymmetric structures with a thinner selective surface combined with a thicker, nonselective supporting layer.

Zeolite membranes are an example of asymmetric membranes, and are typically prepared by in situ hydrothermal synthesis of a thin synthetic zeolite layer onto support tubes or disks of porous stainless steel or alumina.

8.2.2 Membrane module configurations

Symmetric or asymmetric membranes can be prepared either as laminar membranes, as hollow tubes or fibers, or as wafers, and packaged in a variety of different module configurations. The earliest module designs consisted of flat sheets of membrane material held in a frame (plate-and-frame modules) or bundles of relatively large diameter (1–3 cm) tubes (tubular modules). This module design does not in general provide the high surface area to bulk volume ratio required for gas separation applications (Table 8.9), and current commercially available gas separation modules use either spiral-wound or hollow-fiber modules.

Table 8.8 Structural characteristics of membranes

Structure	Characteristics
Symmetric (or isotropic) membranes	Porous or dense membranes with a uniform structure throughout the membrane thickness
Asymmetric (or anisotropic) membranes	Porous or dense membranes with a thin selective layer bonded to a thicker layer that provides mechanical support
Liquid membranes	A liquid selective phase, supported, contained, or immobilized within the pore space of a polymeric or ceramic membrane

Table 8.9 Area-to-volume ratios for various membrane configurations

Module configuration	Typical area-to-volume ratios (m^2/m^3)
Plate and frame	200
Spiral wound	500–1000
Hollow fiber	1500–10,000

Spiral-wound modules

Spiral-wound membrane modules consist of a number of membrane envelopes wound onto a central perforated collecting tube. Each membrane envelope, or leaf, consists of two membranes separated by a permeate spacer and sealed on three sides, leaving one side open for permeate removal. Early designs used a single membrane envelope, but this requires a long spiral path to achieve a high surface area-to-volume ratio, resulting in a high pressure drop along the spiral permeate flow path. Figure 8.12 illustrates the spiral-wound configuration.

The feed gas mixture flows axially along the module through feed spacers held between successive membrane leaves. The permeate passes through the membranes and then flows spirally through the space held open by the permeate spacer, exiting the open edge of the membrane leaf into the central collecting tube.

The construction of spiral-wound modules in commercial production for CO2 removal from natural gas is illustrated in Figure 8.13. Optimization for a specific application requires selection of the number of membrane leaves, module length and diameter, which determine the pressure drop, weight, and cost of the module. Weight and ease of handling are important practical considerations in view of the need to replace modules after performance degradation or failure.

In natural gas processing applications, typically four to six modules are mounted inside a tubular pressure vessel, and many of these tubular

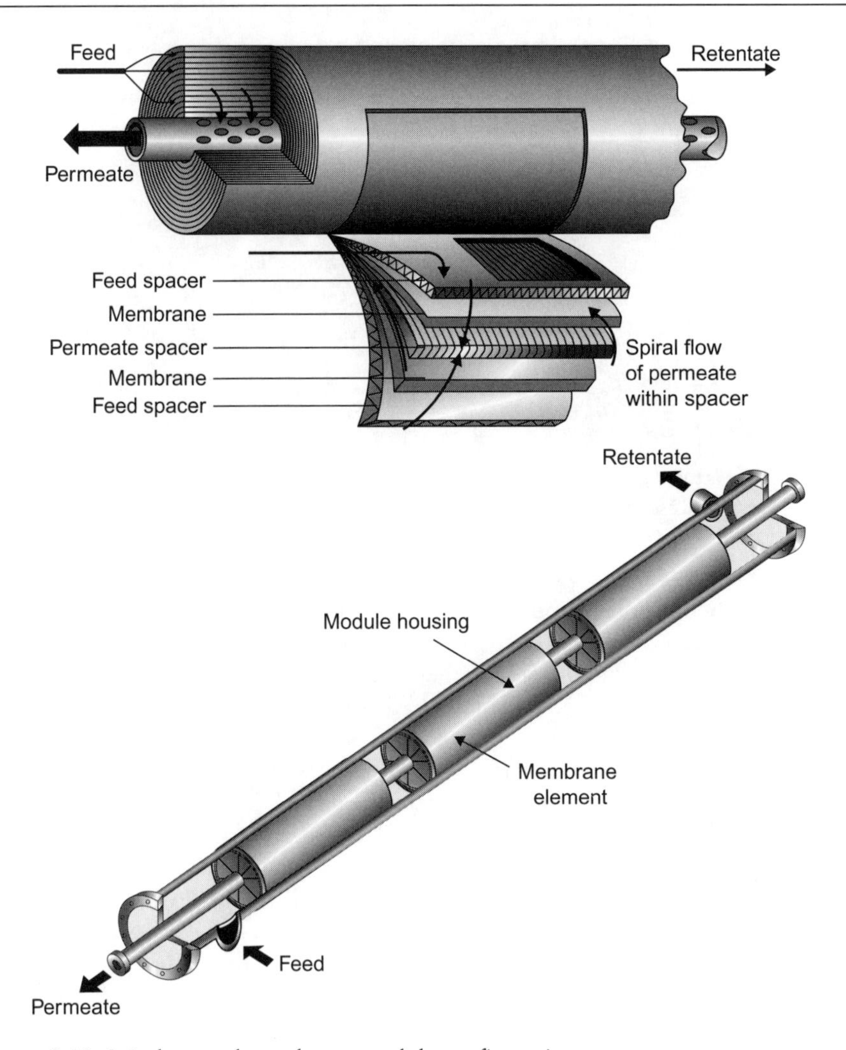

Figure 8.12 Spiral-wound membrane module configuration

elements are mounted in series or parallel to form the complete membrane separation unit. Figure 8.13 shows an example of a natural gas sweetening membrane skid.

Hollow-fiber modules

Hollow-fiber modules are also extensively used in natural gas treatment because of the very high surface to bulk volume ratio noted above. Modules consist of bundles of fibers enclosed in a pressure vessel, with the permeate passing either out of the fibers (bore-side feed) or into them (shell-side feed), as illustrated in Figure 8.14. Fibers are mechanically able to withstand much

Figure 8.13 Natural gas sweetening membrane skid unit (Courtesy; Membrane Technology and Research Inc.)

higher pressures applied externally than internally, so for high-pressure applications the shell-side feed configuration is used.

Optimization of hollow-fiber modules for specific applications involves selection of the fiber diameter, and hence the packing density of fibers in the module, and the flow direction of shell versus fiber—whether these flows are co-current or counter-current.

Hollow-fiber membrane separation units for gas separation are constructed in a similar manner to the spiral-wound skid unit shown in Figure 8.13. Fiber modules are commonly 1 m in length and 0.2–0.25 m in diameter, again for physical handling reasons, and contain $\sim 10^6$ individual fibers with internal and external diameters of ~50 and 100 μm. Hollow-fiber modules for gas–liquid contactors use polymer fibers, which are typically an order of magnitude larger in cross section, with inner and outer diameters of ~600 and 1000 μm.

Hollow-fiber-based separation units are typically smaller than those based on spiral-wound modules, although spiral-wound modules are able to handle higher operating pressures and are more resistant to fouling due to particles in the feed gas stream that can block fine fibers.

Ceramic wafer stack modules

Advanced membrane materials such as metal oxide ion transport membranes require new module structures to cope with the high operating temperatures and pressures. One example is the ceramic wafer stack configuration shown in Figure 8.15, a derivative of the early plate-and-frame type module. This module has been developed for oxygen separation for IGCC and oxyfueling applications by a U.S. Department of Energy National Energy Technology Laboratory-funded consortium led by Ceramatec and Air Products.

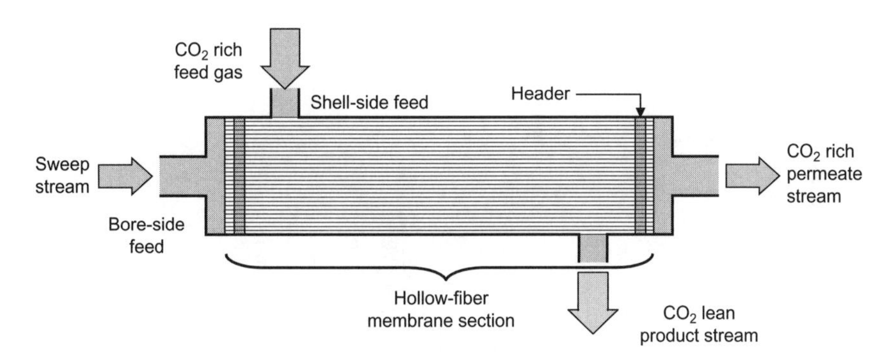

Figure 8.14 Hollow-fiber membrane module structure

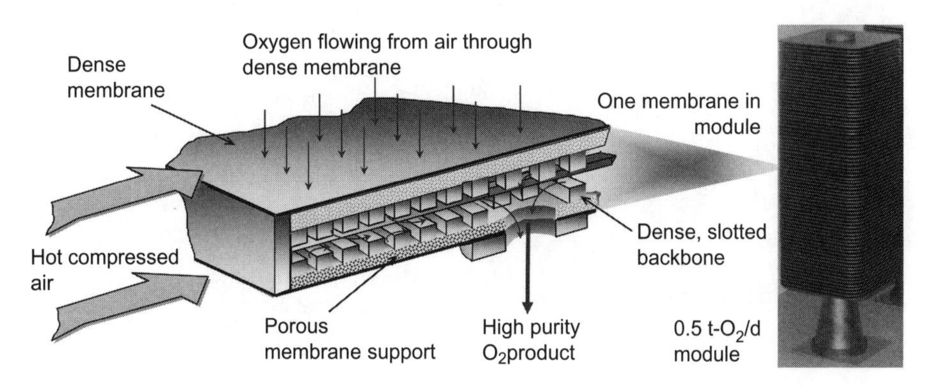

Figure 8.15 Ceramic membrane wafer stack module (Courtesy; Air Products and Chemicals Inc.)

The module consists of a stack of wafers, each consisting of a back-to-back pair of perovskite-based oxygen ion transport membranes. Air is fed into the module at 800–900°C and a pressure of 0.7–2.0 MPa and oxygen is transported through the wafers and collected from the central tube.

8.3 Membrane technology RD&D status

The previous sections illustrate the diverse range of membrane technologies that are available for potential CCS applications. Table 8.10 summarizes the key features of these membrane types, and indicates the current areas of research, development, and demonstration work. The RD&D focus areas listed are not intended to be exhaustive but to give a flavor of the range of topics being investigated for each membrane type.

Table 8.10 Membrane technologies and RD&D summary: Microporous membranes

Membrane technology	Description	Advantages/Disadvantages	RD&D focus areas
Carbon molecular sieve (CMS) membranes	Activated CMS composed of an amorphous, disordered network of graphitic carbon platelets	+ High CO_2 selectivity against N_2 and CH_4 + Durable and high-temperature operation - Low CO_2 permeability - Optimal precursor materials and membrane fabrication are expensive - Brittle nature and thermal expansion make membrane module construction complex	Selection and carbonization of precursor material to control pore structure. Membrane doping to improve mechanical strength and reduce cracking. Techniques to improve regeneration and eliminate aging effects (e.g., due to chemical degradation)
Zeolite molecular sieve membranes	Microporous lattice with well-controlled pore sizes in the range 3–13 Å	+ Perfect selectivity and high permeability in specific molecular sieving applications (e.g., H_2 separation) - Limited selectivity for molecules with similar kinetic diameter (e.g., CO_2, O_2, N_2)	Optimizing adsorption–diffusion by substituting various cations (Li^+, Na^+, Ca^{2+}, K^+) into the zeolite to produce ion-exchanged zeolite membranes

Silica membranes	Microporous membrane operating either by Knudsen diffusion or by molecular sieving and surface diffusion, depending on pore size	+ Chemical, thermal, and structural stability + Structure (pore size and distribution) easily tailored by preparation methods and conditions - Generally low selectivity	Silica deposition methods and conditions to achieve desired, defect-free pore structure. Fine-tuning pore size (e.g., by deposition of alkoxysilane monolayers)
Hybrid membranes (e.g., amine– or polyimide–silica membranes)	Composite membranes with organic groups incorporated into the microporous silica matrix or chemically bonded to the pore surfaces to create so-called functional membrane pores	+ Increased selectivity compared to a nonfunctional ceramic membrane as a result of the enhanced surface diffusion of the permeate + Thermal and mechanical stability of ceramic membranes combined with the selectivity of polymers	Selection of polymers and alternative techniques to uniformly incorporate and activate the functional groups in the ceramic matrix. Optimized pore structure of composite membranes

8.4 Membrane applications in precombustion capture

8.4.1 Oxygen ion transport membranes for syngas production

The formation of syngas ($H_2 + CO$) in the precombustion capture process described in Chapter 3 requires a supply of oxygen for the partial oxidation step. In 1988, Standard Oil Co. (subsequently Amoco and now BP) patented the Electropox (electrochemical partial oxidation) process that integrated oxygen separation, methane oxidation, and methane steam reforming into a single process step. As shown in Figure 8.16, the process is based on a hollow tube-configured ceramic oxygen transport membrane that contains the methane-plus-steam mixture and allows the passage of oxygen ions and electrons. The membrane is composed of a perovskite-like solid metal oxide material, in which the oxidation state of cations in the membrane has been altered to introduce oxygen vacancies into the structure. This allows oxygen ion transport through the membrane by a hopping mechanism between vacancies, described in Section 8.1.4.

The Electropox technology has been the subject of an intensive, collaborative research and development effort since 1997, with the aim of bringing the technology to the stage of a demonstration-scale plant. Table 8.12 summarizes some of the key issues that have been investigated and illustrates the breadth of the R&D effort required to bring such a technology to the demonstration stage.

As an example, optimization of the oxygen conductivity and stability of the membrane material involves investigating the properties of various types of mixed-ion electronic conductors of the general form $A_xA'_{1-x}B_yB'_{1-y}O_{3-\delta}$ where A, A′, B and B′ are metal ions and δ indicates the oxygen nonstoichiometry (i.e., proportion of vacancies in the lattice). Here replacement of a proportion of the A-site ions (e.g., trivalent La^{3+}) by a lower-valency A′ ion (such as divalent Sr^{2+}) increases the oxygen vacancies in the lattice. One such example is $La_{0.7}Sr_{0.3}Co_{0.2}Fe_{0.8}O_{3-\delta}$, which has exhibited high oxygen permeance combined with good chemical and mechanical stability in laboratory investigations.

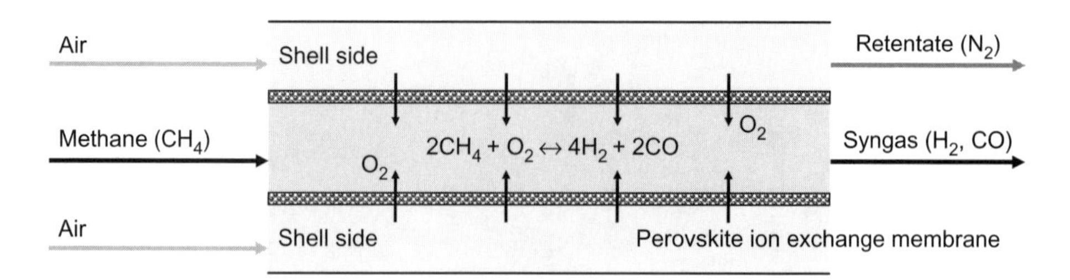

Figure 8.16 Oxygen transport membrane application to methane partial oxidation

A similar application for oxygen ion transport membranes is in the partial oxidation of methane (POM) as an initial step in a gas-to-liquids process. This is essentially the Electropox process but with the steam reforming reaction left out. The partial oxidation reaction:

$$2CH_4 + O_2 = 2CO + 4H_2 \tag{8.19}$$

produces a syngas product stream with the ideal stoichiometric ratio for the Fischer–Tropsch gas-to-liquids process, which is used to convert syngas to liquid hydrocarbons through catalyzed chemical reactions of the form:

$$(2n + 1)H_2 + nCO \rightarrow C_nH_{(2n+2)} + nH_2O \tag{8.20}$$

As an alternative to oxygen supply from cryogenic air separation, which is capital- and energy-intensive, further developments in oxygen ion transport membranes could simplify the gas-to-liquids process and reduce production costs by 20–30%.

8.4.2 Palladium membranes in IGCC applications

Palladium (Pd) or palladium and silver (Pd/Ag) alloy metallic membranes, which are 100% selective for hydrogen, have an important potential application in IGCC power plants. A Pd-based water–gas shift membrane reactor (WGSMR) can be used to separate hydrogen directly from the WGS reaction zone (Figure 8.17), resulting in improved efficiency of the reaction. Removing hydrogen as it is formed results in more complete conversion of CO to CO_2, since it prevents the shift becoming limited by equilibrium between the forward and reverse reaction (see Table 3.3). Alternatively, the reaction temperature to achieve a certain conversion rate can be reduced, thereby reducing the reactor energy requirements.

Water is removed from the retentate by condensation, to leave a near-pure CO_2 stream at a pressure that is close to the high operating pressure of the

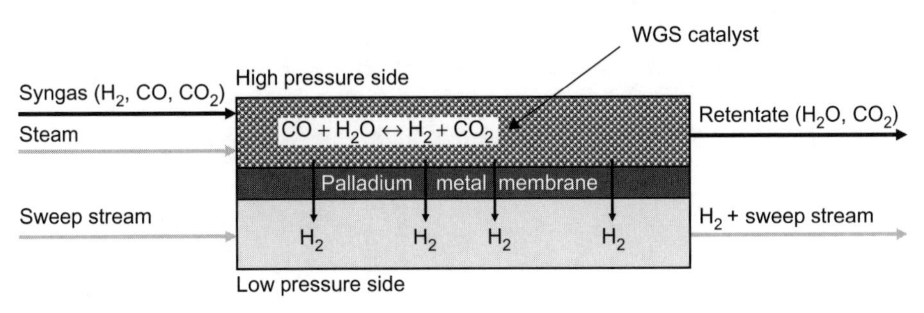

Figure 8.17 Water–gas shift membrane reactor

Table 8.11 Membrane technologies and RD&D summary: Dense (non-porous) membranes

Membrane technology	Description	Advantages/Disadvantages	RD&D focus areas
Polymeric membranes	Predominantly glassy polymer materials, of which polyimides are the most extensively investigated; also copolymer (glassy + rubbery) varieties, combining selectivity and permeability, and hybrid (rubbery + inorganic backbone) to improve mechanical properties	+ Thermal stability, chemical resistance, and mechanical strength of polyimide membranes + High degree of control over permeability and selectivity through control of polymer preparation, and chemical composition, including functional groups - Swelling and plasticization tendency with CO_2 sorption - Limited to relatively low-temperature applications	Polymer selection and preparation methods to improve mechanical strength and chemical stability. High-temperature rubbery and hybrid polymers. Increasing selectivity in rubbery membranes by functional group substitution and in glassy membranes using high CO_2 solubility pore-filling materials
Mixed matrix membranes (MMM)	Glassy or rubbery polymer membranes with molecular sieving particles, such as zeolite or activated carbon, embedded in the polymer matrix	+ Opportunity to combine the high permeability of rubbery polymers with the selectivity of molecular sieves - Formation of microcavities due to poor bonding between polymer and included particles, particularly for glassy polymers	Improved bonding of polymer with sieving particles (e.g., through preparation methods). Selection and optimal size, fraction, and positioning of the sieving phase within the polymer matrix. Preparation and pretreatment of sieving particles
Fixed carrier facilitated transport membranes	Solid polymeric membranes incorporating fixed carriers that react reversibly with the permeate and increase the selectivity and permeability of the solution–diffusion transport mechanism	+ High selectivity and permeate flux + Less carrier loss than mobile carrier membranes - Lack of mobility of the functional group	Selection of polymer and complexing agents (fixed carriers). Methods of membrane fabrication

Immobilized (supported) liquid facilitated transport membranes (ILMs)	A mobile carrier facilitated transport membrane, typically a polymeric or microporous support containing an aqueous solution immobilized in the pore space, including solvent swollen polymer membranes	+ Very high CO_2 permeability and selectivity to N_2 - Loss of liquid through evaporation or leakage - Aging or degradation of the carrier due to chemical instability - Carrier saturation at high permeate pressure	Reaction catalysts and other solution additives. Improved carriers to reduce aging. Fabrication methods and support structures. Low-volatility liquid selection to limit liquid loss through drying
Metal membranes	Hydrogen-selective solution–diffusion membranes formed by depositing a micron-scale metallic layer, such as palladium, onto a porous support	+ 100% hydrogen selectivity - Low permeability - Poor mechanical properties and chemical degradation due to sulfur and chlorine sensitivity	Metal alloys to improve mechanical properties and resistance to chemical degrading. Membrane treatment to improve permeability
Metal oxide ion-exchange membranes	Dense metal oxide membranes, typically perovskite-like minerals, conducting oxygen ions by facilitated transport between oxygen vacancies in the lattice	+ High-temperature operation + High O_2 or CO_2 absorption capacity + Low susceptibility to contaminant gases (SO_x, H_2S)	Scale-up of pilot scale (5 t-O_2 per day) oxygen separation plant for oxycombustion
Dual phase ion-exchange membranes	Ion exchange in a molten carbonate electrolyte, held in the pore space of a metal or ceramic metal oxide structure	+ Very-high-temperature application; close to combustion point in power generation - Support material performance in high-temperature oxidizing environment	Robustness under high-temperature oxidizing conditions

Table 8.12 Electropox R&D focus areas

Key issue	Research tasks
Oxygen diffusion kinetics	Measurement of kinetics of oxygen uptake and transport in ceramic membrane materials under commercially relevant operating conditions
	Measurement of surface activation and reaction rates in ion transport membranes
Grain structure and atomic segregation	Evaluate the effect of defect configuration on ceramic membrane conductivity and long-term chemical and structural stability
	Assess the microstructure of membrane materials to evaluate the effects of vacancy–impurity association, defect clusters, and vacancy–dopant association on the membrane performance and stability
Phase stability and stress development	Evaluate phase stability and thermal expansion of candidate perovskite membranes and develop techniques to support these materials on porous metal structures
Mechanical property evaluation in thermal and chemical stress fields	Determine material mechanical properties under conditions of high temperatures and reactive atmospheres
	Preparation and characterization of dense ceramic oxygen-permeable membranes
Graded ceramic and metal seals	Design, fabrication, and evaluation of ceramic-to-metal seals based on graded ceramic powder and metal braze joints

reactor (>10 MPa). This reduces the energy penalty for compression before transport and storage.

Laboratory-scale WGS membrane reactors have been demonstrated using Pd or Pd/Ag alloy membranes operating in the temperature range 300–600°C, and have achieved 100% hydrogen selectivity and almost complete CO conversion.

Microporous membranes (silica, alumina, carbon, and zeolite), typically asymmetric membranes on mesoporous alumina supports, have also been investigated for WSGMR applications. In this case hydrogen selectivity is the result of molecular sieving and surface diffusion through the selective layer, with pore diameters in the region of 7–20 Å (3–7 Å for zeolite) although these membranes have lower H_2 versus CO_2 selectivity than Pd-based metallic membranes.

8.5 Membrane and molecular sieve applications in oxyfuel combustion

A 500 MW power plant using oxyfuel combustion, operating at currently achievable plant efficiency (45%), requires an oxygen supply of 8000 t-O_2 per day. Cryogenic air separation, discussed in Section 9.3, is the only currently commercial process for delivering oxygen at this rate, and cryogenic air separation units (ASUs) with capacities up to ~4000 t-O_2 per day are installed worldwide. However, with a typical energy requirement of ~200 kWh per t-O_2 to deliver oxygen at 170 kPa for oxyfuel combustion applications, oxygen supply alone would consume >10% of the output of an oxyfueled power plant.

The IPCC threshold of 0.1 Mt-CO_2 per year for defining a large stationary CO_2 source corresponds to a power plant of ~12 MW for current efficiencies. A plant of this size would require ~200 t-CO_2 per day for oxyfuel combustion. This is close to the maximum rate currently achievable using molecular sieve-based ASUs and is in the range of oxygen production rates targeted by current technology development work for membrane-based oxygen production.

8.5.1 Molecular sieves for oxygen separation

The industrial separation of air into oxygen and nitrogen by pressure swing or vacuum and pressure swing adsorption, discussed in Chapter 7, is an established application of carbon and zeolite molecular sieves. Carbon molecular sieves have very similar capacities of adsorption for oxygen and nitrogen, but the rates of adsorption of these two species differ significantly, oxygen adsorption being considerably faster than nitrogen adsorption.

Separation of oxygen and nitrogen by pressure swing adsorption therefore results from the difference between the rates of adsorption of the two species. When a carbon molecular sieve comes into contact with air, the adsorbed phase becomes oxygen-rich while the gaseous phase becomes correspondingly nitrogen-rich. This difference is related to molecular size, the smaller kinetic diameter of oxygen (3.46 Å) compared to that of nitrogen (3.64 Å) resulting in faster adsorption into the activated carbon surface.

The carbon molecular sieve pressure swing adsorption process (CMS-PSA) is widely used for nitrogen generation with, in this case, the adsorbed oxygen being vented back to the atmosphere during the sorbent regeneration part of the PSA cycle. Alternatively, the released oxygen can be captured and the process used for oxygen generation.

Pressure swing absorption units can yield an oxygen stream with 95% purity, with plant capacities up to 200 t-O_2 per day.

8.5.2 Ion transport membranes for oxygen production

Ion transport membranes have been the subject of a U.S. Department of Energy National Energy Technology Laboratory-funded R&D project since 1999, and a 5t-O_2 per day prototype unit (shown in Figure 8.18) started operation in 2005. The perovskite-based metal oxide ion transport membrane modules have a wafer stack configuration (Section 8.2), and multiple modules are housed within the high-pressure membrane vessel. The plant delivers a >99% pure oxygen permeate stream and has successfully demonstrated that this technology can be applied on a commercial scale.

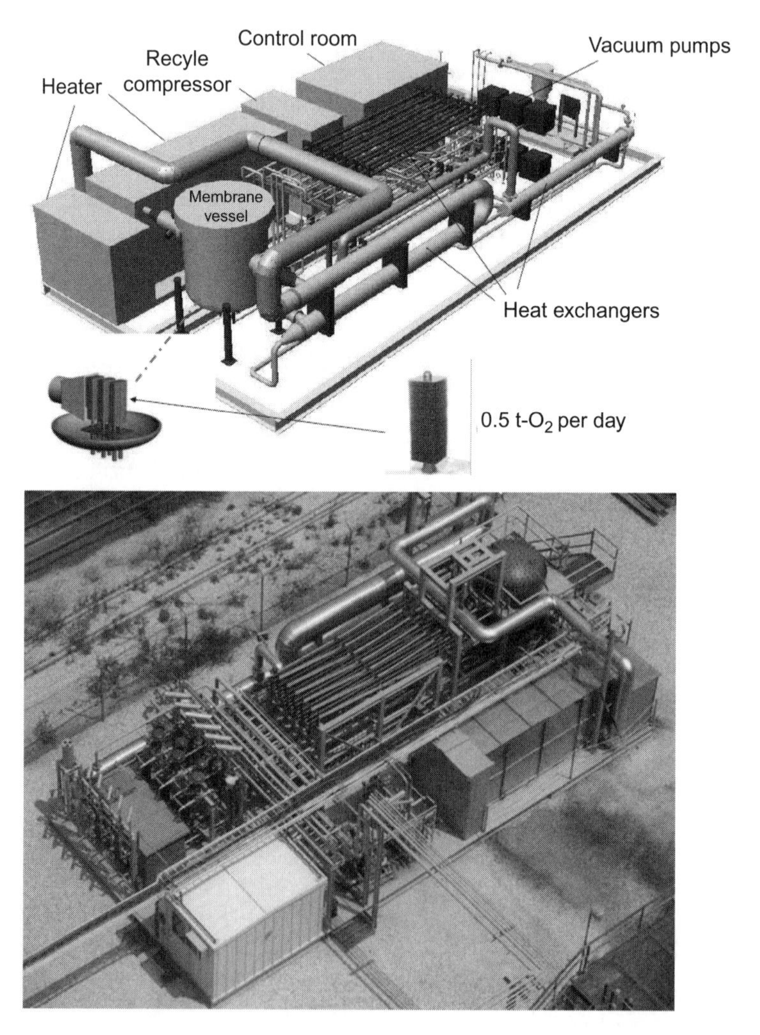

Figure 8.18 Prototype 5t-O_2 per day ITM oxygen separation plant (Courtesy Air Products)

This technology reduces the energy cost of oxygen production by a third compared to cryogenic methods, and future phases of the project aim to scale up the capacity to a commercial-scale 2000 t-O_2 per day by 2012, in what was initially planned to be part of the development of the U.S. Department of Energy FutureGen plant.

8.6 Membrane applications in postcombustion CO_2 separation

Membrane separation of CO_2 from flue gases is a very active research field, which aims to reduce the energy penalty and cost of membranes to enable large-scale demonstration and deployment. The key characteristics that need to be combined in a commercial membrane are:

- Sufficient permeate flux rate to achieve desired operating throughput
- Low space requirements and capital cost, including membrane material cost
- Low sorption capacity for nonselected gases (typically N_2, CH_4, and H_2)
- Chemical resistance to contaminants (SO_x, NO_x, H_2O, H_2S)
- Thermal stability at required operating temperature
- Inexpensive fabrication methods for a variety of module configurations

Achieving a combination of high selectivity, to yield high CO_2 purity in the permeate stream, and high permeability, to give the required flow rate, is the key to unlocking the potential of membranes for CO_2 separation from high-volume flue gases. Current demonstration systems require a large membrane area and multiple separation stages in order for both high flow rate and permeate purity to be achieved.

8.6.1 High-temperature molten carbonate membrane

A dual ion conduction membrane using a molten carbonate electrolyte is under development for high-temperature and -pressure CO_2 separation from flue gas streams close to the point of combustion, or to remove CO_2 from a syngas stream after the production of hydrogen in the WGS reaction. The membrane, illustrated in Figure 8.19, is based on a composite material that combines oxygen ion exchange through a solid metal oxide with CO_2 transport as a carbonate anion (CO_3^{2-}) through the molten carbonate phase.

At the feed surface of the membrane, an oxygen anion from the metal oxide combines with CO_2 to form a carbonate anion:

$$CO_2 + O^{2-} \rightarrow CO_3^{2-} \tag{8.21}$$

This ion is transported through the molten carbonate electrolyte, which is immobilized in the pores of the solid metal oxide. At the permeate side, the carbonate anion releases CO_2 into the gas phase while the oxygen ion enters a lattice vacancy in the metal oxide and is transported back to the feed side, driven by the gradient in vacancies within the metal oxide lattice.

Current R&D efforts focus on varying fabrication methods to control the volume fractions and pore structure of the composite material, and on testing of alternative metal oxide compositions to optimize CO_2 permeability and selectivity.

8.6.2 Facilitated transport membranes

An example of a facilitated transport mechanism that has been developed at the laboratory scale to demonstrate CO_2 separation from flue gas involves the incorporation of amine carriers into a cross-linked polyvinyl alcohol (PVA) polymer membrane. The membrane showed good CO_2 permeability and CO_2 and N_2 selectivity at operating temperatures up to 170°C.

Mathematical modeling studies indicated that a 0.6 m long hollow-fiber module containing $\sim 10^6$ fibers could recover >95% of the CO_2 from a 0.5 m^3 per second flue gas stream into a 98% CO_2 permeate stream, yielding ~12 t-CO_2 per day per module. At this rate, ~800 modules would be required to process the $\sim 1.4 \cdot 10^6$ m^3 per hour of flue gas from a 500 MW coal-fired power plant.

8.6.3 Carbon molecular sieve membranes

CMS membranes hold promise for flue gas CO_2 separation applications because of their high thermal and chemical stability in nonoxidizing environments, essential characteristics for this application. Although high selectivities for CO_2 and N_2 (>100) and for CO_2 and CH_3 (>250) have been demonstrated in laboratory experiments, these have been achieved only at very low permeabilities in the range from 1 to 10 barrer ($3\text{–}30 \cdot 10^{-16}$ mol m^{-1} s^{-1} Pa^{-1}).

Current research efforts for this type of membrane are focusing on:

- Selection, preparation, and pretreatment of the precursor material to reduce the cost of membrane production
- Parameters of the carbonization process (temperature, heating rate, pyrolysis atmosphere) to control the membrane pore structure
- Postcarbonization treatment of the membrane, for example additional carbon deposition by chemical vapor deposition (CVD)
- Membrane doping to improve mechanical strength and reduce cracking
- Techniques to improve regeneration and eliminate aging effects, for example due to chemical degradation

8.7 Membrane applications in natural gas processing

The removal of CO_2 from natural gas or natural gas liquids is important in order to deliver hydrocarbon products to a low CO_2 sales specification and to reduce the potential for corrosion of process equipment, pipelines, and product distribution systems.

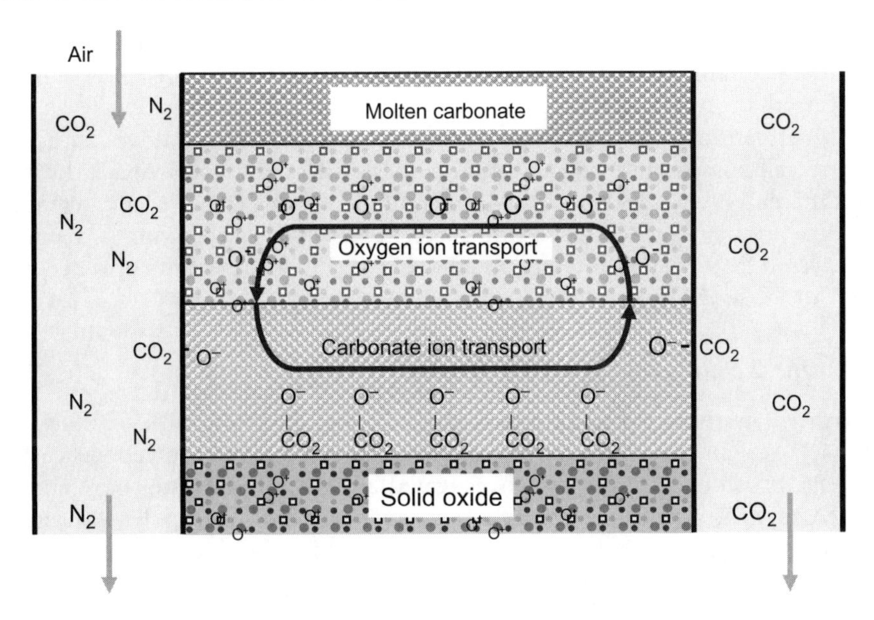

Figure 8.19 Molten carbonate high-temperature dual ion exchange membrane

8.7.1 *Polymeric membranes*

Polymeric membranes have been in use for CO_2 removal from natural gas since the early 1980s and many systems are operating around the world, most commonly using spiral-wound or hollow-fiber cellulose acetate-based asymmetric polymer membranes. Feed gas pretreatment removes heavy hydrocarbon components and particulate contaminants that can cause fouling of the membrane modules.

While CO_2 can also be easily separated from methane and from propane and heavier components (C_{3+}) using traditional distillation columns (Section 9.2), the separation of CO_2 from ethane is problematic due to the closeness of their vapor pressures and due to the formation of an azeotrope (see Glossary) between CO_2 and ethane. This azeotrope has an approximate composition of 70% CO_2 and 30% ethane at ambient temperature and prevents further CO_2–ethane separation by simple distillation.

A pilot scale (~400 m^3 per day) membrane separation plant has been demonstrated by BP for CO_2 separation from liquid ethane. The plant used a commercially available NATCO-Cynara asymmetric cellulose acetate hollow-fiber membrane module, with a shell-side feed of liquid ethane and dissolved CO_2. Methane was used as the sweep gas on the bore-side of the module, necessitating a further CH_4:CO_2 separation stage.

StatoilHydro and UOP have also demonstrated a hybrid distillation-membrane system in the Kårstø Ethane Plant, in which the membrane is used to break the CO_2–ethane azeotrope, maximizing ethane recovery and producing a high-purity CO_2 stream. A feed gas stream comprising 82% ethane, 16% CO_2,

1% methane, and 1% C_{3+} from the ethane plant is distilled in a chilled CO_2 stripping column, yielding an ethane bottom stream and an azeotrope-limited CO_2 + ethane top stream (Figure 8.20).

This stream enters a membrane separator consisting of spiral-wound asymmetric cellulose acetate membrane modules, where the azeotrope is broken by differential transport across the membrane. Both permeate and retentate streams are then purified in second-stage chilled distillation columns, recovering overall 98% of the feed stream ethane in a 95%-pure ethane product, and 78% of the feed CO_2 in a 99.98%-pure CO_2 product.

Polymeric facilitated transport membranes

Polymeric immobilized liquid membranes (ILMs) have also been investigated for CO_2–ethane separation. Diethanolamine (DEA) and polyethylene glycol (PEG) solutions have been investigated as ILM solvents and have shown increased CO_2 permeance and selectivity over nonfacilitated polymeric membranes. One problem with this type of membrane is the saturation of the carrier at higher CO_2 partial pressure, which imposes a fundamental limit on CO_2 transport across the membrane.

8.7.2 Gas–Liquid membrane contactors

Gas–liquid membrane contactors make use of macroporous membranes to provide an interface between a gas and a liquid absorbent and are being developed for CO_2 removal from flue gases and for natural gas sweetening. The relatively large pore diameter of the membrane, typically in the range 10–200 nm, means

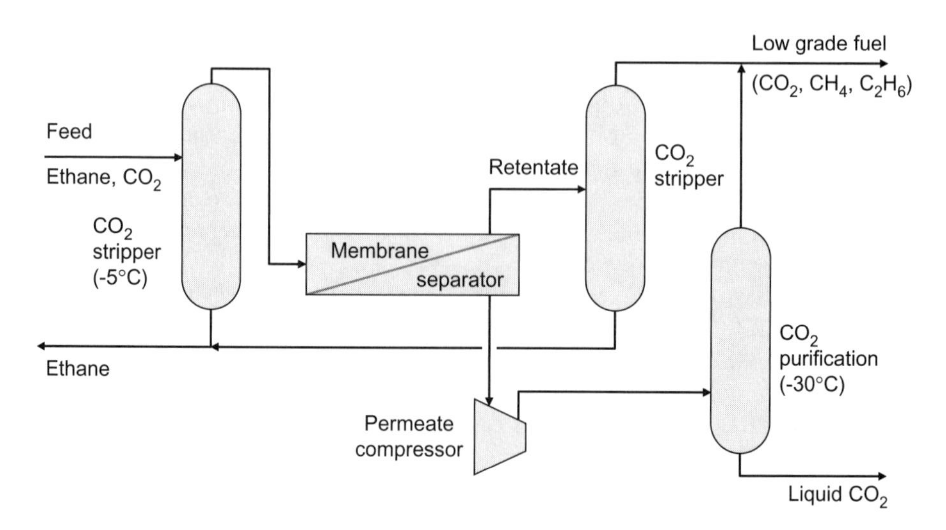

Figure 8.20 Hybrid membrane-distillation process for CO_2–ethane separation

that the membrane itself is nonselective and the selectivity of the contactor results from the physical or chemical absorption process at the permeate side of the membrane.

This hybrid membrane–absorbent configuration offers significant advantages compared to the traditional column-type adsorption contactors, as summarized in Table 8.13.

An important characteristic of the membrane–absorbent combination is that the membrane should not be wetted by the solvent solution, so that the pore space remains fully gas-filled to maximize mass transfer through the membrane. Other solvent requirements include the ability to achieve a high CO_2 loading at the relatively low partial pressures achievable on the permeate side of the membrane, low viscosity to reduce energy requirements for solvent circulation through hollow-fiber modules, and low volatility to reduce solvent losses.

The traditional amine-based solvents have been used in experimental work but require the use of more expensive membrane materials, such as Teflon, to avoid the problem of membrane wetting. Research has therefore focused on the selection of alternative solvents that can be used with commercially available hollow-fiber membrane modules, constructed using inexpensive membrane materials such as polypropylene.

A pilot-scale demonstration unit using Teflon (PTFE) membranes and a conventional amine solvent has been developed by Aker Kvaerner. This unit was able to recover 85% of the CO_2 from the exhaust gas of a 520 kW gas turbine engine, at a rate of 195 kg-CO_2 per hour from 2610 kg per hour of exhaust gas, with a solvent flow rate of 4.5 m^3 per hour.

Table 8.13 Advantages of gas–liquid membrane contactors

Advantage	Description
High area-to-volume ratio	Currently available hollow-fiber membrane modules offer packing densities up to 1000 m^2 per cubic meter, compared to 100–250 m^2 per cubic meter for traditional bubble, packed, and plate column contactors, resulting in significant weight and space reductions
No interpenetration of gas and liquid phases	Ability to independently control gas and liquid flow rates to optimize the absorption process, avoiding operational problems encountered in distillation processes such as liquid entrainment in the gas phase or column flooding

8.8 References and Resources

8.8.1 Key References

The following key references, arranged in chronological order, provide a starting point for further study.

Ward, W. J. and W. L. Robb. (1976). Carbon dioxide–oxygen separation: facilitated transport of carbon dioxide across a liquid film, *Science*, **156** (1481–1484).

Koros, W. J., Y. H. Ma and T. Shimidzu (1996). *Terminology for Membranes and Membrane Processes*. IUPAC Recommendations (see also Web resources).

Mano, H., S. Kazama and K. Haraya. (2003). Development of CO_2 separation membranes (1) Polymer membrane, *Proceedings of the Sixth International Conference on Greenhouse Gas Control Technologies*, J. Gale and Y. Kaya (eds.), Elsevier, Oxford, UK.

Baker, R. (2004). *Membrane Technology and Applications*, John Wiley and Sons, Chichester, UK.

Bredesen, R., K. Jordal and O. Bolland. (2004). High-temperature membranes in power generation with CO_2 capture, *Chemical Engineering and Processing*, **43** (1129–1158).

Cejka, J., H. van Bekkum, A. Corma, and F. Schueth. (2007). *Introduction to Zeolite Molecular Sieves*, Elsevier, Oxford, UK.

Favre, E. (2007). Carbon dioxide recovery from post-combustion processes: Can gas permeation membranes compete with absorption? *Membrane Science*, **294** (50–59).

8.8.2 Institutions and Organizations

Institutions that have been active in the research, development, and commercialization of membranes for CO_2 separation are listed below.

Aker Kvaerner (membrane contactor system development): www.akersolutions.com

Membrane Technology and Research Inc. (development and manufacture of membrane separation systems): www.mtrinc.com

NATCO (membrane systems development and manufacture): www.natcogroup.com

Norwegian University of Science and Technology (NTNU; membrane technology research): www.ntnu.no/engas/labs/mt.htm

Nanoglowa (European Union-funded collaborative project for nanostructured membrane development): www.nanoglowa.com

Research Triangle Institute (R&D on membrane separation systems): www.rti.org/page.cfm/Membrane_Technologies

UOP (membrane systems development and manufacture): www.uop.com

8.8.3 Web Resources

IUPAC Terminology for Membranes and Membrane Processes. Available at http://old.iupac.org/publications/pac/1996/pdf/6807×1479.pdf

9 Cryogenic and distillation systems

Distillation has been applied as a technique for separating a mixture of liquids since the second millennium BC, when it was used in Mesopotamia for the preparation of perfumes, and has been used for alcohol preparation since at least 500 BC. Modern industrial distillation methods were first developed in the 1800s, driven by the need for large-scale alcohol and industrial chemical production, and these techniques have been an area of considerable development since then due to their importance in the petrochemical industry.

Distillation techniques are relevant to CCS in two areas: oxygen production by cryogenic air separation for oxyfuel combustion, and the separation of CO_2 from natural gas either to treat gas to sales specification or to separate CO_2 produced in an enhanced oil recovery (EOR) project for reinjection.

9.1 Physical Fundamentals

The separation of a mixture of liquids by distillation into its components depends upon the difference in the boiling points and volatilities of the components.

Every liquid (and to a lesser extent every solid) has a tendency to evaporate as a result of the escape of molecules from the liquid surface, and the greater that tendency is the more volatile the substance is said to be. When a liquid is in equilibrium with its vapor, the pressure of the vapor is called the equilibrium vapor pressure. The vapor–liquid distribution ratio (or K-value) is used to characterize the volatility of a substance and is defined as:

$$K = C_V/C_L \tag{9.1}$$

where C_V is the concentration of the substance in the vapor phase and C_L is the concentration in the liquid phase, at a given temperature and pressure. A substance with a higher vapor pressure will have a higher concentration in its vapor phase at any given temperature and pressure and will therefore have a higher K-value. In a binary mixture, the relative volatility (α_{ij}) of the two components (i and j) at a given temperature and pressure is given by:

$$\alpha_{ij} = K_i/K_j \tag{9.2}$$

Doi:10.1016/B978-1-85617-636-1.00009-2.

and is therefore governed by the ratio of their vapor pressures. Mixtures with higher relative volatilities are easier to separate by distillation.

As temperature increases, the vapor pressure of a liquid increases until the point is reached at which the vapor pressure becomes equal to the surrounding pressure; this is the boiling point of the liquid. While every pure chemical has a single boiling point, a mixture, such as the binary mixture of water and ethanol, boils over a range of temperatures depending on the ratio of components in the mixture. This is illustrated in Figure 9.1, which shows a schematic boiling point diagram of a binary mixture at a given pressure.

The boiling point of the mixture lies between the boiling points T_i of the pure component i and T_j of component j. Considering a mixture with 25% mole fraction of component i, the boiling point of this mixture is represented by point B, the bubble point curve being the curve joining the boiling points for all mixing ratios.

In the superheated vapor region, no liquid phase is present. As a vapor mixture in this region is cooled (line DC), a liquid phase will start to condense when the temperature drops to the dew point (point C). The dew point curve joins all such points across the mixing ratio.

Considering again the mixture with 25% mole fraction of component i (point A): When this mixture is heated to its boiling point (point B), the vapor produced has an equilibrium composition given by point C, on the dew point curve at this temperature. This mixture has a higher mole fraction of component i, ~70% in Figure 9.1, as a result of the higher relative volatility of this component. The remaining liquid has a reducing mole fraction of this component and, if the original liquid mixture is not replenished, the boiling point of the remaining mixture will rise and boiling will cease.

This illustrates the process of separation by distillation, with a vapor stream that is rich in the lower boiling point component and a residual liquid stream that is lean in that component and rich in the other.

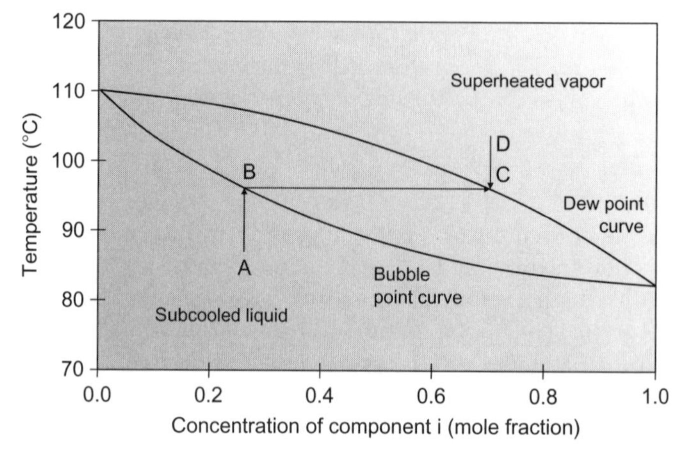

Figure 9.1 Boiling point diagram of a binary mixture

9.2 Distillation column configuration and operation

Industrial-scale distillation employs large distillation columns, in which multiple separations of this kind occur. Figure 9.2 illustrates part of the internal structure of a distillation column (two trays). The column is heated at its base by an internal heater or heat exchanger, or by liquid recycled from an external reboiler, to a temperature close to the boiling point of the required bottom stream. Similarly it is cooled at the top by an internal condenser or cooled reflux stream, to a temperature that is just above the boiling point of the required top-stream product. In separating a binary mixture, these would be the boiling point temperatures of the two pure components.

The input liquid stream is fed into the column and covers the "feed tray." The temperature at this point in the column is at the boiling point of the feed mixture so that vapor that is rich in the volatile component rises up toward the tray above the feed, while residual liquid that is rich in the less-volatile component pours over the weir and drops to the next lower tray.

Rising vapor passes either through holes in a sieve type tray, through lifting valves or, in the example shown, below bubble caps, and bubbles into the liquid in the tray above. This liquid is the condensed phase that is falling from higher, and therefore cooler, trays in the tower. The same separation process occurs here, yielding a vapor that is progressively richer in the volatile component at successively higher trays. This section of the column, above the feed inlet, is called the rectification section.

A similar process occurs in the trays below the feed with the liquid in successively lower, and therefore hotter, trays being progressively richer in the less volatile component. This section of the column is called the stripping section.

The number of trays needed to achieve a specific separation requirement depends on the relative volatility of the components to be separated. For a binary

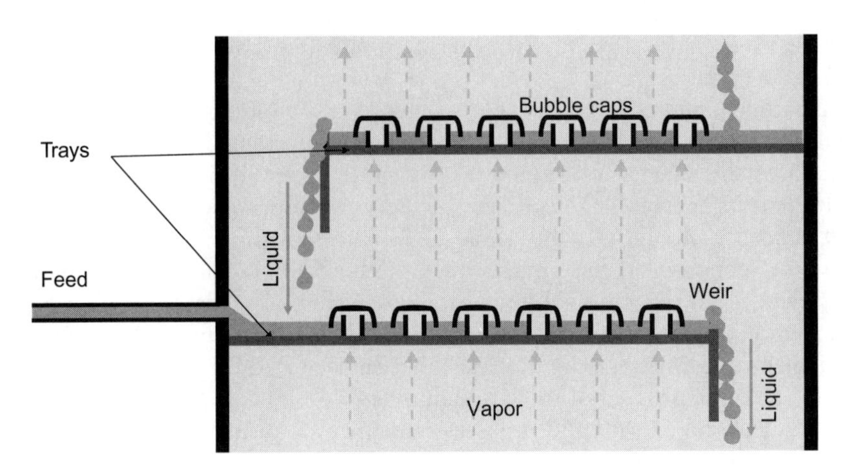

Figure 9.2 Tray structure within a distillation column

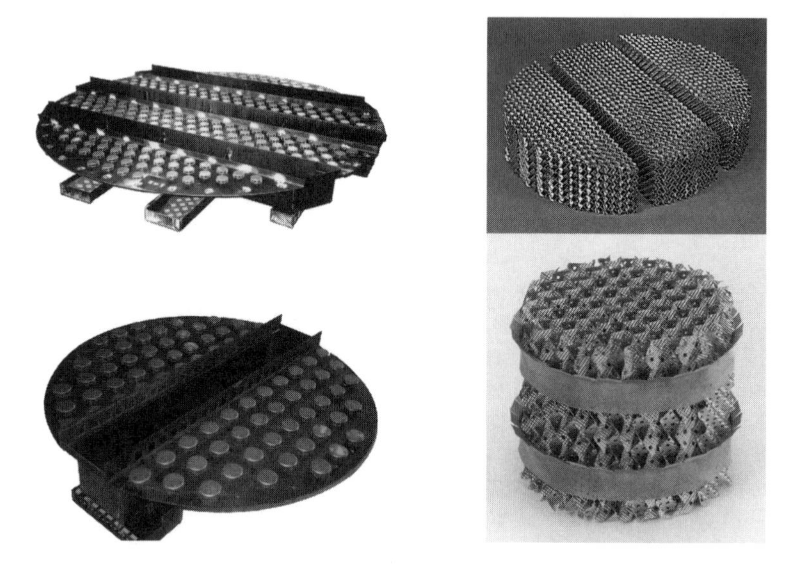

Figure 9.3 Distillation column trays (left) and structured packing material (right) (Courtesy: New Tianjin Corp. (trays) and Sulzer Chemtech (structured packing))

mixture, if α_{ij} is close to 1, the bubble point and dew point lines in Figure 9.1 will be close together, and relatively little change in composition of the mixture will be achieved in each separation stage. In this situation a larger number of trays would be required to achieve a specified purity of the two streams compared to the situation in which the bubble point and dew point lines are far apart.

As well as the vapor–liquid equilibrium behavior, the key operating parameters that determine the number of trays are the reflux rate and the condition of the feed (for example, feed temperature and phase state [liquid, mixed, or vapor]), and various analytical methods are available to calculate the required number of trays for a particular separation task (e.g., McCabe-Thiele diagram or Fenske equation).

A packed column is an alternative to the tray column described above in which the trays are replaced by a continuous packing material, typically a structured sheet metal in the form of corrugated metal plates (Figure 9.3). This configuration increases the contact area between vapor and liquid, since liquid will condense on and wet the whole surface area of the packing material, and allows a reduction in the physical size of the column for a given separation requirement. In this case, design of the column involves determining both the number of trays required and also the so-called height (of packing material) equivalent to a theoretical plate (HETP). The required height of a packed column is then the product of these two numbers.

Figure 9.4 shows a simplified process scheme for a distillation column. As well as cooling the top of the column, the reflux stream also serves to increase the concentration of the volatile component in the upper part of the column, reducing

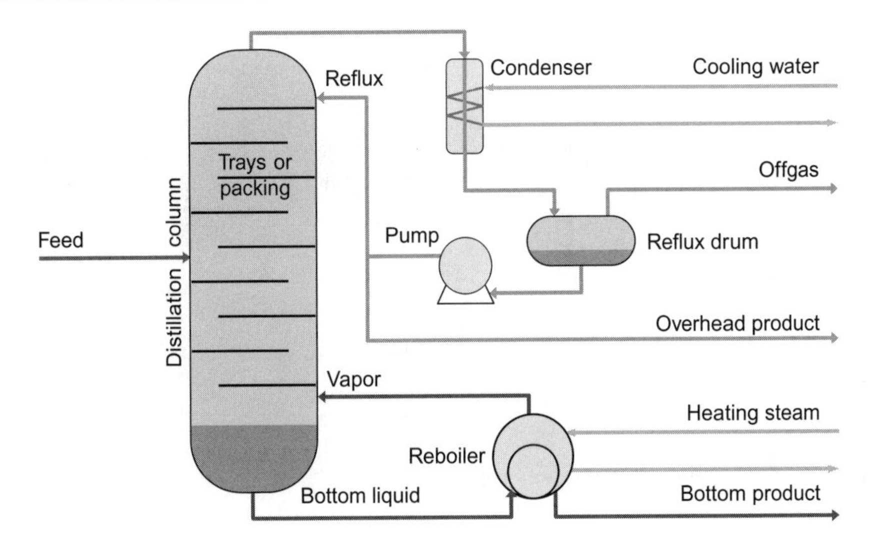

Figure 9.4 Distillation column process flow scheme

the boiling point of the mixture and therefore keeping the less-volatile component in the liquid phase. This increases the efficiency of the separation process and reduces the number of trays required to achieve the required degree of separation.

Practical distillation process design also needs to take into account the range of operating conditions over which column performance must be sustained, for example considering possible variations in feed composition, feed rate, and ambient temperature.

Problems that can reduce separation efficiency include:

- Entrainment: upward transport of excessive liquid as a result of high vapor rates
- Weeping: excessive downward liquid rates due to low vapor pressure
- Foaming: liquid expansion due to the retention of vapor bubbles in the liquid, commonly due to the presence of surfactant components in the feed stream.

9.3 Cryogenic oxygen production for oxyfuel combustion

The separation of oxygen from air for oxyfuel combustion can be achieved by a number of techniques, of which cryogenic systems are the most widely used for large-scale applications (more than ~200 t-O_2 per day).

Cryogenic air separation has been used for oxygen production since 1902, when the first air separation plant was constructed by Carl von Linde. This plant, with a capacity of 120 kg-O_2 per day, used Joule Thomson cooling together with counter-current heat exchange to liquefy air and then separate oxygen and nitrogen in a distillation column. The process is illustrated in Figure 9.5.

Filtered air is compressed, accompanied by heating, and cooled back to ambient temperature in a heat exchanger, with water as the cooling fluid.

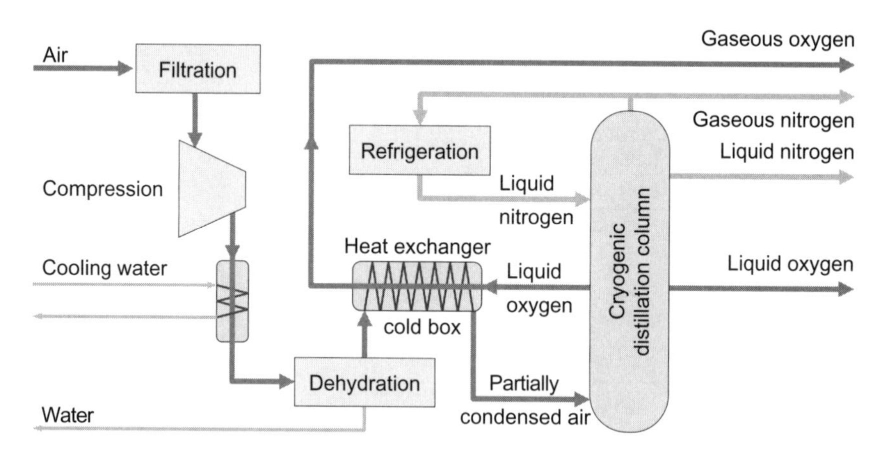

Figure 9.5 Cryogenic air separation process

The cooled high-pressure air is fed through a counter-current heat exchanger in which the returning end product from the cryogenic process is the heat sink. After this second cooling, the air is expanded through a valve and cooled to liquefaction temperature by Joule Thomson cooling.

The liquefied air is then fed into a tray or packed distillation column. Since nitrogen has the lower boiling point, the vapor above each tray will tend to be nitrogen-rich while the liquid will be oxygen-rich, the liquid cascading down the stack of trays while the vapor rises. This simple single-tower process yields a pure liquid oxygen bottom stream and a gaseous nitrogen top stream typically containing 5–10% of oxygen.

An early enhancement was the addition of a second column (the Linde double column) to increase the purity of the nitrogen top stream. In an industrial air separation unit (ASU), water vapor and CO_2 are removed after the first compression using molecular sieve units or absorption beds. A lower operating temperature for this air feed purification stage, achieved by additional cooling, improves CO_2 and water vapor removal and makes overall plant performance less sensitive to seasonal changes in ambient temperature. Figure 9.6 shows a schematic flow scheme for a modern ASU.

As can be seen from Table 9.1, the operating temperature of the distillation tower will be in the range −196°C to −183°C, between the boiling points of nitrogen and oxygen.

Helium and neon will not be liquefied in the process and will stay in the vapor phase with nitrogen, while krypton and xenon, with higher boiling points, will remain in the liquid phase as impurities in the oxygen stream. Argon has a boiling point very similar to that of oxygen, and if a high-purity oxygen product is required (>95% pure) argon is removed at an intermediate point in the distillation column—the so-called argon belly—and can be separated and purified in additional distillation columns to yield a pure-argon product.

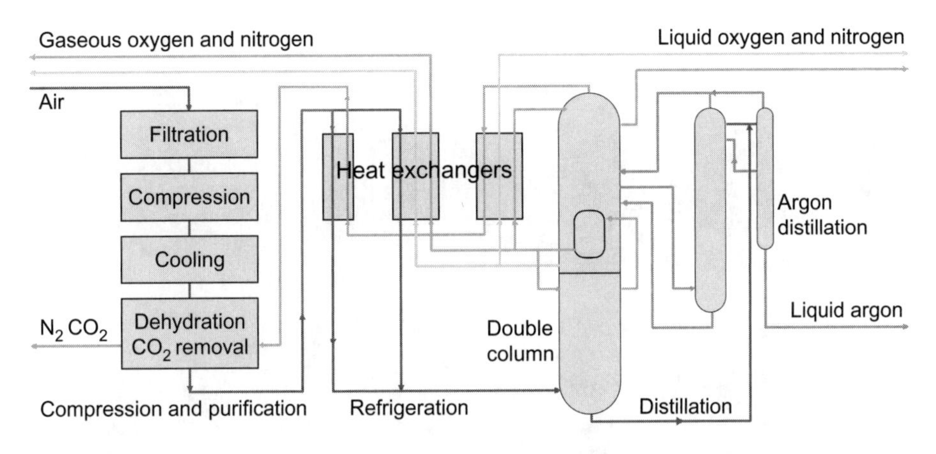

Figure 9.6 ASU flow scheme

Table 9.1 Composition of air and boiling points of its components

Component	Boiling Point (°C)	Volume %
Helium	−268.9	0.0005
Neon	−246.1	0.0018
Nitrogen	−195.8	78.08
Argon	−185.9	0.93
Oxygen	−183.0	20.95
Krypton	−153.2	0.00011
Xenon	−108.0	0.000009
Carbon dioxide	−78.5	0.0384

The principles of cryogenic air separation remain unchanged since Linde's 120 kg-O_2 per day prototype, but ASUs are now routinely constructed with single train capacities of ~4000 t-O_2 per day. The world's largest plant, at 30,000 t-O_2 per day, has been constructed by Linde AG for the Pearl Gas-to-Liquids project in Qatar using eight parallel 3750 t-O_2 per day trains. Figure 9.7 shows the ASU installed by Linde AG at the Schwarze Pumpe oxyfuel pilot plant.

9.4 Ryan–Holmes process for CO_2–CH_4 separation

CO_2 injection is a widely used technology for EOR, a process that will be described in Chapter 11. When CO_2 breaks through into the oil-producing wells, it becomes necessary to separate CO_2 from the gaseous hydrocarbon stream in order to produce sales-quality natural gas and LPG, the CO_2 generally being returned to the field for reinjection.

Figure 9.7 Air separation unit at the Schwarze Pumpe oxyfuel pilot plant (Courtesy; Linde Group)

The separation of CO_2 from light hydrocarbons by distillation is complicated by several factors:

1. At temperatures above –60°C, the vapor pressure of CO_2 lies between those of methane and ethane and very close to that of ethane (Figure 9.8)
2. The risk of solid CO_2 formation in the distillation tower if operated at temperatures and pressures to yield CO_2 in the bottom stream and methane at the top
3. The formation of an azeotrope (see Glossary) between CO_2 and ethane, preventing further CO_2–ethane separation by simple distillation

The Ryan–Holmes process overcomes these problems in a low-temperature distillation process to separate CO_2 from light hydrocarbons by the introduction of additives into two stages of the distillation process, in the form of a chilled C_{5+} hydrocarbon stream. The process flow is shown in Figure 9.9.

The feed gas is first dehydrated, initially by contacting with ethylene glycol in a low-pressure dehydration tower, and finished at higher pressure by molecular sieve adsorption after compression. The dehydrated feed gas enters a propane recovery column, where propane and heavier hydrocarbon components (C_{3+}) are recovered in the bottom stream. The additive acts as an absorbent for propane and increases the separation efficiency of this first stage. This bottom stream is distilled in the additive recovery column, yielding an LPG (propane plus butane) top stream and recovering the C_{5+} additive as the bottom stream.

The top stream from the first stage is compressed and distilled in a CO_2 recovery column, where CO_2 plus ethane is recovered in the bottom stream.

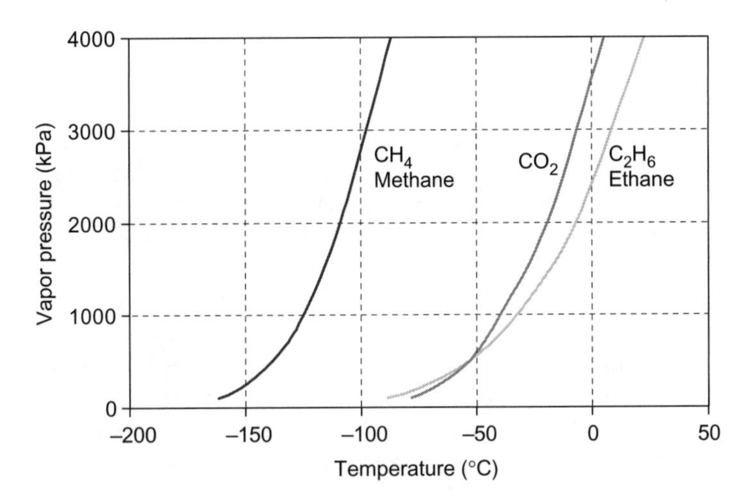

Figure 9.8 Vapor pressure versus temperature for CH_4, C_2H_6, and CO_2

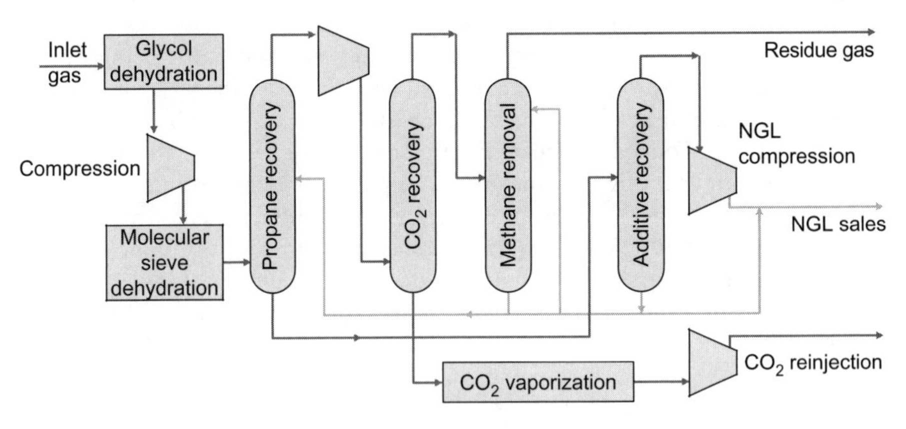

Figure 9.9 Ryan–Holmes separation process

The top stream from this stage is mainly methane, with some carryover of propane and heavier components. This stream is treated in a fourth column, again with the C_{5+} additive stream injected as an absorbent to remove the C_{3+} and yield a methane, sales gas top stream and an additive bottom stream that is recycled.

Although ethane recovery from the bottom stream of the CO_2 recovery column is desirable, simple distillation is ineffective due to the azeotrope formation. Other conventional methods of CO_2 removal such as chemical absorption using amine solutions and adsorption using molecular sieves are capital-intensive, operationally complex, and generally not economically viable. Consequently, the CO_2 plus ethane stream is commonly reinjected into the EOR reservoir, although membrane techniques for CO_2–ethane separation have also been investigated and demonstrated on a pilot scale (Chapter 8).

9.5 RD&D in cryogenic and distillation technologies

Cryogenic and distillation technologies have been applied in industrial-scale processes for over a century and the technology can therefore be considered as very mature. As a result, ongoing RD&D effort is directed at optimizing cost and performance rather than targeting fundamental breakthroughs.

High-performance process control, ensuring optimal plant operation to improve reliability and energy efficiency, is particularly important in reducing the CO_2 penalty of energy-intensive cryogenic separation systems. Approaches to the control problem range from detailed mathematical compositional modeling of the separation process that is used to derive model-based predictive or adaptive control, to model free adaptive (MFA) control systems in which the control system adaptively learns how to optimize product yield and quality, and to maintain operational stability and efficiency.

Increasing single-train capacity is also a development focus, in order to achieve lower unit production costs. Single-train ASU capacity has grown from ~1500 t-O_2 per day in 1990 to ~4000 t-O_2 per day in 2008, achieving significant cost reduction through economies of scale.

Opportunities to achiever tighter process integration are also being investigated in order to improve energy and process efficiency. In one commercially demonstrated example, shown in Figure 9.10, integration with a gas turbine power generator reduced the inlet compression power requirement of the ASU by extracting air from the gas turbine air supply (air integration), while injection of the nitrogen top stream into the turbine combustor enabled increased power output without exceeding the turbine inlet temperature limit (nitrogen integration). NO_x emissions were also reduced as a result of the lower combustion temperature. Preheating of the injected nitrogen was achieved by efficient recovery of low-level heat from the extracted air supply to the ASU.

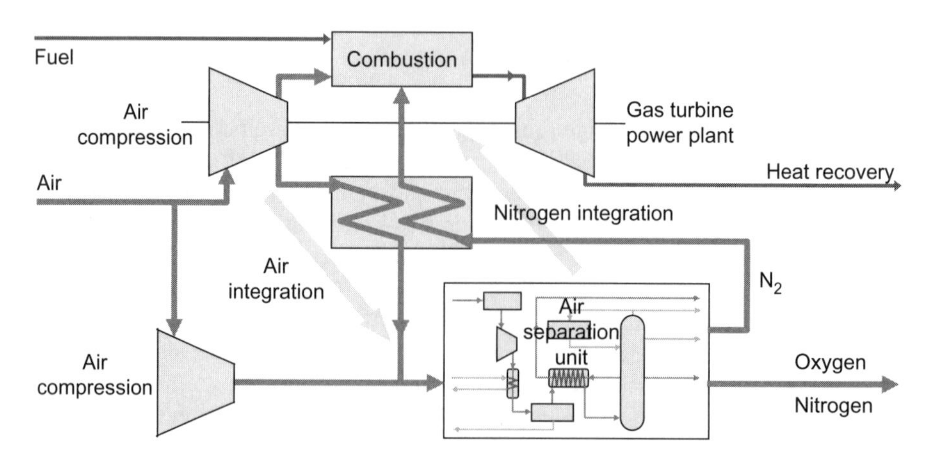

Figure 9.10 ASU process integration with gas turbine

9.6 References and Resources

9.6.1 Key References

The following key references, arranged in chronological order, provide a starting point for further study.

Latimer, R. E. (1967). Distillation of air, *Chemical Engineering Progress*, **63** (35–59).
Holmes, A. S. and J. M. Ryan. (1979). Cryogenic distillate separation of acid gases from methane. U.S. Patent 4,318,732-A.
Meratla, Z. (1997). Combining cryogenic flue gas emission remediation with a CO_2/O_2 combustion cycle, *Energy Conversion and Management*, **38** (Supplement) (S147–S152).
Kidnay, A. J. and W. R. Parrish. (2006). *Fundamentals of Natural Gas Processing*, CRC Press, Boca Raton, FL.
Linde Engineering Division. (2008). *Cryogenic Air Separation; History and Technological Progress*, Linde AG, Pullach, Germany. Available at: www.linde-anlagenbau.de/process_plants/air_separation_plants/documents/L_History_e_100dpi_08.pdf.

9.6.2 Institutions and Organizations

Companies that have been active in the research, development, and commercialization of cryogenic and distillation technology are listed below.

Air Products: www.airproducts.com/products/merchantgases/oxygen
Linde AG: http://linde-anlagenbau.de/process_plants/air_separation_plants/
Praxair: www.praxair.com

10 Mineral carbonation

The aim of the carbon capture methods described in the previous chapters is to produce a pure or near-pure CO_2 stream that can then be stored using one or other of the approaches described in Part II. As the discussion in the following chapters will show, storage as molecular CO_2 has a number of consequences, ranging from the need for monitoring to confirm long-term storage integrity to the potential adverse environmental impacts in the case of ocean storage.

In contrast, approaches based on the processes of mineral carbonation seek to store carbon in the form of products that are chemically stable and relatively benign. These processes mimic the slow, natural processes of weathering of igneous rocks and sequestration into long-term carbon sinks, and in this context "long term" implies storage on a time scale of >100,000 years.

The weathering of rocks and transportation of the products of weathering into the oceans were described in Chapter 1 as key steps in the geological carbon cycle. Mineral carbonation is a capture and storage technology that accelerates this natural process, and involves the reaction of oxides or silicates of magnesium, iron, and calcium with CO_2 to form stable carbonates.

10.1 Physical and chemical fundamentals

The simplest carbonation reaction occurs when a metal oxide, such as magnesium oxide, is reacted with CO_2 to yield magnesium carbonate:

$$MgO + CO_2 \rightarrow MgCO_3 + \text{heat} \tag{10.1}$$

The stability of the carbonate reaction product over geological time scales results from the fact that a carbon atom in a carbonate is in its lowest energy state, and the thermodynamic consequence is that the carbonation reaction is exothermic, the excess energy being released as heat.

Calcium or magnesium oxides would be ideal candidates as mineral feedstocks for carbonation, since their carbonates ($CaCO_3$, $MgCO_3$) have low solubility in water. This is an important consideration if the carbonation products would ultimately be stored on land (for example, back-filling an open cast mine) in order to prevent ground water contamination through leaching. However, as described below, the carbonation reaction typically requires some 2 to 3 tonnes of mineral input per tonne of CO_2 sequestered. Natural abundance is therefore

Doi:10.1016/B978-1-85617-636-1.00010-9.

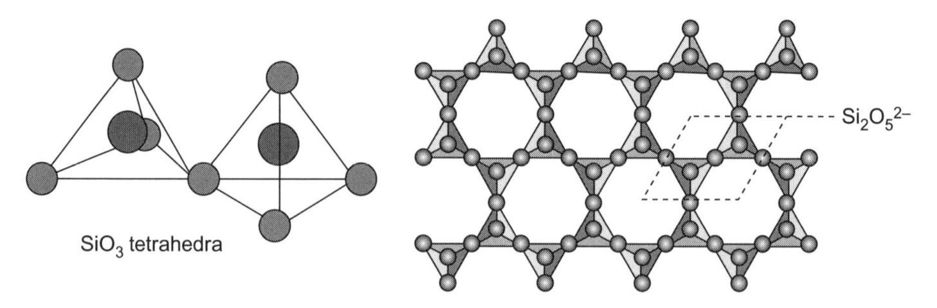

Figure 10.1 Si_2O_5 sheet structure of phyllosilicate minerals

an important characteristic for a suitable feed mineral, and these oxides are rare in nature as a result of their reactivity.

Silicate rocks are an alternative source of metal oxides and make up the majority of the earth's mantle. In particular, mafic and ultramafic rocks are igneous silicate rocks that consist predominantly of magnesium, calcium, and iron minerals. Silicate minerals occurring in these rocks include olivine $((Fe_x, Mg_{1-x})_2SiO_4)$, wollastonite ($CaSiO_3$), and serpentine ($Mg_3Si_2O_5(OH)_4$). In the case of olivine, (Fe_x, Mg_{1-x}) in the chemical formula indicates that the mineral is a mixture (known as a solid solution) of iron and magnesium silicates, with forsterite (Mg_2SiO_4) and fayalite (Fe_2SiO_4) being the pure end members of this range.

The three minerals noted above are representatives of different classes of silicates, which are defined according to the way the silicate ion (SiO_4^{4-}) is bonded in the crystal lattice. Orthosilicate minerals, of which olivine is an example, have the simplest structure and consist of a single SiO_4^{4-} anion bonded to two divalent metal cations (Mg^{2+} or Fe^{2+}). In contrast, pyroxenes form single chains as a result of the sharing of an oxygen atom between two SiO_4 units. Wollastonite is an example of a single-chain silicate. Serpentine is an example of the more complex class of phyllo- (or leaf-like) silicates, with sheetlike structures in which three of the four oxygen atoms from each SiO_4 unit are shared with adjacent units to form hexagonal rings that extend into sheets (Figure 10.1).

In minerals containing hydroxyl ions (OH^-), including serpentine, the OH^- ions bond at the center of the hexagonal ring. Metal cations form bonds with oxygen and hydroxyl to form a second, parallel sheet in which each metal ion is surrounded by eight oxygen or hydroxyl ions. Figure 10.2 shows the resulting lamellar structure for the lizardite form of serpentine. In some cases, including the antigorite and chrysotile forms of serpentine, the combined two-sheet structure takes up a curved form as a result of the longer spacing of the metal ion layer compared to the Si_2O_5 layer. Chrysotile (asbestos) is an extreme example of this curvature in which the structure is rolled up to form fibers. The different structures of these minerals have an impact on the relative ease with which they react with CO_2, necessitating various pretreatment options described later.

The carbonation reactions for a number of minerals that have been considered as possible candidates for CCS are shown in Equations 10.2 to 10.5.

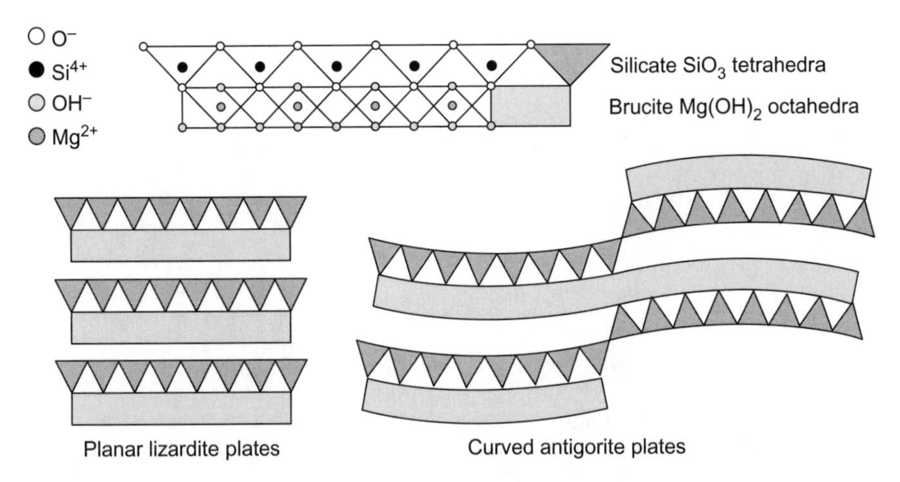

Figure 10.2 Structure of the sheet silicate serpentine: (a) lizardite, (b) antigorite

Olivine (Forsterite)

$$Mg_2SiO_4 + 2CO_2 \rightarrow 2MgCO_3 + SiO_2 \quad (10.2)$$

Olivine (Fayalite)

$$Fe_2SiO_4 + 2CO_2 \rightarrow 2FeCO_3 + SiO_2 \quad (10.3)$$

Wollastonite

$$CaSiO_3 + CO_2 \rightarrow CaCO_3 + SiO_2 \quad (10.4)$$

Serpentine

$$Mg_3Si_2O_5(OH)_4 + 3CO_2 \rightarrow 3MgCO_3 + 2SiO_2 + 2H_2O \quad (10.5)$$

The enthalpy of reaction from carbonation of these minerals is compared in Table 10.1 to the heat initially released in the combustion of carbon to produce a mole of CO_2 ($C + O_2 \rightarrow CO_2$). Clearly the enthalpy of reaction is a significant fraction of the initial heat from combustion, and the effective use of this heat will reduce the energy and carbon penalty of the carbonation process.

Although the carbonation reactions are thermodynamically favored, the reactions shown above proceed at very slow rates at room temperature and pressure. A major focus of research work has therefore been on chemical process optimization to achieve reaction rates that are viable for industrial applications.

Table 10.1 Heat released in carbonation reactions

Mineral	Enthalpy of reaction $-\Delta H^0_r$ (kJ per mol)	% of heat from carbon combustion $(-\Delta H_r/393.8)$%
Calcium oxide (CaO)	179	45%
Magnesium oxide (MgO)	118	30%
Olivine (forsterite)	89	23%
Wollastonite	87	22%
Serpentine	64	16%

Table 10.2 Potential carbonation reaction routes

Reaction route	Options	Suboptions
Direct carbonation	Dry gas–solid reaction	Gaseous CO_2 Supercritical CO_2
	Aqueous reaction	With or without additives pH swing—two-step aqueous route
Indirect carbonation	Acid extraction	Hydrochloric acid Sulfuric acid Acetic acid
	Molten salt extraction Sodium hydroxide	

Several chemical process options have been considered, including methods where CO_2 is reacted directly with the mineral feedstock, and indirect methods where the reactive mineral component is extracted from the feedstock in an initial step to improve the reaction rate. Following the initial extraction step, the reactive component can be carbonated using any of the direct route options as the second step in the indirect reaction route. These options are summarized in Table 10.2 and described below.

10.1.1 Direct carbonation; gas–solid route

The direct carbonation of a solid mineral in gaseous CO_2 (shown, for example, for direct olivine carbonation in Equation 10.2) is the simplest carbonation reaction in terms of the process design and offers the best prospect of effective utilization of the heat released in the reaction. However, using silicate minerals the reaction rate is too slow for industrial-scale application, even at elevated temperatures and pressures (for example, using supercritical CO_2).

The direct gas–solid reaction is however a viable carbonation route if the alkaline–earth metal is first extracted from the silicate feedstock as an oxide or hydroxide, since the carbonation of these compounds is rapid. For example:

$$Mg(OH)_2 + CO_2 \rightarrow MgCO_3 + H_2O \qquad (10.6)$$

The feasibility of the initial extraction step is discussed under the indirect reaction route later.

10.1.2 Direct carbonation; aqueous route

Direct carbonation in an aqueous medium is considered by many research groups as the most promising reaction route and has been the subject of extensive research effort. The presence of water speeds up the carbonation reaction, which then proceeds in three steps; for example, for forsterite olivine:

1. CO_2 dissolves in water to form carbonic acid containing protons (H^+) and bicarbonate ions (HCO_3^-)

$$CO_2 + H_2O \rightarrow H_2CO_3 \rightarrow H^+ + HCO_3^- \quad (10.7)$$

2. The presence of free protons (H^+) enables the release of Ca and Mg from the mineral matrix:

$$Mg_2SiO_4 + 4H^+ \rightarrow 2Mg^{2+} + SiO_2 + 2H_2O \quad (10.8)$$

3. The metal and bicarbonate ions combine to precipitate carbonate:

$$Mg^{2+} + HCO_3^- \rightarrow MgCO_3 + H^+ \quad (10.9)$$

Without the use of additives, the rate-limiting step in this reaction is the release of metal ions from the mineral matrix, and the speed of the reaction can generally be increased by decreasing the mineral particle size, for example by fine grinding, thereby increasing the surface area available for the reaction. The use of additives, such as sodium chloride (NaCl) and sodium bicarbonate ($NaHCO_3$), can also enhance the release of metal ions from the silicate. Under these conditions the dissolution of CO_2 in water (Step 1 above) can become the rate-limiting step.

The first step in the aqueous carbonation route (Equation 10.8) requires a low pH for dissolution of the metal ion from the mineral feedstock, while a higher pH is optimal for the precipitation step (Equation 10.9). This has led to the so-called pH swing option in which the pH is varied by the introduction of additives, as the reaction proceeds, to achieve optimal conditions for each step.

These and other aspects of reaction optimization are discussed in Section 10.2.

10.1.3 Indirect carbonation; acid extraction

Several indirect carbonation options have been proposed that are based on the application of acids to extract the alkaline–earth metal from the mineral feedstock, and either precipitate this as a more reactive hydroxide for carbonation in a second step, or carbonate the metal ion directly in solution. The use of hydrochloric acid (HCl) was the first such approach proposed and, with a serpentine feedstock, proceeds according to the following steps:

1. HCl is used to extract magnesium as magnesium chloride:

$$Mg_3Si_2O_5(OH)_4 + 6HCl + H_2O \rightarrow 3MgCl_2 \bullet 6H_2O + 2SiO_2 \quad (10.10)$$

2. HCl is recovered by heating the solution to ~250°C:

$$MgCl_2 \bullet 6H_2O \rightarrow MgCl(OH) + HCl + 5H_2O \quad (10.11)$$

3. Water is introduced to reform MgCl(OH) to $Mg(OH)_2$ which is precipitated:

$$2MgCl(OH) \rightarrow Mg(OH)_2 + Mg_2Cl_2 \quad (10.12)$$

The precipitated hydroxide is then carbonated via the direct carbonation step in Equation 10.6. The energy cost of Step 2 in this extraction routes is very high and has been estimated as four times the energy released by carbon combustion for the quantity of CO_2 eventually sequestered. Clearly, without a substantial reduction in energy requirements this is not a feasible sequestration process.

A similar route has been proposed using sulfuric acid, either forming magnesium hydroxide for carbonation as in the HCl route, or carbonating the magnesium sulfate directly in solution according to:

$$Mg_3Si_2O_5(OH)_4 + 3H_2SO_4 \rightarrow 3MgSO_4 + 5H_2O + 2SiO_2 \quad (10.13)$$

followed by:

$$3MgSO_4 + 3H_2O + 2SiO_2 + 3CO_2 \rightarrow 3MgCO_3 + 2SiO_2 + 3H_2SO_4 \quad (10.14)$$

Although in principle these reaction routes result in recovery of the acids, some make-up volume of acid is likely in practice. This would be required, for example, if the feedstock contains alkali metals, which would react with HCl to form soluble chlorides, consuming chlorine.

The sulfuric acid process does not require the energy-intensive dehydration step that was necessary for HCl extraction, but would also require acid make-up. The use of acids would also increase the construction cost of a carbonation reactor in view of the need to use more corrosion-resistant materials.

Acetic acid extraction has also been proposed, using wollastonite as a feedstock to produce calcium carbonate. This has the advantage of being a less corrosive medium and therefore puts less stringent constraints on reactor material selection, although laboratory trials indicate that recovery of the acetic acid may be problematic.

10.1.4 Indirect carbonation; molten salt extraction

The use of a molten salt, such as magnesium chloride, as an extraction medium can partially reduce the energy cost of the dehydration step in the HCl extraction option. In its simplest form, the carbonation reaction (Equation 10.5) takes place in a single direct step in a molten salt medium ($MgCl_2 \cdot 3.5H_2O$) at ~200°C. Alternatively, a multistep process similar to HCl extraction can be used, with the intermediate precipitation, separation, and carbonation of magnesium hydroxide.

Once again, the energy cost and corrosive nature of the extraction medium are challenges for this option.

10.1.5 Indirect carbonation; sodium hydroxide extraction

The use of sodium hydroxide has also been proposed as an alternative to acid extraction. In this option calcium or magnesium is released from the mineral matrix and precipitated as hydroxide, with sodium taking the place of the extracted metal ion in the mineral. For example, with plagioclase (calcium feldspar) as a feedstock, the overall reaction is:

$$3CaAl_2Si_2O_8 + 8NaOH + 4CO_2 \rightarrow 3CaCO_3 + Na_8(AlSiO_4)_6\ CO_3 \cdot 2H_2O + 2H_2O \quad (10.15)$$

Since sodium hydroxide is consumed in the reaction, substantial quantities (1.8 t-NaOH per t-CO_2) would be required if this reaction route were to be applied on an industrial scale.

10.2 Current state of technology development

A technology development road map for mineral carbonation is shown in Table 10.3. The technology is currently at the development stage, with a wide range of applied research studies being conducted, focusing on increasing the carbonation reaction rate.

The following sections summarize some of the key issues addressed and outcomes to date.

10.2.1 Identification of preferred mineral feedstock

A number of important characteristics can be identified that would need to be fulfilled by the optimal feedstock for mineral carbonation:

- Low cost of supply and preparation
- High reaction rate, either natural or achievable with reaction optimization
- Availability and global distribution
- High molar fraction of alkaline–earth metal
- Low reaction energy requirements

Table 10.3 Technology development road map for mineral carbonation

Research
Identification of preferred mineral feedstock
Assessment of global availability of preferred mineral resources
Development
Optimization of direct vs. indirect, and gas-solid vs. aqueous reaction options
Assessment and optimization of mineral pretreatment and activation options (mechanical, thermal, chemical, electromagnetic)
Increasing reaction rate through optimization of reaction parameters (CO_2 partial pressure, reaction temperature, solution additives)
Assessment of in situ versus ex situ carbonation options
Use of alkaline solid waste as a feedstock and integration with industrial processes
Assessment of disposal and reuse options for carbonation end products
Design and development of a demonstration-scale carbonation reactor
Assessment of economic feasibility
Demonstration (possible example demonstrator projects)
Steel mill slag carbonation
Coal gasification, liquid fuel synthesis with integrated mineral carbonation
Power plant, precombustion CO_2 capture and integrated mineral carbonation
Deployment
Integrated mineral carbonation

Low cost of supply and preparation is essential, since typically 2–3 tonnes of mineral feedstock are required to capture 1 t-CO_2. Costs therefore cannot exceed a few U.S. $ per tonne. Supply and preparation steps include mining, transportation, grinding, and removal of any impurities or components that would impede the carbonation reaction or cause problems in storing the reaction products (e.g., components that could leach out of mine back-filling). Many olivines and serpentines contain 5–20% by weight of iron oxides, a valuable by-product that can reduce the net cost of the feedstock.

A high reaction rate, either natural or through reaction optimization, is essential to accelerate the carbonation reaction from a geological to an industrial time scale. While the carbonation reaction is faster for the alkali metals such as sodium and potassium than for the alkaline–earth metals calcium and magnesium, the higher reactivities of sodium and potassium also result in their carbonates being soluble in water. This would pose problems for storage of reaction products. Of the alkaline–earth metals, the carbonation reaction proceeds faster for calcium than for magnesium, as is indicated by the greater heat

Table 10.4 Relative mineral feedstock quantities and conversion efficiencies

Mineral	R_{CO_2} (kg per kg)	R_C (kg per kg)	R_x (%)
Olivine (forsterite)	1.6–1.8	5.9–6.6	60–80
Olivine (fayalite)	2.3–2.8	8.4–10.3	61
Wollastonite	2.6–2.8	9.5–10.3	40–80
Serpentine (lizardite)	2.1–2.1	7.7–7.7	9–40
Serpentine (antigorite)	2.1–2.5	7.7–9.2	60–90

release in the exothermic carbonation of CaO ($\Delta H^0_r = -179$ kJ per mol) compared to MgO ($\Delta H^0_r = -118$ kJ per mol).

For a given carbonation feedstock, the efficiency of the carbonation reaction is expressed by the parameter R_x, which measures the percentage conversion of available metal cations into carbonate. The conversion efficiency depends on the specific reaction conditions (temperature, pressure, time) and mineral pretreatment. The R_x values shown in Table 10.4 for each mineral illustrate the range of values achieved in various experiments, the lower figure without and the higher figure with activation, either by ultrafine grinding or by heat treatment in the case of serpentine.

Availability and global distribution is clearly very important for a feedstock if mineral carbonation is to be applied at a scale that would have a material impact on global CO_2 emissions. Of the alkaline–earth silicates, olivine and serpentine occur in nature in vast quantities. The igneous rocks that contain these minerals originated from upwelling magma on ancient ocean floors, and several countries, including the United States and Oman, have sufficient deposits to store the CO_2 generated from combustion of a greater part of the world's coal reserves. Studies of deposits within the United States have shown a good match on a regional scale between the distribution of these resources and the regional CO_2 emissions from coal-fired power plants. In other countries, cost-effective exploitation of the available mineral resources may require the development of a CO_2 pipeline infrastructure to transport captured CO_2 to carbonation plants located at major mineral deposits.

A high molar fraction of alkaline–earth metal in the feedstock will minimize the quantity of rock that needs to be mined per t-CO_2 captured. Two ratios commonly used to compare the amount of candidate feedstocks required to capture a certain quantity of CO_2 are:

- R_{CO_2}: the quantity of mineral required to capture 1 t-CO_2
- R_C: the quantity of mineral required to capture the CO_2 produced by combustion of 1 t-C (equal to R_{CO_2} times 44/12, the ratio of the molar weights of CO_2 and C)

These ratios are shown in Table 10.4 for some potential mineral feedstocks. R_C and R_{CO_2} are determined purely by the mineral feedstock composition and, in the case of a pure mineral, by the ratios of molecular weights. The table shows two R_{CO_2} values for each mineral, first the theoretical value for a pure mineral and secondly a typical value for an actual prepared mineral sample.

Table 10.5 Ranking of potential mineral carbonation feedstocks

Characteristic	Olivine	Serpentine	Wollastonite	Alkaline industrial waste
Cost of supply and preparation	=	=	=	++
Reaction rate	=	−	+	++
Availability	++	++	−	−−
Global distribution	+	+	−	+
High molar fraction of alkaline–earth metal	+	+	−	=
Low reaction energy requirements	=	−	=	+

Note: =represents the basis for comparison. +and ++ imply better or much better performance against a particular characteristic; −and − − imply poorer performance.

The R_{CO_2} values show that the magnesium minerals are more efficient in terms of quantities required, due to the lower molar weight of magnesium (24) versus calcium (40). Olivine occurs naturally as a solid solution of magnesium and iron silicates (forsterite and fayalite) with the general composition $(Fe_x, Mg_{1-x})_2SiO_4$. R_{CO_2} for pure olivine will be below that for pure serpentine provided the iron mineral fraction in the olivine is <70%. (i.e., x <0.7).

Low reaction energy requirements will keep down the cost of the carbonation process and also contribute to its net capture efficiency by reducing CO_2 emissions attributed to the process energy requirements. Olivine and wollastonite are slightly favored over serpentine due to the higher exothermic heat of reaction noted earlier, although the practicality of usefully applying this reaction heat in an integrated carbonation process is a subject for further investigation. If an indirect acid extraction method is used (Section 10.1.3), the carbonation reaction heat can be easily used in acid recovery (equation 10.11).

Table 10.5 compares a number of potential feedstocks for mineral carbonation using these characteristics.

Based on these considerations, most development work has been focused on either olivine or serpentine as the likely feedstocks for large-scale application. Alternative feedstocks may also be important at the technology demonstration stage and for niche applications in specific industries, as discussed in the following section.

10.2.2 Alternative feedstocks and industrial integration

Many industrial processes produce alkaline waste that could be used as a feedstock for mineral carbonation. Although these wastes are not available on the Gt scale that would be required for a carbonation feedstock on a global scale, the quantities are significant for the industries involved and their use as a carbonation feedstock could have advantages over the minerals considered earlier.

The candidate alkaline wastes include:

- Municipal waste incinerator (MWI) ash
- Ash from coal combustion
- Slag from steelmaking
- Ash and waste cement from cement production
- Waste concrete from building demolition
- Asbestos mine tailings

These materials contain a significant proportion of primarily calcium oxides and hydroxides (10–15% for concrete waste, 40–50% for steel slag) and can be carbonated using any of the reaction routes applied to mineral feedstock, direct aqueous carbonation being the front-runner. The use of these materials in a direct air capture system has also been considered and is described further in Section 14.1.1.

Many of these wastes (e.g., steel slag and waste concrete) have a reuse value in the $5–10 per tonne range that is comparable to or exceeds the expected mined cost of mineral feedstocks. However, carbonation can in many cases be effectively integrated with the reuse process, adding value to the end product. Examples are the production of precipitated calcium carbonate (Section 14.1.1) or the reuse of steel slag as a cement additive. This process requires the slag to be crushed and finely ground, and these are also necessary preparation steps for carbonation. Integration of a carbonation step in the recycling process therefore would not incur additional preparation costs.

A particular advantage of the use of asbestos (chrysotile serpentine) mine tailings is that the hazardous nature of the waste is effectively remediated by the carbonation reaction, which consumes the asbestos fibers. The quantity of available asbestos mine tailings in the United States and Canada is estimated at 170 Mt, sufficient to sequester ~70 Mt-CO_2. While this is a relatively small quantity (<1 day of global CO_2 emissions!), other environmental benefits may be an important enabler for a demonstration project using this feedstock.

10.2.3 Reaction optimization, including mineral pretreatment and activation

Investigations of the aqueous carbonation route have resulted in the identification of the optimal carrier solution for the reaction as a sodium bicarbonate plus sodium chloride solution with strength 0.64 M for $NaHCO_3$ and 1 M for NaCl[1]. The addition of sodium bicarbonate increases the concentration of HCO_3^-, accelerating the carbonation reaction (Equation 10.9). CO_2 solubility is also increased by a factor of 20 in the optimal carrier solution (20 g per liter vs. <1 g per liter in distilled water), allowing the reaction to take place at lower CO_2 pressure. Finally, the bicarbonate–salt mixture buffers the pH of the carrier solution to a slightly alkaline 7.7–8.0, aiding precipitation of the carbonate.

[1] A 1M, or Molar, solution contains 1 gram mole per litre of water. Thus a 0.64M $NaHCO_3$ solution contains 53.8 g/l and a 1M NaCl solution contains 58.4 g/l NaCl.

Table 10.6 Serpentine pretreatment options

Pretreatment option	Description
Dry and wet grinding	Grinding to increase the specific surface area of the mineral feedstock is the first basic pretreatment step. Early investigations showed that reduction of average particle size increased the extent of conversion for a given residence time in line with the increase in specific surface area. Attrition grinding, in which the mineral particles are vigorously stirred together with hard balls, can further increase reactivity by creating dislocations in the mineral lattice structure, which increases the accessibility of the metal ion. Wet attrition grinding has also been investigated, but although it produces smaller particles, the equivalent dry process is more effective at disrupting the lattice structure and produces a more reactive product. Mineral feedstock is typically reduced to a mean particle diameter of $<75\,\mu m$, a further stage reduction to $<38\,\mu m$ being a trade-off between additional energy costs (~80 kWh per tonne) and increased mineral conversion.
Heat treatment	Heating serpentine results in progressive removal of the hydroxyl ion (for lizardite, 45% residual OH at 580°C, 10% at 650°C) and eventually, at ~1100°C, to the crystallization of forsterite and enstatite ($MgSiO_3$). Above 580°C the dehydroxylation results in breakup of the layered serpentine structure shown in Figure 10.2 into an amorphous phase, which exhibits a substantially increased carbonation rate compared to the untreated mineral (the crystallized form resulting at higher temperature is less reactive). However, the energy required for pretreatment is substantial (specific heat to 580°C plus enthalpy of dehydroxylation = ~300 kWh per tonne for lizardite feedstock) and can exceed the energy released from carbon combustion that generates the volume of CO_2 captured as carbonate, resulting in a net negative sequestration efficiency! Heat treatment seems likely to be feasible only when waste process heat is available at low cost by integration with another industrial process.
Chemical and steam activation	Sulfuric and nitric acids and steam have been investigated as options to increase the specific area and hence reactivity of a ground mineral feedstock. These surface activation techniques can increase specific surface area by a factor of 40, equivalent to reducing mean particle diameter by a factor of 6 through additional grinding. Disadvantages are the consumption of acids and the reduction in alkaline–earth metal content of the feedstock as a result of leaching.
Ultrasonic and microwave activation	Speculative investigations have been conducted that attempt to achieve mineral activation by ultrasonic treatment or microwave irradiation but have failed to achieve an increase in reactivity.

The optimal reaction temperature and CO_2 pressure established in laboratory studies are for olivine T = 185°C and P_{CO_2} = 15 MPa, and for heat-treated serpentine (see below) T = 155°C and P_{CO_2} = 11.5 MPa.

Although early studies have identified serpentine as an attractive mineral feedstock for industrial-scale mineral carbonation, the layered lattice structure of the phyllosilicates, including serpentine, causes the magnesium atom to be relatively inaccessible, resulting in a slow reaction rate. A variety of pretreatment options have been investigated with the aim of disrupting or breaking down the lattice structure to increase the reaction rate, as summarized in Table 10.6.

Investigations aimed at optimizing the carbonation reaction have spawned additional basic research into the molecular-level controls on the reaction process. Examples of this work are X-ray diffraction (XRD) and scanning electron microscopy (SEM) studies of the dehydroxylation process in heat treatment of serpentine and of the surface morphology of mineral particles during the carbonation reaction. This is an example of the nonsequential nature of the technology development process. The latter work has led to insights into the development and exfoliation of a surface reaction "passivating" layer and the accelerating exfoliation of surface layers by the addition of abrasive, nonreactive particles such as quartz, further improving the reaction rate. Any additives used will need to be efficiently recovered to avoid an economic penalty.

Although heat pretreatment of serpentine has been shown to be prohibitively energy-intensive, a possible route to mitigate this high energy cost is through integrating mineral carbonation into a coal or biomass gasification plant. The considerable process heat that is available from the initial partial combustion step in the gasification process would be used to heat-treat the serpentine feedstock, and CO_2 produced from the WGS reaction (Section 3.1.2) would be captured in a carbonation reaction with the heat-treated serpentine. A fuller description of this process integration option is given in Section 10.3.

10.2.4 Disposal and reuse options for carbonation end products

As noted above, the mineral carbonation process generates in excess of 2.5 t-reaction product per t-CO_2 sequestered. If mineral carbonation is ultimately applied on a material global scale for CO_2 sequestration, Gt of reaction products including magnesium carbonate, silica, and other compounds will have to be dealt with annually. The eventual fate of these products, including potential reuse options, is therefore a significant issue warranting research at the technology development stage. As well as mine back-filling, options that have been proposed include reuse as an additive in cement or asphalt, or for soil enhancement. However, the quantities that could be reused in these applications would be dwarfed by the amounts available from large-scale mineral carbonation.

One intriguing possibility is to follow the same route as described in Section 12.4 for accelerated weathering of limestone (AWL). The carbonation reaction would be prevented from proceeding as far as the precipitation of the

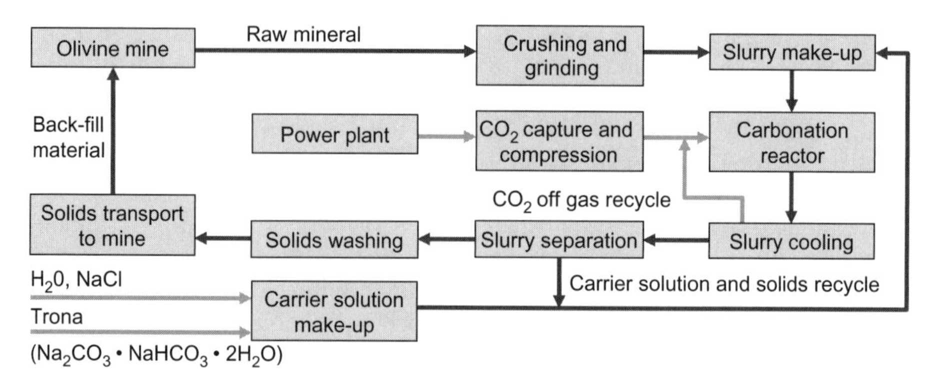

Figure 10.3 Simplified process for direct aqueous carbonation of olivine

carbonate, and instead a solution containing the metal (Mg^{2+}) and bicarbonate (HCO_3^-) ions plus orthosilicic acid (H_4SiO_4) would be discharged into the ocean. Orthosilicic acid occurs abundantly in nature and is an important nutrient for diatoms. These microscopic single-celled algae metabolize the silica from orthosilicic acid and use it to construct their cell walls. Diatoms and other phytoplankton drive the biological CO_2 pump in shallow ocean waters (Section 12.2.2), fixing CO_2 from surface water by photosynthesis, and ultimately contributing to the long-term carbon sink in ocean floor sediments. Diatom growth depends on the supply of nutrients, including silicon in the form of orthosilicic acid. Each spring algae bloom as surface waters warm, the bloom typically ending when the silicon supply is exhausted. Observations have shown that the bloom is extended, resulting in greater CO_2 conversion into oceanic biomass, if there is a new supply of silicate from upwelling deep ocean water. Discharge of the solution containing carbonation products could therefore play a further role in CO_2 sequestration by extending the annual bloom of diatomaceous algae.

10.2.5 Design and development of a demonstration-scale carbonation reactor

Figure 10.3 shows a simplified process for direct aqueous carbonation of olivine.

A number of different carbonation reactor designs have been proposed, including continuous-flow stirred autoclaves and pipeline reactors, although none has yet been constructed beyond a laboratory scale. A scaled-up version of the laboratory autoclave would consist of a series of stirred and heated high-pressure reactor vessels, with a total volume determined by the feedstock slurry flow rate and the residence time required to achieve the required conversion to carbonate.

Figure 10.4 shows a flow-through pipeline reactor proposed by the National Energy Technology Laboratory Albany Research Center that has been successfully

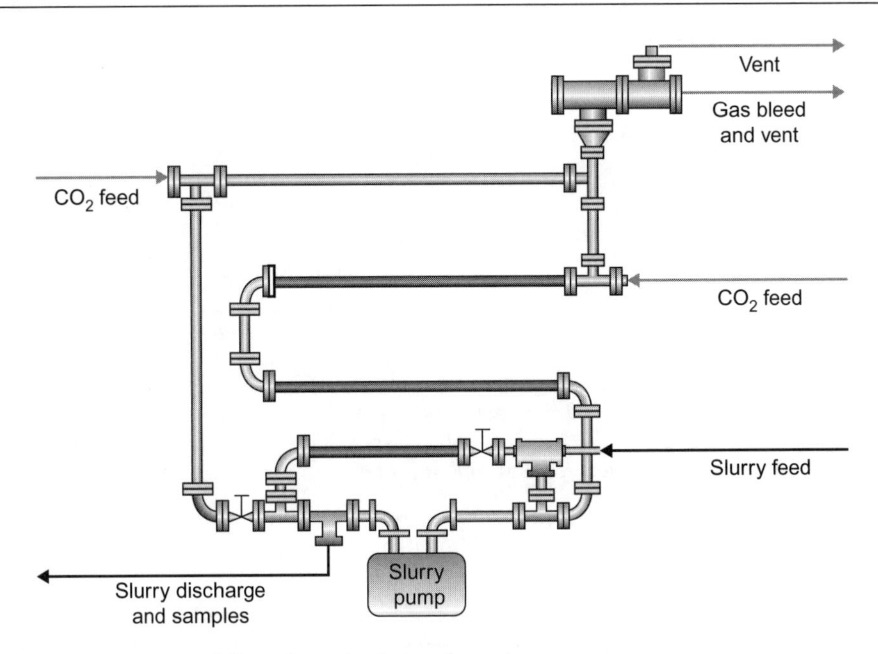

Figure 10.4 Proposed flow through pipe carbonation reactor

developed and operated at a laboratory scale in a flow-loop configuration. The reactor achieved higher carbonation conversion (R_x) than had been achieved in batch autoclave experiments.

Pipe reactors offer a potential advantage of lower capital cost, although the volume requirement to achieve the required slurry throughput rate and residence time will translate into large-diameter or long pipelines. As an example, a carbonation unit to sequester CO_2 from a 100 MW power plant, based on wollastonite feedstock, would require a minimum reactor volume in the region of 200 m^3, equal to 1.6 km of pipe with a 40 cm internal diameter. Pipe reactors may also be less effective operationally since, in the absence of an active agitation device, slurry settling during operational shutdowns would be problematic.

10.3 Demonstration and deployment outlook

Deployment of mineral carbonation as a carbon sequestration technology on a material global scale will involve the construction and integration of a number of large-scale industrial operations. For example, sequestration of 1 Gt CO_2 per year would involve:

- Capture of 1 Gt-CO_2 per year, from approximately one hundred 1 GW_e coal-fired power plants (assuming 35% thermal efficiency)
- Transportation of captured CO_2 via pipeline infrastructure to the nearest suitable mineral locations

Table 10.7 Breakdown of estimated CO_2 sequestration cost by mineral carbonation

Cost element	Relative cost	Comment
Feedstock cost	53%	Wollastonite assumed in this example
Electricity cost	26%	Of which 18% for feedstock grinding
Capital cost	10%	Depreciation basis
Other costs	11%	Maintenance, staff, etc.

- Mining of >2 Gt per year of mineral feedstock. With a typical mine producing ~100 kt per day, this would require 60 such operations to achieve the required scale.
- Mine back-filling (100 kt per day) plus disposal of excess (50–100 kt per day) carbonation products generated at each location

The cost of CO_2 sequestration using a large-scale mineral carbonation infrastructure has been estimated at roughly $120 (100 €) per t-$CO_2$ avoided, which is made up as shown in Table 10.7.

The best prospects for cost reduction are therefore in reducing the cost of feedstock or of energy for mineral pretreatment. As a result, early-stage mineral carbonation demonstration projects are likely to be more closely integrated to specific industries in contrast to the more stand-alone type of operation outlined above. Possible examples are:

- Steel mill with integrated mineral carbonation, eliminating feedstock cost
- Coal gasification power plant with integrated mineral carbonation, reducing pretreatment energy cost
- Steel mill with mineral carbonation of slag by direct air capture, eliminating feedstock cost and reducing capital cost

The first two of these examples are described below, while the third is covered in Section 14.1.1.

10.3.1 Steel mill with integrated mineral carbonation

There are a number of synergies that would result from the integration of mineral carbonation with steel production, including:

- Similar requirements for mining and preprocessing of iron ore and carbonation feedstock
- Potential use of peridotite (a magnesium- and iron-rich mixture of olivine and pyroxene) as a common feedstock for iron ore and magnesium silicate
- Sequestration of CO_2 produced both from iron ore reduction and from the rest of the steelmaking process, totaling ~1.7 t-CO_2 per t-steel produced
- Carbonation of steel slag with added value as a sellable product

Figure 10.5 shows the concept of an integrated steel mill with CO_2 sequestration by mineral carbonation that has been investigated by the American Iron and Steel institute together with Colombia University.

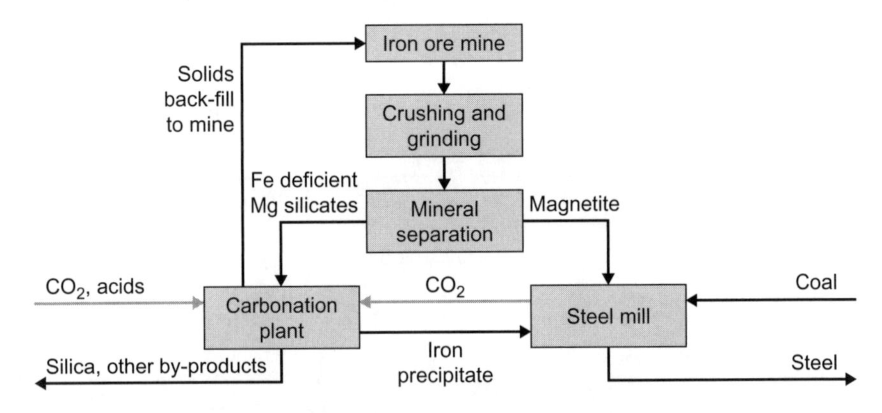

Figure 10.5 Integrated steel mill with CO_2 sequestration by mineral carbonation

10.3.2 Coal gasification power plant with integrated mineral carbonation

Gasification of hydrocarbon, coal, or biomass feedstock, described in Chapter 3, is a process that can be integrated into power generation, or the production of synthetic liquid fuels or hydrogen, with high levels of energy efficiency. The synthesis gas (syngas) produced is a mixture of carbon monoxide and hydrogen and is generated by combustion of the feedstock in an oxygen-poor environment, leading to partial oxidation.

Depending on the application, some or all of the carbon monoxide in the syngas can be converted to CO_2 using a WGS reaction, typically carried out at a temperature of 300–500°C:

$$CO + H_2O \rightarrow CO_2 + H_2 \tag{10.16}$$

For example, if the syngas is to be used for liquid fuel synthesis using the Fischer–Tropsch process, ~35% of the syngas needs to be shifted in order to achieve the required CO to H_2 ratio. For a typical syngas composition this results in a post-WGS syngas containing ~60% CO_2 by weight.

Syngas leaves the partial oxidation step at a temperature of ~1500°C, and the heat available in cooling the gas to the desired temperate for the WGS reaction (~1.47 MJ per kg of syngas) can be used for heat pretreatment of serpentine to create a more active feedstock for carbonation. Each kg of syngas cooled can heat-activate ~1.1 kg of serpentine (1.32 MJ per kg) and yields 0.75 kg-CO_2 if 35% of the CO is water-shifted. Since the R_{CO_2} of heat-activated serpentine is 2.1 kg per kg-CO_2, 0.52 kg-CO_2 can be sequestered via carbonation, which is 70% of the CO_2 generated in the shift reaction.

Figure 10.6 shows a schematic diagram of a coal gasification, liquid fuel synthesis plant with integrated CO_2 sequestration by mineral carbonation.

Additional process heat integration is shown, with the cooled syngas exiting the fluidized bed reactor being used to preheat the incoming crushed and

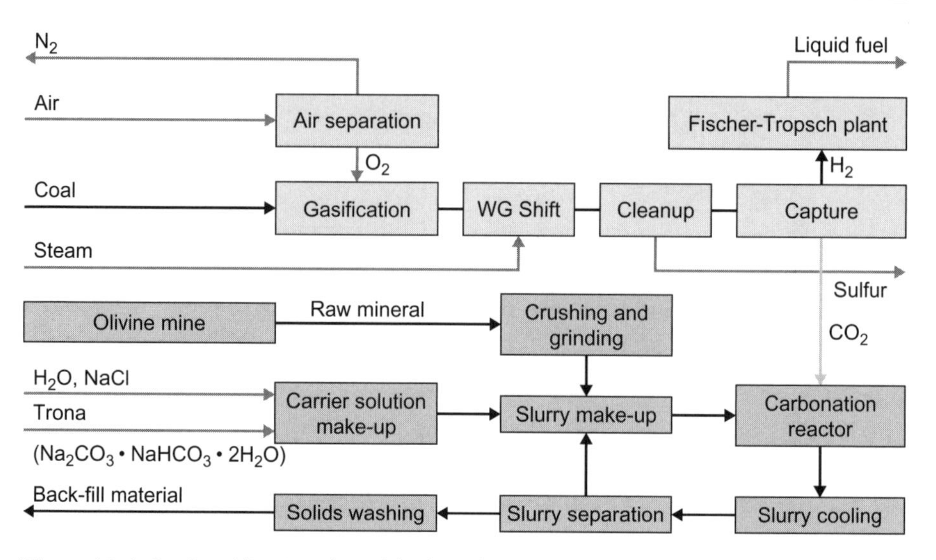

Figure 10.6 Coal gasification, liquid fuel synthesis plant with integrated mineral carbonation

coarsely ground serpentine feedstock. Fine grinding of the mineral is not required because larger particles, up to a few millimeters, will fragment due to the expansion of steam released during dehydroxylation.

10.4 References and Resources

10.4.1 Key References

The following key references, arranged in chronological order, provide a starting point for further study.

Seifritz, W. (1990). CO_2 disposal by means of silicates, *Nature*, **345** (486).

Lackner, K. S., *et al.* (1995). Carbon dioxide disposal in carbonate minerals, *Energy*, **20** (1153–1170).

IPCC. (2005). *Special Report on Carbon Dioxide Capture and Storage, Chapter 7–Mineral Carbonation and Industrial Uses of Carbon Dioxide*, Cambridge University Press, Cambridge, UK.

IEA Greenhouse Gas Programme. (2005). *Carbon Dioxide Storage by Mineral Carbonation,* IEA Greenhouse Gas Programme Report 2005/11.

Stolaroff, J. K., G. V. Lowry, and D. W. Keith. (2005). Using CaO- and MgO-rich industrial waste streams for carbon sequestration, *Energy Conversion and Management*, **46** (687–699).

Gerdemann, S. J., W. K. O'Connor, D. C. Dahlin, L. R. Penner, and H. Rush (2007). Ex situ aqueous mineral carbonation, *Environmental Science & Technology*, **41** (2587–2593).

Huijgen, W. J. J., R. N. J. Comans, and G.-J. Witkamp. (2007). Cost evaluation of CO_2 sequestration by aqueous mineral carbonation, *Energy Conversion and Management*, **48** (1923–1935).

10.4.2 Institutions and Organizations

The institutions that have been most active in the research and development work on mineral carbonation are listed below, with relevant web links.

U.S. Department of Energy National Energy Technology Laboratory (NETL), notably the Albany Research Center (NETL-Albany): www.netl.doe.gov
Energy Research Centre of the Netherlands (ECN): www.ecn.nl
Los Alamos National Laboratory: www.lanl.gov
Arizona State University, Center for Solid State Science and Science and Engineering of Materials Graduate Program: www.asu.edu/clas/csss
University of British Colombia, Department of Earth and Ocean Sciences (EOS): www.atsc.ubc.ca/research
Columbia University, New York, USA: www.seas.columbia.edu/earth/index.html
ETH Zürich, Switzerland: www.ipe.ethz.ch/laboratories/spl
Åbo Akademi University, Finland: http://web.abo.fi/fak/tkf/vt/Eng/research.htm

10.4.3 Web Resources

U.S. Department of Energy National Energy Technology Laboratory (NETL), mineral carbonation subscription list: http://listserv.netl.doe.gov/mailman/listinfo/mineral-carbonation
U.S. Department of Energy National Energy Technology Laboratory (NETL), mineral carbonation archive: http://listserv.netl.doe.gov/mailman/private/mineral-carbonation

Part III

Storage and monitoring technologies

11 Geological storage

11.1 Introduction

The injection of CO_2 into permeable rock formations—geological storage—is the only method of carbon storage that has been applied on a commercial scale to date. The mature technologies that have been developed and deployed worldwide in the oil and gas industry are applicable with minor modifications to the selection of suitable injection sites, the injection of CO_2, and the prediction and monitoring of the movement of the subsurface CO_2 plume.

In this chapter the key oil and gas industry concepts and technologies that are important for geological storage of captured CO_2 are introduced, and they provide the foundation for a discussion of the geological storage in producing or depleted oil and gas fields, including enhanced recovery aspects, or in saline aquifers. Storage in unmineable coal seams, with enhanced coal-bed methane recovery, is also described.

11.2 Geological and engineering fundamentals

11.2.1 Sedimentary reservoir rock properties

Sedimentary processes

The rock formations that contain gas, oil, or water in the subsurface were formed over thousands and millions of years by the gradual deposition of sediments—predominantly either silicate or carbonate grains—in a wide range of sedimentary processes such as the deposition of sand bars in meandering rivers, the gradual buildup of sediments on a flood plain or of a delta system at a river mouth, the subsea "rain" of carbonate skeletal material, or the deposition of wind-blown sand grains in dune systems, to name but a few.

Postdepositional burial results in compaction of the sediments and commonly the cementation of grains into a consolidated rock. If the space between the grains remains open the rock will be porous, with a capacity to hold fluids, and if these pores remain connected, the rock will also be permeable, allowing fluids to flow through the rock.

Doi:10.1016/B978-1-85617-636-1.00011-0.

Sedimentary processes also transport and deposit organic material in sediments, and subsequent burial and heating result in thermal decomposition of this material to produce hydrocarbons. Buoyancy forces usually result in the migration of these fluids away from the point of origin (descriptively known as the "kitchen") until the migration path is impeded by faults or impermeable formation and the fluids are trapped, forming an oil or gas accumulation.

Porosity and permeability

The porosity and permeability of sedimentary rocks are typically correlated by a logarithmic relationship of the form:

$$\Phi = a + b\log(k) \tag{11.1}$$

where Φ is the porosity and k is the matrix permeability, and a and b are constants that will differ for individual formations. In some cases, particularly in carbonate rocks where a variety of geochemical processes may have occurred since deposition, this type of relationship may be very approximate or absent. Porosity and permeability are measured directly on rock samples recovered during well drilling (core samples), and these data are used to determine porosity and permeability for uncored formations by calibrating downhole measurements of bulk density or acoustic velocity. The permeability of a formation can also be determined directly by performing a flowing or injection test on a well and evaluating the results using an appropriate form of Darcy's equation (Section 11.2.4).

Seals and caprocks

In fine-grained sediments with a high content of clay particles, compaction and dewatering of the clays will result in impermeable rocks that can provide a shale seal, or caprock, if deposited above a permeable formation. Caprocks can also be formed from other impermeable material such as salt and other evaporites-minerals deposited from evaporating bodies of water.

Injection of fluids, such as water to improve oil recovery or CO_2 for storage, will result in increasing pressures close to the injection well that exceed the initial fluid pressure within the rock pores. In order to maintain the integrity of the caprock, pressures must not be allowed to exceed the caprock fracture pressure, which depends on the in situ stress regime and the geomechanical properties of the caprock. This limit, the fracture pressure, can be determined by the measurement of rock mechanical properties on core samples, or by performing a "minifrac" test to fracture the formation in situ, and knowledge of this limit will be a requirement for detailed characterization of potential storage sites.

Heterogeneity

In attempting to characterize a sedimentary rock formation and thereby delineate its potential either for oil and gas exploitation or for CO_2 storage, the most significant problem faced by the geologist is heterogeneity. The vertical and areal variability of key parameters such as porosity, permeability, sand thickness, and shale content are relevant to such an evaluation on all length scales from the pore scale (10^{-4}m) to the whole field (typically tens of km). Permeability heterogeneity will be the most important factor determining the position and shape of an injected CO_2 plume.

11.2.2 Structural and stratigraphic traps

After deposition, deformation and faulting of the sediments commonly occurs due to tectonic and gravitational forces, and uplift may also result in erosion of previously deposited sediments. These processes result in the formation of various structural features; a number of common types are illustrated in Figure 11.1.

A structural trap is formed when deformation, faulting, or an erosion surface (known as an unconformity) prevents the upward movement of a buoyant fluid, such as gas, oil, or supercritical CO_2. A trap may also occur as a result of the geometric form of a permeable sedimentary body, for example a marine sediment that gradually becomes more shaly away from the sediment source. This form of trap, which depends on the sedimentary layering of the formation, is known as a stratigraphic trap, and is illustrated in Figure 11.2.

Depending on the properties of the rock when deformation and faulting occurs, these forces may also cause fracturing. Over geological time these fractures may become closed, by further deformation, deposition of minerals, etc., or may remain open and contribute an additional fracture permeability for

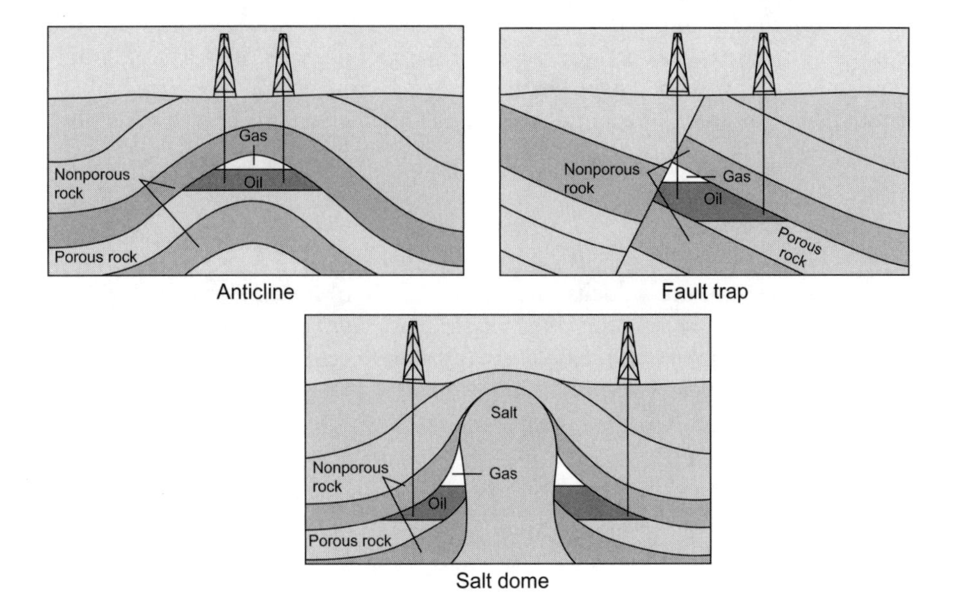

Figure 11.1 Structural features and traps

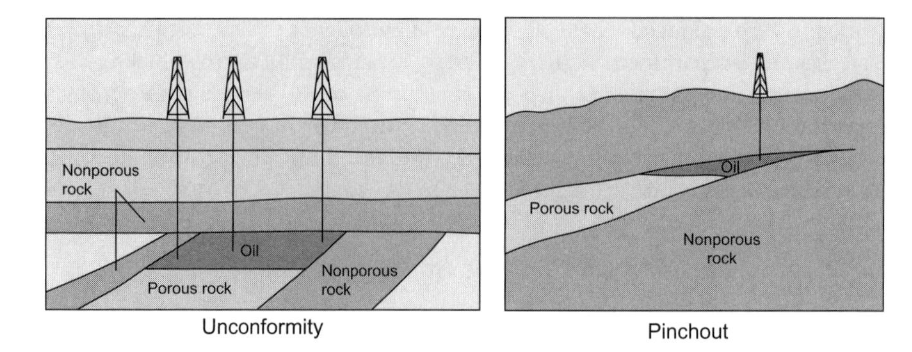

Figure 11.2 Stratigraphic trapping

fluid flow. In extreme cases, intense fracture networks may provide the only significant permeability in an otherwise tight (low-permeability) rock.

11.2.3 Seismic surveying

Although gravitational and magnetic surveys can provide information on subsurface rock types and structuration, seismic surveying is the key technology used to generate high-resolution 3D images of the subsurface. Reflection seismology was developed in the 1920s as a tool for oil exploration and relies on the reflection and refraction of acoustic energy from interfaces between formations with differing acoustic impedances.

In conducting a land-based survey, an array of acoustic sensors known as geophones are first arranged in a series of parallel lines on the surface. A pulse of seismic energy with a well-defined wave characteristic is emitted by the dropping or vibration of a large weight, and the acoustic energy travels down into the earth. The acoustic impedance (Z) of a rock layer (a) is given by:

$$Z_a = v_a \rho_a \tag{11.2}$$

where v is the acoustic velocity (meters per second) and ρ is the rock density (kg per m^3) of the layer. As the wave passes from layer a to layer b, with impedance Z_b, a fraction of the incident energy is reflected. For normal incidence, this reflection coefficient is given by:

$$R = (Z_b - Z_a)/(Z_b + Z_a) \tag{11.3}$$

The reflected energy is detected by the geophone when it reaches the surface after a two-way travel time:

$$t = 2D/V \tag{11.4}$$

where D is the depth below surface of the reflecting layer and V is the average acoustic velocity of rock down to that depth. The received energy at each geophone is then recorded for a period of several seconds after each "shot" of the seismic source, and several shots are made at each location to improve the signal-to-noise ratio by averaging the acquired data. The geophone array is moved to the next location and the process is repeated. For 3D surveying, geophones are typically located on an x-y grid at a spacing of 12.5 or 25 m over an area that may cover hundreds or thousands of square kilometers.

The acquisition process is similar although logistically simpler offshore, with arrays or streamers of hydrophones pulled along by one or more vessels while an air gun is used to generate the source pulse. Accurate GPS positioning of the geophones at each shot point is critical to accurately locate identified subsurface features, and strong currents or poor surface weather conditions can therefore disrupt the offshore acquisition process.

The apparent simplicity of Equations 11.2 to 11.4 belies the computational complexity of the processing that follows from the acquisition stage. Individual geophone traces are "gathered" to improve signal-to-noise ratio, deconvolution using the waveform of the source pulse (known as the wavelet) reveals the impedance contrasts causing reflection, and "migration" of these events in 3D corrects for the source-to-geophone geometry to correctly locate these events in the subsurface space.

High-resolution 3D seismic can reveal features on a 20 to 25 m scale at depths of several kilometers, including the internal structure of sedimentary bodies, and can often indicate the fluid fill and pressure in a formation as a result of the impact of these parameters on impedance. Acquisition and processing parameters can be adjusted to achieve the best resolution for a particular target depth, so that surveys acquired to evaluate oil and gas reservoirs may need to be reacquired or reprocessed to achieve optimal results for other targets such as shallower aquifers.

4D seismic involves the repeated acquisition of a 3D survey over a period of time, typically with an interval of a few years between surveys. When a repeat or "monitor" survey is acquired using the same acquisition geometry as the original "baseline" survey, the processed results can be subtracted to identify changes in impedance during the intervening period. This change can be used to identify the movement of fluids of different density, for example oil replaced by water in an oil recovery process, or water replaced by CO_2 in a storage process.

11.2.4 Fluid flow in porous media

Darcy's law

The flow of water through sand beds was first studied by French engineer Henry Darcy in the 1850s, and the resulting Darcy equation remains the foundation of hydrology and of oil and gas reservoir engineering today. The empirically derived relationship allows the flow rate of a fluid to be determined, as follows:

$$q = -(kA/\mu)dP/dL \tag{11.5}$$

where q is the flow rate (m^3 per second), k is the permeability of the porous medium (m^2), A is area (m^2), μ is the fluid viscosity (kg/m.s or Pa.s), and dP is the pressure drop (Pa) over length dL (m). With a knowledge of formation pressure and permeability and of fluid viscosity, this relationship enables the reservoir engineer to determine the expected production rate of an oil or gas well, or the required bottom hole pressure to achieve a desired rate of CO_2 injection.

Wettability and relative permeability

When two fluids such as oil and water are present in the pore space, one of the fluids will preferentially wet the rock grains and spread over the grain surfaces. This depends on the relative interfacial tensions between the fluids and the grain material, and in most oil- and gas-bearing formations water is the wetting phase. The nonwetting phase tends to occupy the open pore spaces, and in this situation the ability of each phase to flow will depend on the relative proportions, or saturations, of the two phases within the pore space. An example of this relative permeability behavior for an oil–water system is shown in Figure 11.3.

The degree of curvature of the relative permeability curves is a measure of the degree to which the presence of one fluid in the pore space inhibits the flow of the other fluid; the more concave the curves are, the greater the interference. Conversely, straight-line relative permeability lines would be applicable if there were no interaction, for example in the case of segregated flow in a very high permeability or vertically fractured reservoir or of fully miscible flow of two fluids.

The curves show that as the saturation of either phase reduces, a point is reached at which the permeability for that phase becomes essentially zero. The

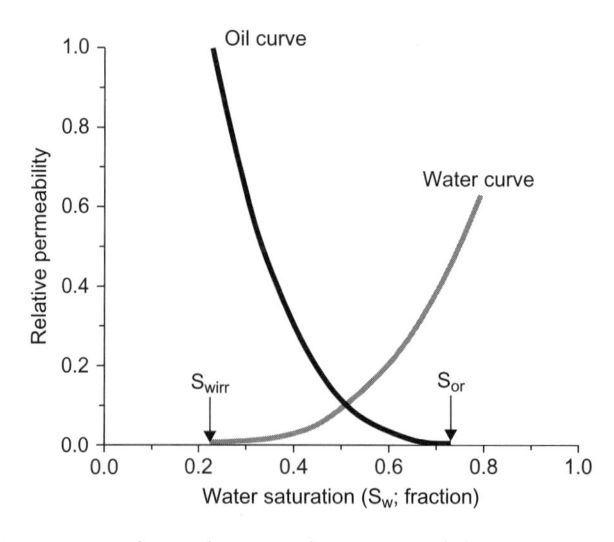

Figure 11.3 Generic two-phase oil–water relative permeability curves

saturation at this point is called the residual saturation (e.g., S_{or} for residual oil), and the residual fluid is trapped in the pore space. Further reduction of this residual saturation requires a modification of the capillary forces within the pore, for example by the injection of a surfactant to reduce interfacial tension or injection of a fluid that is miscible with the residual phase. The relative permeability behavior of two- or three-phase systems including CO_2 has not been widely investigated to date.

11.2.5 Hydrocarbon fluid phase behavior

The phase behavior of a hydrocarbon reservoir fluid under varying pressure and temperature is an important factor in oil and gas field exploitation. Figure 11.4 shows the phase diagram for a typical oil reservoir fluid.

The liquid phase in the reservoir may be initially undersaturated (point P'_U), in which case the fluid will remain in a liquid phase until the pressure drops to P_b, or saturated (if initial pressure is on the bubble point curve), in which case any drop in pressure will result in the liberation of dissolved (solution) gas. As production proceeds and the reservoir pressure declines, liberation of solution gas will cause shrinkage of the remaining oil in the reservoir and, if the gas flows to a well and is produced, will result in unwanted high gas production at the surface, as well as a further loss of reservoir pressure. As the reservoir fluid travels to the well and up the production tubing (Section 11.2.7), pressure and temperature will decline to the wellhead pressure and temperature (P_{WH}, T_{WH}), and gas and oil will finally be separated into two product streams in the surface processing facilities (at P_S,T_S).

The situation is simpler for lean hydrocarbon gas, composed typically of >90% CH_4, as shown in Figure 11.5.

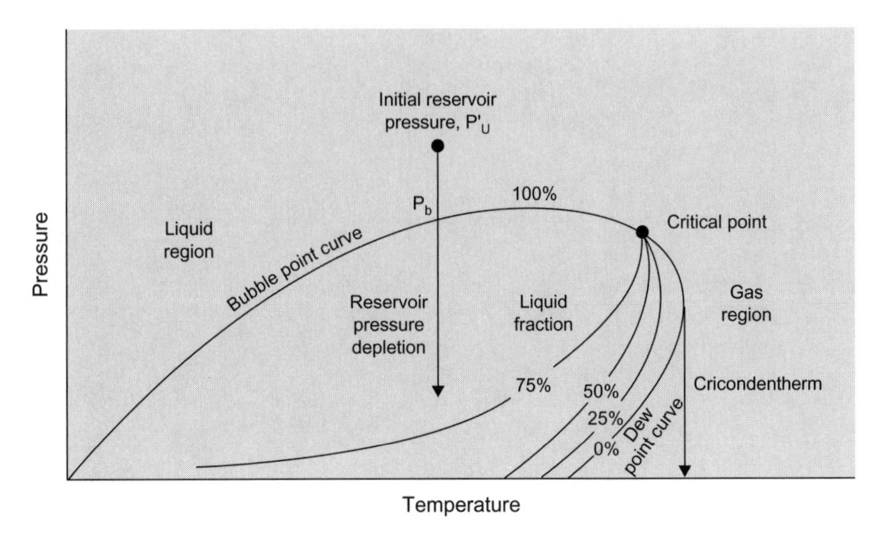

Figure 11.4 Phase diagram for a typical oil reservoir

Here the fluid remains in the vapor phase until the dew point is reached, and this may occur in the upper part of the production tubing or in the surface facilities. Condensing heavy hydrocarbons are separated in the surface facilities.

For rich gas reservoirs, which contain a significant fraction of heavier (C5+) hydrocarbons, the phase behavior in the reservoir becomes more complex. In the example shown in Figure 11.6, with a reservoir temperature below the cricondentherm for this fluid, depletion from the initial pressure P_i to below the

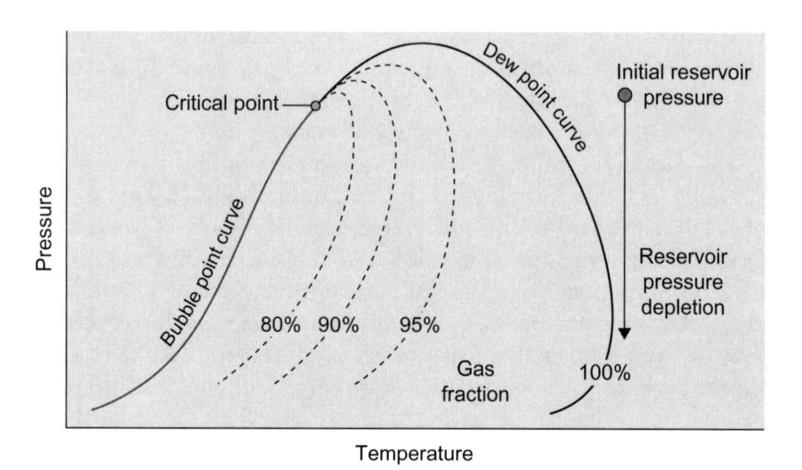

Figure 11.5 Phase diagram for a lean gas reservoir

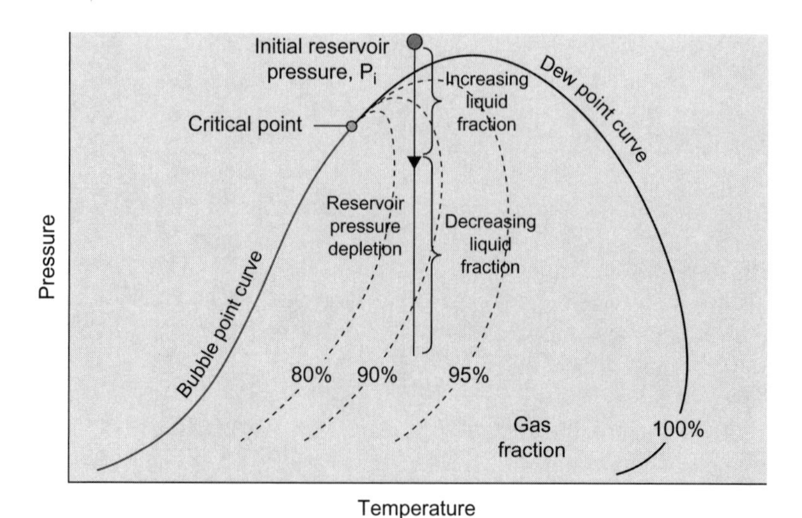

Figure 11.6 Retrograde condensation in a rich gas reservoir

dew point pressure results in condensation, resulting in the formation of a liquid phase within the reservoir. This is known as retrograde condensation—retrograde since here condensation is occurring as a result of a pressure decrease rather than a pressure increase as is the case for most vapors.

This liquid phase could be completely revaporized if the reservoir pressure subsequently drops back below the dew point curve, but more commonly a liquid phase will remain in the reservoir. If the liquid saturation does not build up to the critical value at which the liquid becomes mobile, this liquid will remain trapped and will not be recoverable.

The impact of a CO_2 component on the phase behavior of the initial reservoir fluids is an important aspect in the design of enhanced recovery schemes based on CO_2 injection. This is typically addressed using laboratory pressure–volume–temperature (PVT) measurements that are used to calibrate mathematical equation of state (EOS) models.

11.2.6 Reservoir modeling and monitoring

Rock and fluid data acquired during well construction and operation (see next section) are used in combination with the interpreted results of seismic surveys to build 3D numerical models of oil and gas reservoirs. The construction of these models is strongly guided by geological understanding of the original environment in which the sediments were deposited, in particular the impact of depositional and postdepositional processes on rock characteristics and heterogeneity.

These models are then used to simulate the dynamic behavior of reservoirs under various production or injection scenarios. For oil and gas field developments, reservoir models are built in order to maximize oil and gas recovery, including assessing the effectiveness of water or gas injection as a "secondary recovery" method, or CO_2 injection as an enhanced oil recovery (EOR) method. Reservoir models are validated by matching the model output to actual production and injection performance (history matching), with adjustments of the value of uncertain parameters being used to improve the match between predicted and actual performance. The validated model can then be used in prediction mode as the basis for forecasts and decision making.

Reservoir management is an important aspect of the exploitation of oil and gas fields and involves the gathering of monitoring data, such as production and injection measurements, pressure surveys in producing and injecting wells, postproduction seismic surveys, etc., and the comparison of these data with the behavior predicted by the reservoir model. Refinement of the model based on these performance data underpins confidence in the validity of forecasts.

This type of monitoring and verification is an essential aspect of CO_2 storage projects and will build on these oil and gas industry approaches. An important additional challenge will be the need to ensure the validity of model predictions on a centennial and millennial time scale—not a common requirement in oil and gas field practice.

11.2.7 Oil and gas well drilling and production technology

Well construction

The construction of an oil or gas well starts with the drilling of a surface hole, typically 76cm (30 inches) or 91cm (36 inches) in diameter, using a rotary drilling bit on the end of a string of steel drill pipes. Rock cuttings are lifted out of the well by circulating a viscosified fluid known as drilling mud through the drill pipe, out of the bit and up the annulus around the drill pipe.

When drilling of the hole section has reached the required depth and all cuttings have been removed, a steel casing is constructed from threaded ~10m long joints and lowered into the well. The casing is typically 5–10cm smaller in diameter than the drilled hole size and is cemented into the rock by pumping cement down the inside of the casing and up the annulus. Effective casing cementation is essential to eliminate the potential flow path between the casing and the face of the drilled formation, which could otherwise conduct fluids from deeper, high-pressure formations to the surface. In a CO_2 storage project, this well integrity issue is an important consideration if old wells are present in the region that will be invaded by a mobile CO_2 plume.

This process of drilling and casing, illustrated in Figure 11.7, is repeated a number of times with progressively smaller drill bit and casing diameters, until the target formation is reached and a final casing is set. For CO_2 injection service, or for wells that penetrate a storage formation and may be exposed to high $[CO_2]$, material selection for casings and cement will be important in order to maintain well integrity in the corrosive environment.

The well construction geometry can be varied depending on the specific well objectives, as illustrated in Figure 11.8. Where surface access is unimpeded,

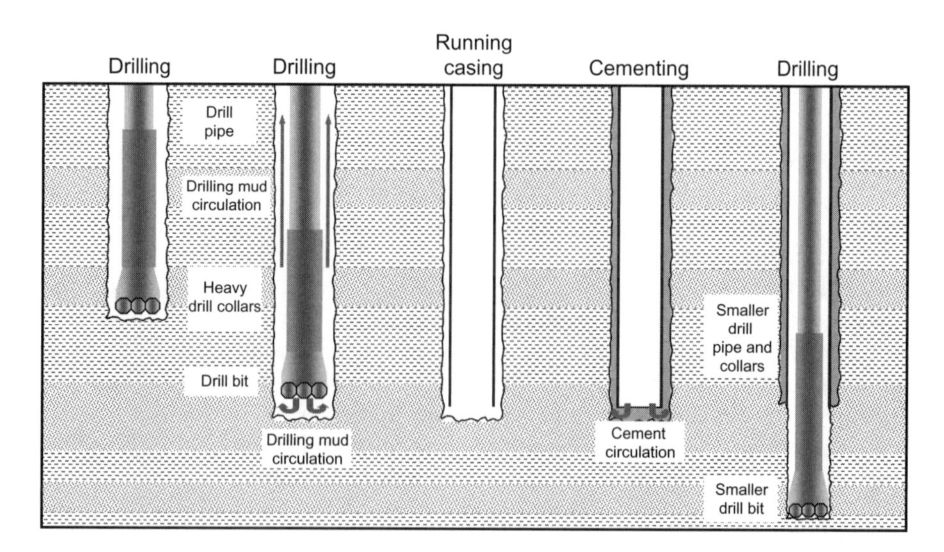

Figure 11.7 Drilling and casing a hole section during well construction

vertical wells are the cheapest option, while deviated, horizontal, and multilateral wells can be drilled to overcome more complex surface and subsurface constraints, or to increase well productivity or injectivity as a result of a larger contact area with the target formation.

A horizontal injection well has been used in the StatoilHydro Sleipner project to inject CO_2 into the Utsira aquifer, as described in Section 11.4.5.

Data acquisition in wells

The construction of a well provides an opportunity to gather data on the subsurface formations penetrated by the bit, both the target formation and those in the shallower sections of the well—the overburden.

Once wells are in service, either for production or injection, additional data can be gathered from instruments permanently installed in the well or by running instruments into the well to conduct surveys. The types of data that can be acquired are summarized in Tables 11.1 and 11.2 for the construction and operating phases respectively.

Well completion and control

A well completion consists of two parts, the sand-face completion that provides the interface between the reservoir rock and the wellbore, and the upper completion, comprising equipment installed within the last casing string to control and monitor the well flow and ensure well safety.

The sand-face completion is required to protect the well from the ingress of formation material should the rock fail under production or injection conditions. Rock strength measurements on core samples, the planned well geometry (particularly the deviation from vertical), and production or injection pressures and flow rates are used to decide on the appropriate sand-face completion type.

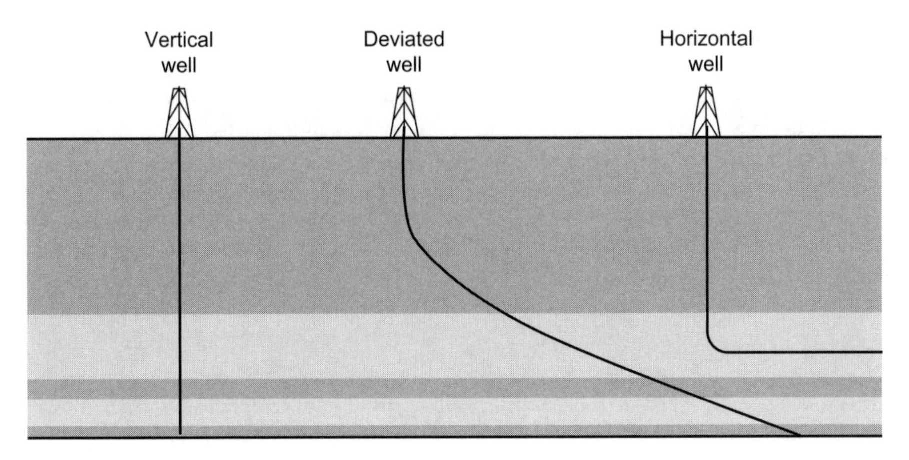

Figure 11.8 Schematic overview of the Sleipner CO_2 storage project (Courtesy; StatoilHydro)

Table 11.1 Data gathering during well construction

Data gathering technique	Description	Data types
Open-hole logging	Lowering of measuring sondes into the borehole either on an electrical cable (wireline logging) or as an integral part of the drill string (logging-while-drilling)	Radioactivity, density, and porosity inferred from nuclear measurements, acoustic velocity, electrical resistivity, nuclear magnetic resonance response
Coring	Cutting of rock samples either from the wall of an already drilled hole (side-wall coring) or while drilling using a special coring drill bit and core collection barrel	Grain size distribution and mineralogy, radioactivity, rock strength, porosity and permeability by laboratory analysis, fluid content, relative permeability, and capillary pressure behavior
Formation fluid sampling	Lowering a fluid sampling tool into the well on wireline or as part of the drill string	Formation pressure and permeability, recovery of fluid samples, downhole analysis of fluid composition
Vertical seismic profile (VSP) logging	Use of downhole geophones and surface-induced vibrations to generate high-resolution seismic data for the formations near the wellbore	High-resolution seismic data

Table 11.2 Data gathering during well operations

Data gathering technique	Description	Data types
Cased-hole logging	Lowering of measuring sondes into the cased hole on an electrical cable (wireline logging) or on coiled steel tubing during a well intervention	Fluid density and flow rate, static and flowing temperature profiles, and several of the open-hole data types (radioactivity, density, and porosity inferred from nuclear measurements, nuclear magnetic resonance response)
Permanent downhole instrumentation	Installation of instrumentation as part of the well completion to allow continuous well and reservoir performance monitoring	Static and flowing pressure, spot temperatures, and full well temperature profile

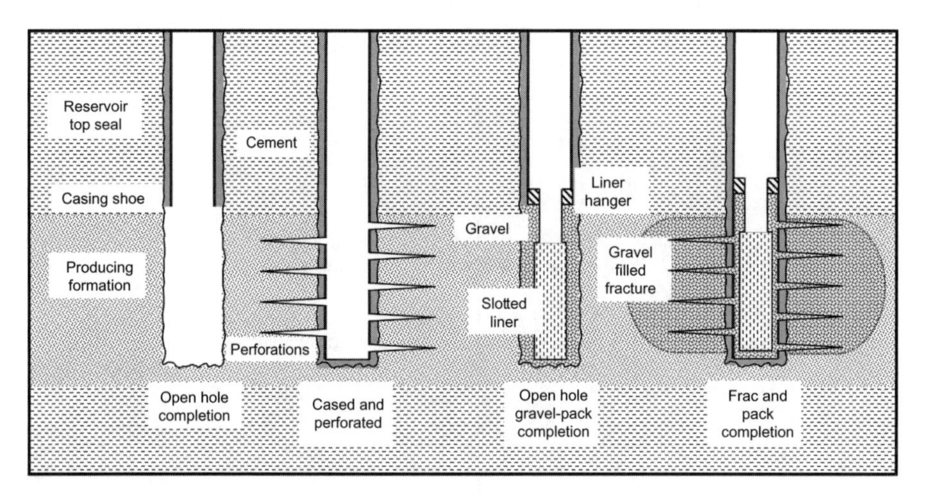

Figure 11.9 Types of sand-face completion

For the most competent rock, the bottom part of the well may be left uncased, resulting in an open-hole completion (Figure 11.9). The most common completion is a "cased and perforated" completion, in which many holes or perforations, typically 1–2 cm in diameter and 40–60 cm in length, are shot through the final casing and surrounding cement using explosive shaped charges mounted in a perforating gun. Formations that are expected to experience mechanical failure are completed using wire-wrapped screens or gravel packs in order to prevent ingress of formation material following sand failure.

The upper completion comprises a number of components, as illustrated in Figure 11.10. A tubing string contains the production or injection flow and, together with the production packer, protects the inner casing from corrosion. Mechanically operated sliding sleeves or electrohydraulic control valves may be installed together with additional packers to allow operational control over which formations are open for production or injection.

Close to the surface, a safety valve is installed in the tubing string to prevent the escape of fluids from the tubing in the event of a loss of containment at the surface, for example due to damage to surface equipment. At surface, the well is capped by a so-called xmas tree, which houses a master control valve and other valves to control production and allow access to the various tubing and annular spaces. Electrical and hydraulic feeds through the body of the tree provide for downhole control and monitoring signals.

Selection of suitable materials is an important aspect of well and completion design, with corrosion resistance being a key design factor. This in turn depends on the nature of the produced or injected fluids, and will be an important consideration in wells for CO_2 service, especially if there is a risk of free water presence resulting in carbonic acid corrosion.

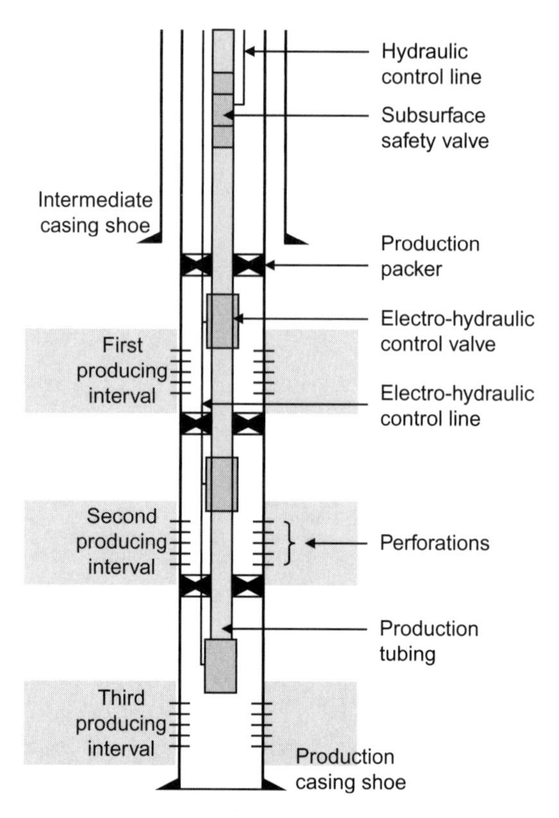

Figure 11.10 Upper completion components

11.2.8 Oil and gas reservoir exploitation

The fraction of the oil or gas initially in place that can be economically recovered from a given reservoir depends on a wide range of factors, including the fluid and rock properties, the areal and vertical variability of rock properties, and the presence of an active aquifer connected to the formation, which may provide pressure support when hydrocarbons are withdrawn.

Primary oil recovery results from production under pressure depletion, in which the reservoir pressure declines as fluids are withdrawn, or with pressure support from an aquifer. In this phase, the recovery of oil can range from 10–20% of the original oil in place (OOIP) in the case of pure pressure depletion up to 50–60% or occasionally higher for reservoirs with pressure support from a natural aquifer. In the absence of pressure support from an aquifer, water or gas injection may be used to maintain reservoir pressure. These so-called secondary recovery methods can increase primary depletion recovery to 30–50% as a result of sustained production levels and improved sweep of oil out of the rock pores as the injected water or gas pushes the oil saturation down toward the residual value (Figure 11.3). Since the influx of a natural aquifer occurs along the whole oil–water boundary while injection takes place only at a number of discrete points, the former is generally more efficient at reaching the parts of the reservoir that secondary recovery methods cannot reach, resulting in higher recovery efficiencies.

In gas reservoirs, the higher mobility of gas compared to oil and the drive energy available from expansion of the compressed gas result in primary recoveries that can reach 90–95%. Gas recovery is generally higher if the reservoir pressure is not maintained since, from simple gas law considerations ($pV = znRT$), the amount of unrecovered gas remaining in a given pore volume declines as the abandonment pressure is reduced. Rather than injecting fluids into the reservoir, gas recovery is usually maximized by installing compressors at surface, so that the gas can be produced down to the lowest possible abandonment pressure.

However, if the reservoir fluid in a gas reservoir is susceptible to retrograde condensation (Section 11.2.3), maintenance of reservoir pressure above the dew point will increase the recovery of the high-value heavier hydrocarbon components. This is typically achieved by an initial stage of gas recycling, in which lean gas is reinjected into the reservoir after condensate has been stripped out at the surface. Once condensate production drops as a result of the breakthrough of the recycled lean gas, reinjection will cease and the reservoir will be produced under pressure depletion.

11.3 Enhanced oil recovery

When primary and secondary recovery schemes have run their course, oil recovery can be further enhanced by the application of tertiary or enhanced recovery (EOR) methods, of which the most common are steam injection, which is applied to heavy viscous crude oils, or CO_2 injection.

EOR by CO_2 flooding is applicable to reservoirs deeper than ~800 m, at which depth hydrostatic pressure reaches the CO_2 critical pressure of 7.38 MPa, and for crude oils with a density of less than ~0.9 at 15°C. Under these conditions, dissolution of supercritical CO_2 increases the mobility of the residual oil in the formation by increasing its volume, and saturation, and reducing its viscosity. From a CCS perspective, depths >800 m also enable high storage efficiency by storing CO_2 as a dense supercritical fluid. For pressures >10–15 MPa, depending on oil composition and temperature, CO_2 and crude oil will become fully miscible, resulting in improved mobilization and sweep of remaining oil toward production wells, as illustrated in Figure 11.11. The produced CO_2 plus crude oil mixture is separated in surface facilities, yielding oil, hydrocarbon gas, and, after breakthrough, CO_2 product streams.

EOR by miscible CO_2 flooding has been extensively applied in the Permian Basin in west Texas and southeast New Mexico, primarily using CO_2 produced from natural CO_2 reservoirs. The incremental oil recovery covers a wide range from 5% to 15% of OOIP, depending on reservoir characteristics and the recovery efficiency of the preceding secondary recovery phase.

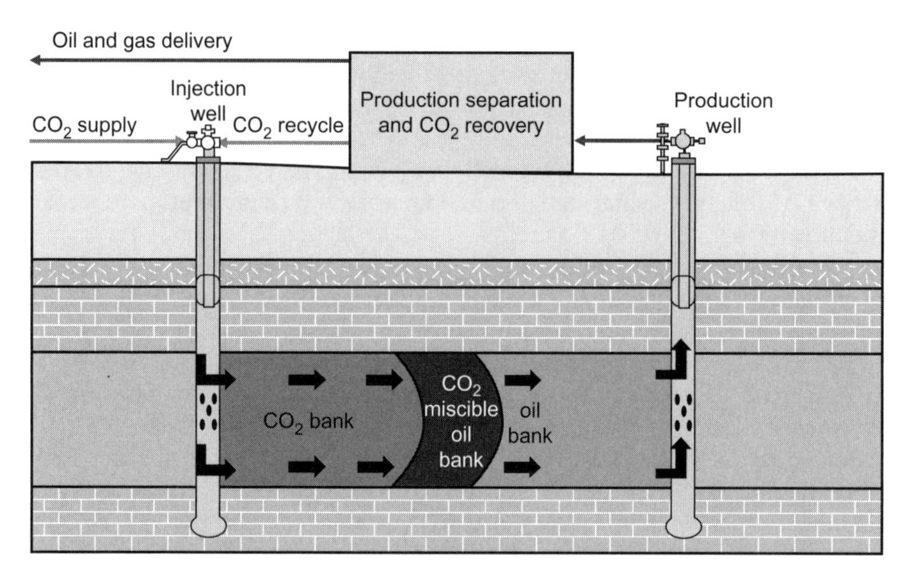

Figure 11.11 Configuration of EOR by miscible CO_2 flooding

If CO_2 for an EOR flood is purchased, project economics are generally improved by alternating injection of CO_2 and water in a so-called water-alternate-gas (WAG) scheme, with the alternating period being anything from days to months. The volume of gas purchased is also minimized by separating CO_2 from the produced fluids and recycling this to the injection wells. If an EOR flood will eventually be used for CO_2 storage (after the end of economic oil production), recycling of early breakthrough gas would also be required and storage capacity would be maximized by using continuous CO_2 injection rather than WAG, depending on the available volume of CO_2 for storage.

11.3.1 EOR case study: EnCana Weyburn field

The largest EOR project to date using anthropogenic CO_2 is the EnCana Weyburn project, which began injecting CO_2 from a synfuels gasification plant in September 2000, together with the Apache Canada-operated Midale field, which began injection in 2005. The combined project involves the transport of >2.2 Mt-CO_2 per year (6000 t-CO_2 per day) of 96% pure, supercritical CO_2 from a coal gasification plant in North Dakota through a 325-km pipeline to the Weyburn and Midale fields in Saskatchewan, Canada. The CO_2 is injected into the Midale fractured carbonate formation at a depth of 1400–1500 m.

Prior to the start of EOR operations, 25% of the Weyburn field's 220 million cubic meters of OOIP had been produced by primary recovery and secondary waterflooding, and it is estimated that the EOR project will result in an additional recovery of 9% of OOIP, as well as storing 22 Mt-CO_2.

The Weyburn and Midale units were developed using a so-called inverted nine-spot pattern, illustrated in Figure 11.12, in which each injection well is

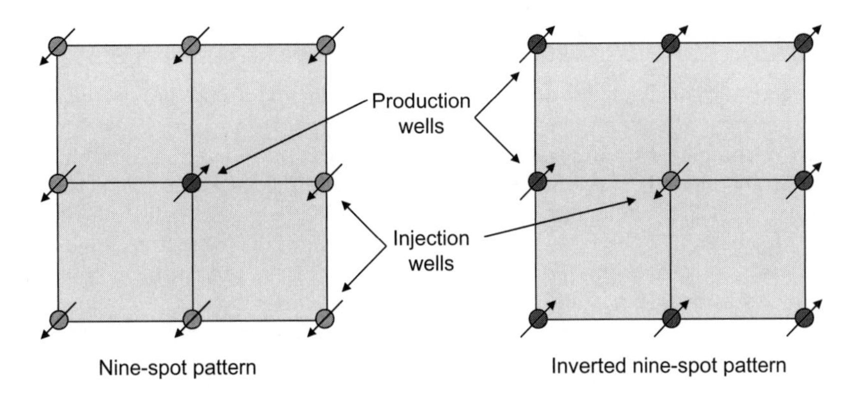

Figure 11.12 Inverted nine-spot and modified inverted nine-spot well patterns

surrounded by eight producers at a grid spacing of ~150 m between producing wells. In an extended grid using this configuration, the ratio of producing to injecting wells is 3 to 1. The Weyburn patterns had been modified by the drilling of horizontal wells within the pattern (known as "infill" wells).

The first phase of the EOR project involved CO_2 injection into 18 such patterns, extending in subsequent phases to 75 patterns, covering a total area of close to 700 ha.

Through the involvement of the International Energy Authority (IEA), the Weyburn project has become an important demonstrator for geological storage in general, not limited to EOR applications. A comprehensive research and study program was launched in 1999, covering four key research themes:

- Geological characterization of the site
- Prediction, monitoring, and verification of CO_2 movement after injection
- Prediction of the technical and economic limits to storage capacity
- A long-term risk assessment for the site.

Geological characterization

Geological characterization of the site includes both the target injection formation and the overburden layers and provides the basis for the construction of an integrated 3D geological model that is used to predict and monitor the performance of the project, both during the injection phase and in the long term.

The geological characteristics that must be assessed for a potential site to be considered suitable for storage are summarized in Table 11.3, together with the assessment techniques that can be applied.

Since a significant proportion of injected CO_2 will eventually be trapped by dissolution in formation water, flow in a dynamic aquifer is a mechanism that could transport CO_2 outside the desired boundaries. For this reason the hydrological regime, which is rarely a major issue in oil field operations, is an important consideration for CO_2 storage.

Table 11.3 Aspects of geological characterization assessed for the Weyburn project

Required geological characteristic	Description and assessment techniques
Presence of competent sealing boundaries Effective trapping in the target formation Absence of vertical conduits through open faults or fractures Isolation from surface-connected aquifers A suitable hydrodynamic regime	Regional structural and stratigraphic mapping of the target formation and overburden up to surface Assessment of the regional tectonic activity and the presence and recent activity of any related faulting Stratigraphic interpretation of cores and well logs Rock porosity, permeability, and mineralogy from core samples Integration of all data and interpretations into a 3D reservoir geological model for use in risk assessment and performance prediction and monitoring Identification and evaluation of manmade conduits such as abandoned boreholes Hydrological assessment and distribution of shallow aquifers

Prediction, monitoring, and verification of CO_2 movement

To provide assurance of long-term storage it is necessary to be able to predict the lateral and vertical movement of CO_2 after injection, and then to monitor the actual movement in order to verify these predictions. This verification from monitoring data provides a degree of assurance for the predictions of storage capacity and longer-term CO_2 movement.

In the case of an EOR storage project, this task will be aided by the subsurface geological and dynamic performance data gathered during the primary and secondary recovery phases, and by integrated numerical simulation models that will have been constructed as tools to manage these phases. During CO_2 flood EOR operations, this database will be supplemented by a range of additional surface and subsurface monitoring data. The types of data that may be acquired and their application are summarized in Table 11.4.

Prediction of the technical and economic limits to storage capacity

Prediction of the technical limit to the storage capacity of an EOR site is an extension of the numerical reservoir simulation work performed during the monitoring and verification stage. The aim of this work is to predict the long-term distribution of CO_2 within the reservoir, long after the end of the injection period, and thereby to establish the maximum CO_2 volume that can

Table 11.4 Description and application of CO_2 flood monitoring data

Monitoring data	Description and application
Pressure and volume measurements during injection and production	Basic data to perform material balance calculations and assess the dynamic response of the reservoir to injection and production
Geochemical composition of produced fluids, including carbon isotopes	Tracing the movement of injected CO_2 and identification of reservoir geochemical processes taking place on the time scale of injection, such as carbonate mineral dissolution
Time-lapse (4D) seismic monitoring surveys	Starting with a baseline survey before the start of injection, repeated monitor surveys allow tracking of CO_2 saturation changes that occur at the flood front.
Soil gas sampling or laser-based CO_2 detection systems	Comparison with off-site control measurements to detect CO_2 that may have leaked from the reservoir

be accommodated while maintaining this long-term distribution within specified boundaries.

Valid long-term predictions can be derived only from a reservoir model that is grounded in a detailed understanding of the physical and chemical processes occurring in the reservoir. Laboratory measurements of rock and fluid properties and interactions, including potential changes in these parameters through time, provide the basis for understanding the mechanisms that will trap CO_2, and simulation of these processes allows the long-term distribution to be predicted.

Figure 11.13 illustrates the distribution of CO_2 during the injection phase of an EOR project, for both miscible and immiscible conditions. Mixing or dissolution of CO_2 in the oil phase will dominate in the flood front, while CO_2 will be present as a separate mobile or residual phase behind the front, depending on whether a WAG scheme is employed.

When injection ceases, mixing will continue under gravitational and diffusive forces. Free supercritical CO_2 will continue to dissolve in both oil and formation water, and density and concentration differences will drive the redistribution of fluids. Geochemical processes involving the removal of CO_2 by mineral precipitation (mineral trapping) can also occur over a long time scale (Figure 11.14). Over a millennial time scale, and provided that primary containment in the subsurface is maintained, all free CO_2 will eventually be trapped by one of these mechanisms—dissolution in oil or water or mineral trapping. In the case of the Weyburn project, the predicted outcome of these

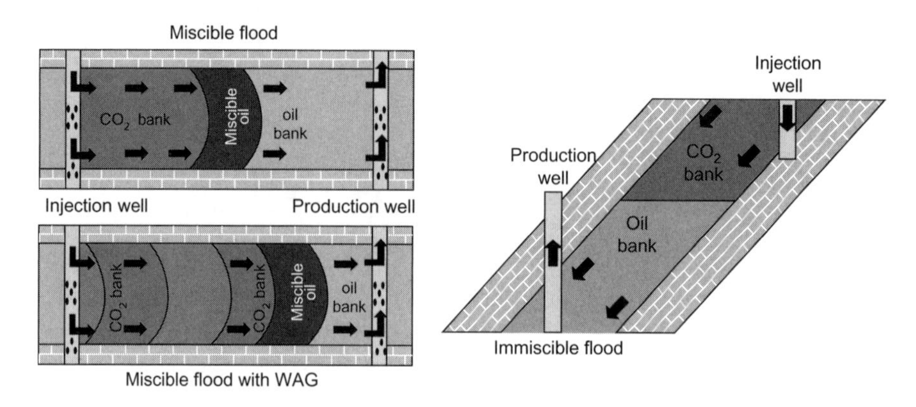

Figure 11.13 CO_2 distribution during EOR injection phase

trapping processes after 5000 years is that ~45% of CO_2 will be dissolved in the oil phase, with the remainder distributed equally between trapping in water and mineral trapping.

In parallel with the technical modeling work, an economic model is constructed to integrate the technical, economic, and fiscal aspects of the project, including the effects of oil price and CO_2 costs or credits. This model is used to determine the point at which CO_2 injection is no longer economical, for example as a result of declining oil production, which in turn will determine the economic storage limit of the project. In the Weyburn example, a technical storage limit of >45 Mt-CO_2 was estimated, while the economic limit of the EOR project is expected to be reached after injection of roughly half this quantity. CO_2 credits would be required to make the remaining storage capacity economically viable.

Modeling predictions are subject to many uncertainties, from measurement uncertainties in the basic data to subsurface heterogeneities on a range of length scales, which are captured using various approximations in the reservoir model. The standard approach to dealing with these uncertainties is to perform sensitivity studies using a number of alternative models to capture the possible ranges of uncertain parameters. The resulting range in predicted results provides one of the key inputs to the final theme—risk assessment.

Long-term risk assessment

The objectives of the long-term risk assessment are three-fold;

- To identify the mechanisms that could lead to the leakage of CO_2 from the target storage reservoir into the biosphere- comprising shallow potable aquifers, surface soil, and the atmosphere
- To assess the likelihood and consequences of leakage as a result of these mechanisms

Table 11.5 Typical risk-assessment process input

Characteristics	Events	Processes
Primary caprock properties and extent	Start of CO_2 injection	Pressure-driven flow of injected CO_2
Presence and abandonment status of old boreholes	Earthquake occurrence	Pressure-driven flow of any dynamic aquifers, including deep saline and fresh groundwater
Presence, potential conductivity, and recent activity of nearby faults	Fault reactivation	Density-driven convective flow of fluids
Presence and extent of saline aquifers above the primary caprock	Construction of new boreholes	Dissolution or precipitation of minerals
Hydrology of groundwater system	Loss of integrity of production and injection wells or other boreholes	Diffusion of CO_2
	Maintenance (workover) activity on existing wells	Microbial activity

- To identify any mitigating actions that may be needed to reduce these risks to an acceptable level.

Formal risk-analysis tools are applied, which build on an inventory of the relevant characteristics, events, and processes to define a range of possible scenarios for the future performance of the system. A nonexhaustive example of a risk-assessment input listing is shown in Table 11.5.

Quantitative risk assessment (QRA), applying probabilistic methods, can be used to estimate the probability of containment within or movement out of the target reservoir, including the volume and probability range for CO_2 leakage into the biosphere. In the Weyburn risk assessment, it was estimated with 95% confidence that 98.7–99.5% of the CO_2 in place at the end of the EOR injection phase would remain stored in the geosphere and that the mean leakage through abandoned wells would be $<10^{-5}$ of the stored volume.

The approaches outlined above, which were developed as part of the IEA Weyburn project, provide a framework for assessing the long-term storage aspects of CO_2 flood EOR projects, and are also for the most part applicable to other storage options such as saline aquifer storage described later.

11.3.2 Planned EOR sequestration projects

With incremental oil revenues providing a helpful economic incentive, a number of EOR-based CCS demonstration projects are currently in the planning stage with start-up dates in the 2010 to 2014 period, as shown in Table 11.6.

These projects will demonstrate a range of capture technologies, as well as providing opportunities to further develop and test subsurface modeling and monitoring techniques.

Table 11.6 Planned EOR projects storing anthropogenic CO_2

Project; operator and partners	Project description	Planned start-up
Williston Basin; Plains CO_2 Reduction Partnership (PCOR), Encore, Basin Electric Power	Retrofit postcombustion capture of 0.5–1.0 Mt-CO_2 per year from Antelope Valley coal-fired power plant, with injection for EOR in deep, high-pressure (33 MPa) oil reservoirs	end 2010
Northeastern; American Electric Power (AEP), Alstom	Retrofit postcombustion capture of 1.5 Mt-CO_2 per year from 200 MW coal-fired power unit at AEP's Northeastern Station in Oklahoma, with injection for EOR	2012
Masdar; BP Alternative Energy, Rio Tinto	Hydrogen production from natural gas (steam reforming) to fire a 420 MW power plant, with up to 1.7 Mt-CO_2 per year captured for EOR	2012
California (DF2); Hydrogen Energy	Gasification plant, fueled with refinery residue (petroleum coke), producing hydrogen for a 500 MW power plant. 4–5 Mt-CO_2 per year captured for EOR	2014

11.4 Saline aquifer storage

11.4.1 Introduction

The vast majority of deeply buried sedimentary rock formations do not contain hydrocarbons, but instead are saturated with high salinity formation brine, which is unsuitable for consumption, irrigation, or other uses. In the absence of nearby economic EOR opportunities, saline aquifers provide a geological storage option that is widespread and will be easily accessible to many capture sites.

Saline aquifers have a number of other advantages and disadvantages when compared to hydrocarbon-bearing reservoirs as potential CO_2 storage sites, as summarized in Table 11.7.

11.4.2 CO_2 trapping mechanisms

The storage potential of a formation depends on the effectiveness of a number of trapping mechanisms that control the fate of injected CO_2. These processes operate on different time scales, as illustrated in Figure 11.14, and a full assessment of the long-term distribution of injected CO_2 requires the integration of all of these processes into a reservoir simulation model.

Table 11.7 Advantages and disadvantages of saline aquifers versus hydrocarbon reservoirs as geological storage sites

Saline aquifer	**Oil or gas reservoir**
Disadvantages	Advantages
Increased site selection and proving requirements due to a relative lack of data for geological characterization	A wealth of geological data is usually available from the hydrocarbon exploitation stage.
Lack of established methods to establish site suitability, long-term integrity, and storage capacity	Effective structural or stratigraphic trapping is already demonstrated by the presence of hydrocarbons.
Lack of economic boost from enhanced oil or gas recovery	More mature legislative and regulatory framework and greater public acceptance of CO_2 injection for EOR projects
Storage capacity limited by water compressibility and aquifer volume	Storage capacity created by hydrocarbon withdrawal
Advantages	Disadvantages
More widespread and therefore more accessible to capture sites, reducing or eliminating transportation costs.	Existing boreholes provide potential leak paths if their long-term integrity is not ensured.
Typically fewer well penetrations, reducing the risk of leak paths.	

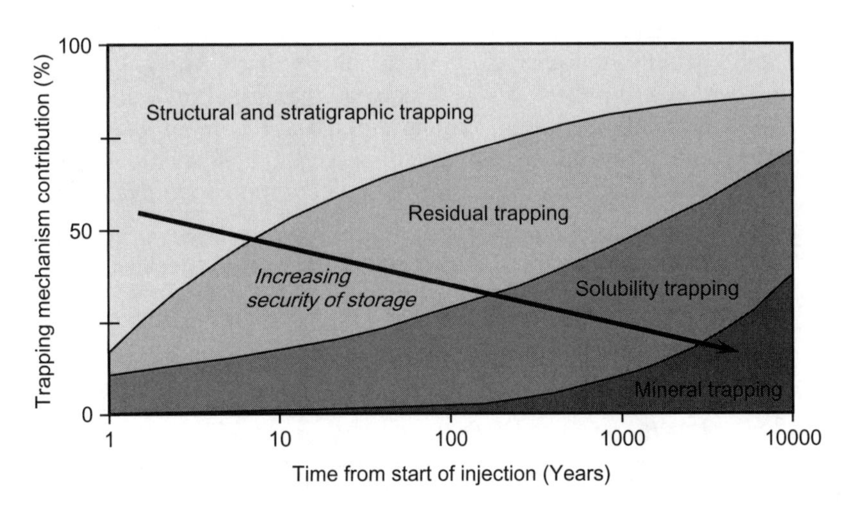

Figure 11.14 Relative importance of trapping mechanisms through time

Structural and stratigraphic trapping

The buoyancy of injected supercritical CO_2 will result in physical trapping of CO_2 that migrates into a structural trap beneath a competent caprock. This typically occurs on a 1- to 100-year time scale, after which physically trapped CO_2 will be gradually dispersed as a result of dissolution or geochemical reactions with the caprock (see mineral trapping).

In the absence of a large-scale structural trap, such as those typical in oil and gas fields, physical trapping can still occur under smaller trapping elements, such as irregularities in the caprock surface or small-scale impermeable layers within the aquifer. Detailed geological assessment during the planning phase will provide the basis for estimating the physical trapping capacity of the target formation.

Residual trapping

As a plume of injected CO_2 rises through a water-saturated rock, the buoyant CO_2 will initially expel water from the pore space. Behind the rising plume, water will again be drawn into the pores by capillary imbibition. As described in Section 11.2.4, as the saturation of a mobile phase reduces, the permeability of the rock to that phase declines, eventually approaching zero at the residual saturation. In this case the whole rock volume swept out by the rising plume will initially retain a residual saturation of CO_2.

Over time, this residual saturation will decline as CO_2 is dissolved in the formation brine and diffuses into the surrounding unsaturated aquifer, a process that will be considerably accelerated if the aquifer is hydrodynamically active.

Solubility trapping

Dissolution of CO_2 in formation water is likely to be the major trapping mechanism in saline aquifer storage and is active on a 1- to 1000-year time scale. Solubility will depend on the geochemical composition of the formation brine, which should be fully characterized in the planning phase. Analysis will include ionic composition and pH, as well as dissolved organic and inorganic carbon.

Depending on the brine composition and pH, and the mineralogy of the target reservoir, solubility trapping may be enhanced by the formation of bicarbonate ions and other ionic complexes, resulting in additional ionic trapping.

Mineral trapping

The reaction of dissolved CO_2 with Ca-, Fe-, or Mg-containing minerals in the rock matrix can result in the precipitation of carbonates in the pore space. This is the slowest potential trapping process, operating over a decadal to millennial time scale, but is the one process that leaves CO_2 in a completely immobile state in the reservoir.

Mineralogical reactions may also occur when CO_2 contacts the caprock above the injection formation. Geochemical analysis for the Sleipner project (Section 11.4.5) indicated that silicates present in the caprock would react with formation brine and CO_2 to form carbonates. The resulting increase in mineral bulk volume reduces porosity and permeability and leads to enhanced caprock integrity over time.

Similar reactions will occur on a much shorter time scale (weeks to years) for CO_2 in contact with basalt, as a result of the high alkaline mineral content in this type of rock. CO_2 injection into or below laterally extensive flood basalts has been proposed as a geological storage option that would exploit this in situ mineral carbonation trapping mechanism.

11.4.3 Site selection

Site selection for deep saline aquifer storage is less constrained than the identification of an EOR-related storage option and is best approached as a two-stage process. A preliminary feasibility stage using site visits and existing geological (structural and stratigraphic) data can provide a regional assessment of storage potential and identify a ranked list of possible sites for more detailed evaluation. A set of qualifying criteria such as those shown in Figure 11.15 can be used in this phase to screen possible sites.

The second evaluation and selection phase will in most cases require the acquisition of additional data and test results so that a preferred site can be selected from the short list of qualifying sites as the basis for the project design. Acquisition of a 3D seismic survey may be required in order to assess the subsurface, including the presence of possible structural traps and the location and intensity of faulting. If subsurface data are not already available, exploratory drilling and the acquisition of well logs and cores will also be required before final site confirmation, in order to characterize the target formations as well as the primary and any secondary seals. A mini-frac test on the primary seal and injectivity tests into the target formation may also be needed to validate laboratory measurements and confirm achievable injection rates, as this is likely to be required to determine the number of wells needed to achieve the planned CO_2

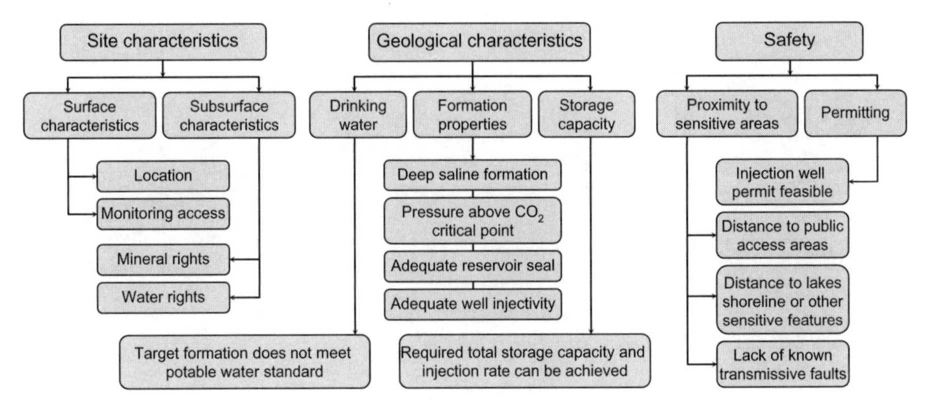

Figure 11.15 Geological storage site qualification criteria (after FutureGen)

injection rate. Aquifer water sampling and analysis will be required to assess the impact of CO_2 dissolution on water geochemistry.

Whether this more detailed and costly evaluation is performed sequentially or concurrently on several short-listed sites will depend on a cost-versus-schedule trade-off specific to the project. Figure 11.16 shows a set of selection criteria that can be used in this phase for site ranking and selection.

11.4.4 Monitoring saline aquifer storage

The ability to track the movement of injected CO_2 is an important component of the risk-management strategy for a storage project, and aims firstly to establish whether injected CO_2 is fully retained within the target formation and secondly to identify any side effects of injection, such as geomechanical disturbance in the formation or at the surface. Early identification, particularly of leakage to shallower formations, provides the opportunity to take remedial action in order to avoid or mitigate undesirable consequences.

A major focus of monitoring activities in demonstration projects to date has also been the development and validation of modeling tools to aid in the prediction of plume behavior. In the longer term, as tools and techniques become well established and as emissions trading frameworks mature, this is likely to give way to a verification role, since measurement, monitoring, and verification (MMV) will be an essential part of these schemes. As a result, monitoring activities for commercial-scale storage projects are likely to be less comprehensive than has been typical for demonstration projects, and program design will become more risk-based, with appropriate techniques selected to address the main risks specific to each particular storage site.

Similarly, as the main risks of a project change through its life cycle, the monitoring objectives, activities, and frequency will be adjusted to match the changing risk profile. For example, during the injection phase, risks associated

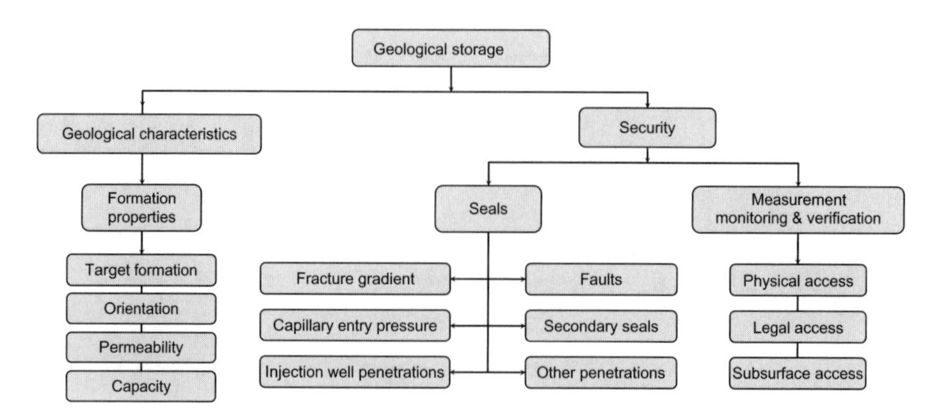

Figure 11.16 Geological storage site selection criteria (after FutureGen)

with increasing reservoir pressure may dictate monitoring for geomechanical deformation and seismic surveys to assess possible leak paths along reactivated faults or fractures (either natural or induced), while, after injection has ceased and reservoir pressure is declining, the focus of monitoring may shift toward plume redistribution and the implications for the relative importance of different trapping mechanisms, the latter on a far longer time scale.

The wide range of physical and chemical monitoring techniques that are available are summarized in Table 11.8, and the most important of these are described in the following sections.

Time-lapse seismic monitoring

The use of time-lapse (4D) seismic is the most important monitoring technique developed to date, and its effectiveness has been well established in a number of cases, notably in the Sleipner project described in the next section. This technique relies on the acoustic impedance contrast between rock containing water, supercritical CO_2, or hydrocarbons, and the comparison of successive surveys indicates areas in the subsurface where one fluid has replaced another or where significant pressure changes have occurred. An initial (baseline) survey acquired before the start of injection is desirable to provide a baseline for the interpretation of the first monitor survey.

As well as imaging plume movement within the target formation, time-lapse seismic will also be able to detect gas migration paths in the overburden if CO_2 migrates above the depth limit for supercritical conditions.

Gravity and electromagnetic surveys

Other surface geophysical monitoring techniques that have been applied include gravity and electromagnetic surveys. These measurements are unable to match the resolution of 4D seismic but may have a role to play in longer-term monitoring programs, long after the end of the injection phase of a storage site, as a trigger for more extensive and expensive monitoring if anomalous behavior is identified.

Surface sampling and seabed monitoring

Although monitoring aims to identify anomalies well before any leakage to the biosphere occurs, a variety of techniques are also available to assess surface leakage. On land, groundwater sampling and atmospheric monitoring can be used, while isotopic analysis can be used to discriminate injected from background CO_2.

Offshore, sonar measurements of the water column have been shown to successfully detect gas seepages from natural gas reservoirs. Subbottom profiling is a sonar technique that images shallow sediments up to tens of meters below the sea bed and can indicate gas migration paths in the shallow subsurface.

Figure 11.17 shows an example of a subbottom profiling survey indicating the presence of hydrocarbon gas in shallow sediments.

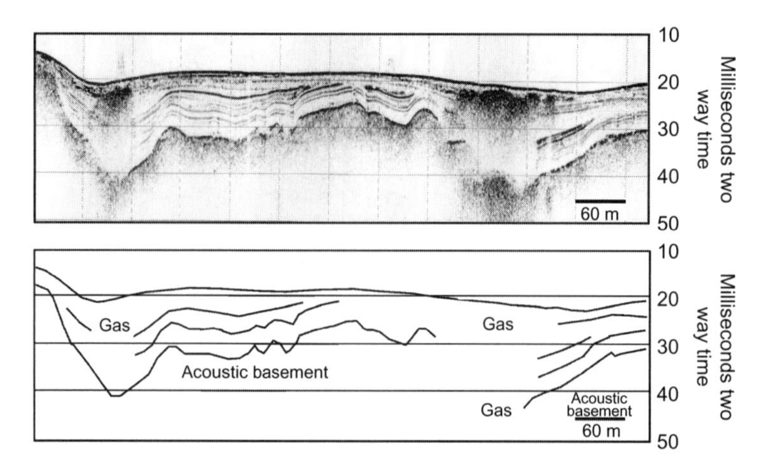

Figure 11.17 Subbottom profiling survey indicating hydrocarbon gas in shallow sediments (Courtesy; Jonathan Bull and Justin Dix, National Oceanography Centre, University of Southampton.)

Table 11.8 Monitoring techniques for geological storage

Monitoring technique	Description
Injection well monitoring	
Injection well pressure and temperature	Standard measurement of well pressures, volumes, and temperatures; used as input to dynamic reservoir modeling and to identify injection well problems
Plume location and movement	
Time-lapse seismic	Typically acquired on a 3D grid, but also possible on 2D lines if sufficient. May also be acquired using permanently installed sources and geophones, particularly offshore.
Vertical seismic profiling	Seismic survey from a surface source to geophones located in a well, generating a one-way vertical profile that can be used to calibrate 3D data and provide near-well imaging of reservoir properties and CO_2 movement
Cross-well seismic	Seismic survey with source and geophones located in two boreholes, generating a profile between the wells that can be used to calibrate 3D data
Microseismic	Measurement of microseismic noise generated by fracturing events as a result of injection, using either surface or downhole-mounted geophones
Electrical resistance tomography	Electrical resistance tomography (ERT) images the subsurface resistivity with measurements between ground-based or borehole-mounted electrodes; long electrode ERT uses the casing of a well as an electrode to extend this technique to greater depth.

(Continued)

Table 11.8 Continued

Monitoring technique	Description
Cross well resistivity	A variant of ERT using electrodes mounted or moving in two wells, and providing a 2D profile of the resistivity distribution between the wells
Gravity surveys	Measurement of the local gravitational field using surface or airborne gravimetry, providing a low-resolution 2D areal image of changes in pore fluid density due to plume movement
Ground movement detection	
INSAR	Interferometric synthetic aperture radar (INSAR) is a technique for processing radar images acquired by satellites to detect surface deformation down to millimeter scale over time scales of months to years
Tiltmeters	Shallow borehole-mounted ($<10\,m$) sensors that can detect microradian-scale geomechanical deformation, equivalent to 1 mm of deflection over a 1 km range
CO_2 leak detection	
LIDAR	Light detection and ranging (LIDAR) is a laser-based monitoring technique that can detect the location and movement of CO_2 in the atmosphere with ppb sensitivity
Hyperspectral remote sensing	Also known as imaging spectroscopy, hyperspectral remote sensing uses space-borne sensors that collect a spectrum at each image point, rather than an amplitude at a single sensed wavelength, to detect atmospheric gases and surface minerals
Eddy covariance measurement	A technique for determining the vertical flux of a gas species such as CO_2 by statistical analysis of the vertical velocity and fluctuating concentration of turbulent eddies
Soil gas sampling	Collection and analysis of gas samples from the soil or shallow boreholes. Isotopic analysis can distinguish leakage of injected from biosphere-generated CO_2
Fluid sampling	Collection and analysis of fluid samples from drinking water aquifer wells, and any observation wells located above the primary seal, to detect elevated CO_2 concentration

11.4.5 Saline aquifer storage case study: Sleipner

In the Norwegian sector of the North Sea, StatoilHydro and partners have been producing natural gas from the Sleipner East and West fields since August 1996. The Sleipner gas contains 4.0–9.5% CO_2, which had to be reduced to <2.5% to meet the European Union pipeline gas delivery specification. In the absence of an economic EOR opportunity, the overlying Utsira aquifer was identified as a CO_2 disposal reservoir. Figure 11.18 shows a schematic of the project.

Similar to the Weyburn EOR example, an international multidisciplinary research program was established to collect and analyze data from the project and to build on this to provide guidance for similar projects in the future. The Saline Aquifer CO_2 Storage (SACS) and subsequent SACS2 programs ran from 1998 to 2003 and culminated in the publication of a Best Practice Manual (Section 11.6.1) based on the knowledge gained and lessons learned in this project.

Site characterization and selection

The Utsira formation is an unconsolidated water-bearing sandstone with high porosity (30–42%) and permeability (1–3 $10^{-12}\,m^2$ [1–3 Darcy]). Initial characterization of the aquifer, which is >400 km long and 50–100 km wide, was based on an extensive regional 2D seismic dataset, as well as data from some 300 well penetrations, of which 30 were within 20 km of the selected injection location.

At the Sleipner Field the Utsira formation is at a depth of 800 m, deep enough for CO_2 to remain in a supercritical state, and forms an almost flat structure with a gentle dip to the south and southwest. This type of structure poses specific problems for predicting plume movement, since minor changes in the topology of the caprock can have a significant impact on the migration direction of the injected CO_2. In this situation accurate depth mapping of the top structure is essential, based on dense 2D or 3D surveys tied to adequate well data. This is less of an issue in formations with high-relief structural or stratigraphic traps, where plume movement will be more constrained by these features.

To minimize the risk from potential leak paths, the injection site was chosen to ensure that predicted migration of the plume would take it away from the gas production wells, and a small structural trap in the aquifer was selected as the

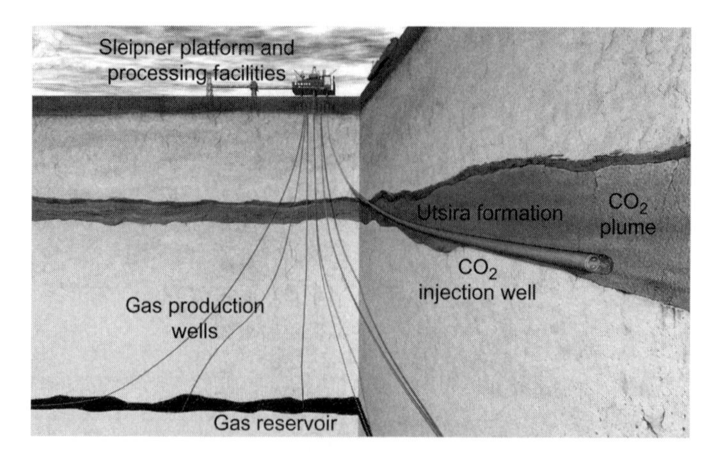

Figure 11.18 Schematic overview of the Sleipner CO_2 storage project (Courtesy; StatoilHydro)

injection target. Preliminary reservoir modeling work indicated that a plume with a maximum lateral extent of 3 km would be formed after 20 years of injection.

Injection operation and monitoring

Capture of ~1.0 Mt-CO_2 per year using amine stripping and injection into the single disposal well commenced in September 1996 and has continued without operational problems since then. A 3D survey acquired in 1994 was used as the baseline for monitoring, and repeated time-lapse (4D) surveys acquired in 1999, 2001, 2002, 2004, 2006, and 2008 provided a detailed picture of the evolution of the plume over the first decade of injection.

The survey results, Figure 11.19, show that the vertical movement of CO_2 was impeded by small-scale impermeable shale layers that were below the resolution of the seismic. These layers caused lateral spreading of the buoyant fluid, which then spilled around the edge and rose to the next thin layer (Figure 11.20). The 4D seismic results could be well matched in the reservoir simulation once these fine-scale details were incorporated into the geological model.

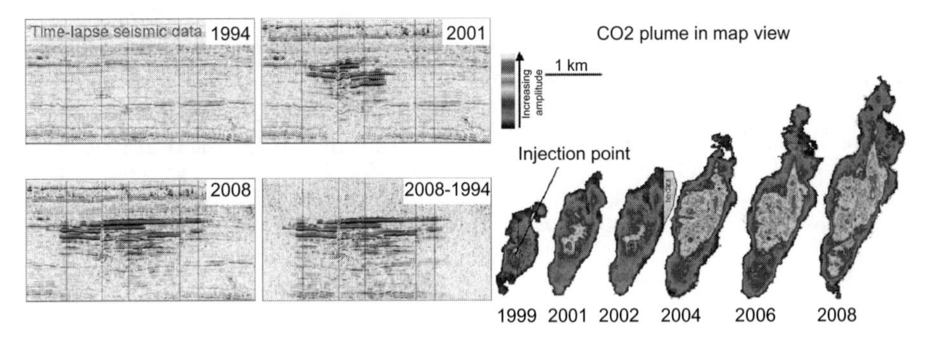

Figure 11.19 Sleipner time-lapse seismic results (Courtesy; StatoilHydro)

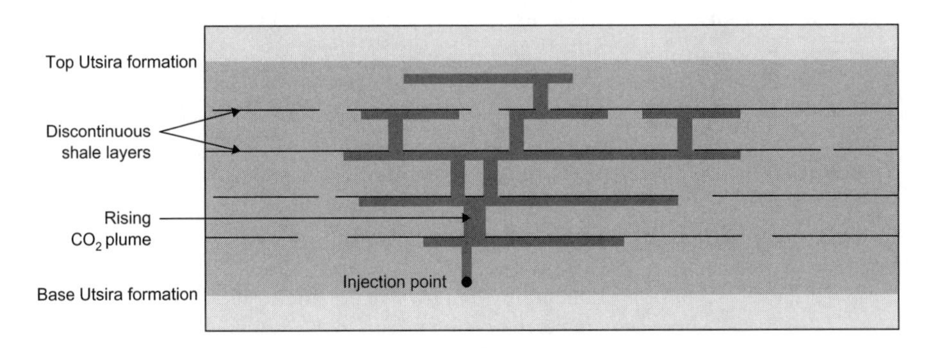

Figure 11.20 Schematic plume evolution controlled by thin shale layers

In the longer term (>50 years) all mobile CO_2 is expected to migrate to the top of the formation, and the caprock topology will be the geological factor controlling further evolution of the plume.

These results demonstrated the central role of time-lapse seismic in monitoring geological storage operations, as well as the potential for plume dynamics to be affected by small-scale geological heterogeneities that may not be fully represented in the initial geological dataset.

11.4.6 Planned saline aquifer storage projects

A number of CCS demonstration projects with saline aquifer storage are currently in the planning stage with start-up dates in the 2011 to 2017 period, as shown in Table 11.9.

Smaller-scale demonstration projects include a 3 kt CO_2 injection pilot into the Columbia River basalt formation in Washington State as part of a Pacific Northwest National Laboratory (PNNL) program investigating in situ mineral carbonation in basalt formations.

11.4.7 RD&D for saline aquifer storage

Although the technologies developed in the oil and gas industry have enabled saline aquifer storage projects to proceed rapidly to the demonstration stage, many of the geomechanical and geochemical processes and monitoring requirements for these projects represent a step-out from oil and gas field practice and

Table 11.9 Planned saline aquifer CO_2 storage projects

Project; operator and partners	Project description	Planned start-up
Fort Nelson; Plains CO_2 Reduction Partnership (PCOR), Spectra Energy	Storage of 1.2–1.5 Mt-CO_2 per year captured from Spectra Energy's Fort Nelson natural gas processing plant with storage in a saline carbonate aquifer	2011
Kimberlina; WESTCARB, Clean Energy Systems, Schlumberger	Capture of 0.25 Mt-CO_2 per year from a newly built 170 MW_{th} oxyfuel power plant (ZEPP-1) with injection into the Vedder aquifer	2012
ZeroGen; ZeroGen Pty. Ltd., Shell, Australian Coal Association	Stage 1: 120 MW_{th} demonstrator IGCC plant with 75% CO_2 capture and aquifer storage. Stage 2: 200 MW_{th} commercial scale-up with 90% CO_2 capture and aquifer storage	2012 2017

are the subject of ongoing RD&D work. These RD&D themes are summarized in Table 11.10.

11.5 Other geological storage options

11.5.1 Enhanced gas recovery

Gas reservoirs share many of the advantages enjoyed by oil reservoirs as potential CO_2 storage sites, particularly in relation to the availability of reservoir characterization data and the presence of a demonstrated caprock. In gas reservoirs this is more strongly demonstrated than in oil reservoirs as a result of the higher overpressure acting against the caprock that arises from the greater density difference between gas and water, as illustrated in Figure 11.21.

As discussed in Section 11.2.8, maintaining pressure by fluid injection into gas reservoirs generally does not increase recovery, although pressure maintenance by gas recycling may be applied in order to avoid a loss of condensate recovery if the reservoir fluid is subject to retrograde condensation in the reservoir.

However, as a gas reservoir approaches abandonment, CO_2 injection can yield enhanced gas recovery (EGR) by repressurizing and sweeping residual

Table 11.10 RD&D themes for saline aquifer storage

RD&D theme	Description
Subsurface imaging	Improvements in imaging technologies to quantify plume location, saturation, and volumetrics, to monitor caprock integrity and enable leak detection; improvements in imaging resolution to identify important small-scale heterogeneities
Geochemical processes	Reaction of CO_2 with reservoir rocks and seals, including self-sealing potential of the plume; mineral trapping, including in situ mineral carbonation in flood basalts; processes that dissolve and mobilize heavy metals or other potential contaminants
Geomechanical processes	Impact of injection pressures and temperatures on reservoir and caprock, faults, and fracture networks; long-term impact of CO_2 on caprocks
Operational technologies	Development of efficient tools and automation technologies for real-time integration of subsurface and surface monitoring data, and comparison against baseline surveys
Subsurface modeling	Incorporate improved process knowledge into mathematical simulation models

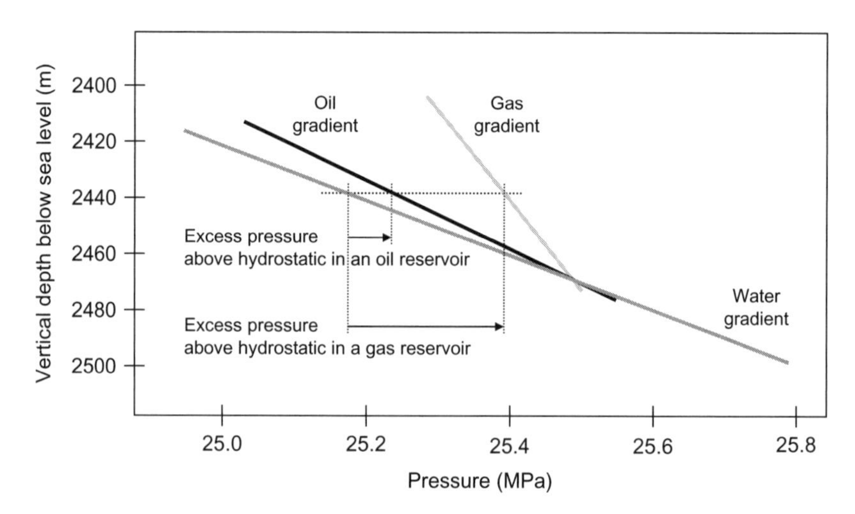

Figure 11.21 Pressure gradients in oil and gas reservoirs

gas toward production wells. Due to the density and viscosity differences, the sweep of methane by supercritical CO_2 will be stable so that, depending on the configuration of injecting and producing wells, it should be possible to fill the greater part of the reservoir volume with CO_2 before breakthrough into the producing wells occurs.

Active EGR pilot project: K12-B field, North Sea, Dutch sector

A pilot-scale EGR project was started in the K12-B gas field in the Dutch sector of the North Sea in 2004. The field, operated by Gaz de France, was discovered in 1981 and contained 14.4 Gm^3 of gas, with 13% CO_2 content, in fluvial and aeolian Rotliegend sandstones at a depth of ~3900 m. CO_2 is separated at surface by amine stripping and from 1987, when the field was brought on stream, until 2004 all produced CO_2 was vented.

During this period 11.5 Gm^3 of gas was produced and reservoir pressure dropped from 40 to 4 MPa. An EGR opportunity was identified in 2001, and detailed geological characterization and model building studies were completed as part of the feasibility and pilot planning phase. A particular focus of these studies was evaluation of the impact of the large historical reservoir pressure drop on formation injectivity and on the sealing capacity of the Zechstein evaporite caprock. Feasibility studies concluded that the field was suitable for CO_2 storage, and a small-scale pilot consisting of two injection tests was initiated in 2004, as summarized in Table 11.11.

Although there was no clear demonstration of EGR from these tests, a further phase of monitoring and investigation under the MONK project (Monitoring K12-B) was commenced in 2007.

Table 11.11 K 12-B pilot injection test scope

Pilot test	Description	Test objectives and outcomes
Test 1	Single injection well test, 11 kt-CO_2 injected from May 2004 to January 2005	CO_2 injectivity confirmed; CO_2 wellbore phase behavior and reservoir response confirmed to be within the range of model predictions
Test 2	EGR test, two producing gas wells plus one CO_2 well, injecting 23 kt-CO_2 from February 2005 to August 2006	Evaluate EGR potential and economics using perfluorocarbon tracer injection and simulation studies to monitor reservoir displacement process; no clear demonstration of enhanced recovery

The final phase of the project provides for injection ramp-up to 480 kt-CO_2 per year, limited by the capacity of the amine stripping plant, with a total of 8 Mt-CO_2 potentially to be injected over a 20-year project life.

Planned EGR demonstration project: Hewett Reservoir, North Sea, UK sector

The Hewett gas field, discovered in 1966 and located in the UK sector of the southern North Sea, is a candidate EGR storage site that is one component of the RWE npower, Peel Energy entry to the UK CCS Demonstration competition. The project is based on CO_2 capture from a 1.6 GW (2 × 800 MW) SC coal-fired power plant that is planned to be built by RWE npower at Tilbury in Essex, UK, with storage in the depleted Hewett gas field.

The field, currently operated by Tullow Oil plc, contained an estimated 98 Gm^3 of gas initial in place (GIIP), and came into production in 1973 from two high-permeability sandstone reservoirs (the Upper and Lower Bunter sands) at depths of 800–1250 m. By 2014, when the field is expected to cease production, ~95 Gm^3 of gas will have been produced. The two Bunter sands in the Hewett field are considered to be an ideal storage site in view of their depth, size, and high permeability, and are estimated to have a combined storage capacity of >0.35 Gt-CO_2, sufficient to store 30 years of captured CO_2 from the Tilbury plant.

An unusual aspect of the proposed project is that marine transport of CO_2 from Tilbury to the field location is being considered, alongside pipeline transport (Section 15.2.2).

11.5.2 Enhanced coal bed methane recovery

Coal is an effective sorbent for methane, CO_2, and other gases, and coal bed methane (CBM) is economically exploited from unmineable coal beds in a number of areas including the United States, Canada, and Australia. CBM is released from coal by reducing the pore pressure within the coal bed, a variant of pressure swing desorption (Section 7.2.2). This is achieved by drilling wells into the bed and pumping off water to reduce the hydrostatic pressure. As the pressure is reduced, liberated methane is transported to the well through the system of fractures, or cleats, in the coal bed. The key requirements for CBM recovery are:

- High initial methane content, which increases with pressure and depth, and can be up to 20–25 m^3 per tonne
- Adequate permeability for gas flow from a well-developed, open cleat system, which typically decreases with increasing depth due to cleat closure under the increasing confining stress.

These opposing requirements dictate a depth range of ~300 m to ~1000 m for economic CBM exploitation.

Injection of CO_2 into coal beds can result in enhanced CBM recovery as a result of the competitive adsorption between these two gases (Section 7.1.1). Due to its higher heat of adsorption, CO_2 will be preferentially adsorbed onto the coal surface. This is illustrated in Figure 11.22, which shows the Langmuir isotherms for methane and CO_2.

Preferential adsorption of CO_2 results in the desorption of methane, which is then able to flow through the cleat system to a producing well. Although this process has been demonstrated in a number of pilot-scale trials (see CO2CRC inventory in Section 11.6.3), it suffers from a number of technical and economic challenges, notably:

- Reduction of coal permeability due to swelling of the coal matrix on CO_2 adsorption, potentially leading to an increase in the number of injection wells required. In

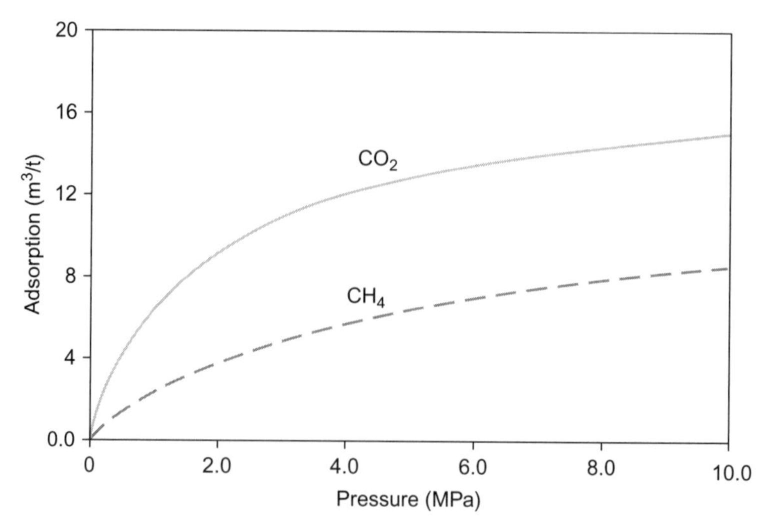

Figure 11.22 Langmuir isotherms for CO_2 and methane adsorption onto coal

some pilot tests this loss of permeability has been partially overcome by increasing injection pressure to maintain the rate of CO_2 injection.

- Permeability reduction as a result of coal softening, which occurs when the adsorption of CO_2 reduces the glass transition temperature (see Glossary) to below the formation temperature, resulting in an in situ transition from a glassy phase to a soft rubbery phase.
- Low ratio of incremental methane recovery to CO_2 injected volume. Pilot results have yielded incremental methane recoveries between 15 and 50% of the injected CO_2 volume.

Active ECBM pilot project: PCOR, North Dakota

Characterization studies conducted by the Plains CO_2 Reduction Partnership (PCOR) have shown that low-rank coal seams in North Dakota have the capacity to store up to 8 Gt-CO_2 and suggested that $>500\,Gm^3$ of methane could be produced from these seams. The studies have been followed by a field validation test, commencing in 2007 with the construction of a five-spot test pattern, with one injection well surrounded by four monitoring wells. Injection of a planned 400 t-CO_2 into a 3 m thick lignite seam at 330 m depth started in March 2009. The project is the first field trial conducted to test the ability of lignite coal seams to store CO_2.

In view of the high volumetric CO_2 requirement, further development of enhanced CBM beyond the pilot stage is likely to require the availability of zero- or negative-cost CO_2 that will follow the introduction of financial incentives for carbon storage.

11.6 References and Resources

11.6.1 Key References

The following key references, arranged in chronological order, provide a starting point for further study.

Marchetti, C. (1977). Geoengineering and the CO_2 problem, *Climate Change*, **1** (59–68).

StatoilHydro. (2003). *Best Practice Manual from SACS: Saline Aquifer CO_2 Storage Project*. Available at www.co2store.org, SACS project page.

Wilson, E. J., T. L. Johnson, and D. W. Keith. (2003). Regulating the ultimate sink: Managing the risks of geologic CO_2 storage, *Environmental Science & Technology*, **37** (3476–3483).

Benson, S. (2005). Overview of geological storage of CO_2, *Carbon Dioxide Capture for Storage in Deep Geological Formations*, **2** (665–672).

Preston, C., *et al.* (2005). IEA GHG Weyburn CO_2 monitoring and storage project, *Fuel Processing Technology*, **86** (1547–1568).

van der Meer, B. (2005). Carbon dioxide storage in natural gas reservoirs, *Oil & Gas Science and Technology—Revue de l'Institut Français du Pétrole*, **60** (527–536). Available at http://ogst.ifp.fr

Bachu, S. (2007). Carbon dioxide storage capacity in uneconomic coal beds in Alberta, Canada: Methodology, potential and site identification, *International Journal of Greenhouse Gas Control*, **1** (374–385).

IEA GHG. (2007). *Storing CO_2 underground*, IEA Greenhouse Gas R&D Programme 2007. Available at www.co2net.eu/public/reports/storingCO2%20sml.pdf.

U.S. Department of Energy National Energy Technology Laboratory. (2008). *Methodology for Development of Geological Storage Estimates for Carbon Dioxide*. Available at www.netl.doe.gov/technologies/carbon_seq/refshelf/methodology2008.pdf.

11.6.2 Institutions and Organizations

The institutions and organizations that have been most active in the research and development work on geological storage are listed below, with relevant web links.

Apache Canada (operator of Midale field EOR project): www.apachecorp.com
CO_2 DeepStore (sequestration in depleted gas fields): www.co2deepstore.com
Gaz de France (K12-B injection project): www.k12-b.nl
DNV (developing standards to qualify sequestration sites): www.dnv.com/industry/energy/segments/carbon_capture_storage
EnCana Corporation (operator of the Weyburn EOR project): www.encana.com
StatoilHydro (operator of the Sleipner CO_2 storage project): www.statoilhydro.com

11.6.3 Web Resources

CO2CRC (global inventory of operating and planned CO_2 storage projects): www.co2crc.com.au/demo/worldprojects_pro.html

IEA GHG (interactive tool for design of monitoring programs for geological storage): www.co2captureandstorage.info/co2tool_v2.1beta/index.php

Kansas Geological Survey (online tools to evaluate saline aquifers for CO_2 storage): www.kgs.ku.edu/PRS/publication/2003/ofr2003-33/toc0.html

U.S. Department of Energy 2008 Carbon Sequestration Atlas II (documents >3500 Gt-CO_2 storage potential in oil and gas reservoirs, coal seams, and saline formations in the United States and Canada): www.netl.doe.gov/technologies/carbon_seq/refshelf/atlasII

12 Ocean storage

12.1 Introduction

As described in Chapter 2, the world's oceans contain an estimated 39,000 Gt-C (143,000 Gt-CO_2), 50 times more than the atmospheric inventory, and are estimated to have taken up almost 38% (500 Gt-CO_2) of the 1300 Gt of anthropogenic CO_2 emissions over the past two centuries.

Options that have been investigated to store carbon by increasing the oceanic inventory are described in this chapter, including biological (fertilization), chemical (reduction of ocean acidity, accelerated limestone weathering), and physical methods (CO_2 dissolution, supercritical CO_2 pools in the deep ocean).

12.2 Physical, chemical, and biological fundamentals

12.2.1 *Physical properties of CO_2 in seawater*

The behavior of CO_2 released directly into seawater will depend primarily on the pressure (i.e., depth) and temperature of the water into which it is released. The key properties are:

- The liquefaction pressure at a given temperature: the point at which with increasing pressure, gaseous CO_2 will liquefy
- The variation of CO_2 liquid density with pressure, which determines its buoyancy relative to seawater
- The depth and temperature at which CO_2 hydrates will form

Saturation pressure

At temperatures of 0–10°C, CO_2 will liquefy at pressures of 4–5 MPa, corresponding to water depths of ~400–500 m, with a liquid density of 860 kg per m^3 at 10°C and 920 kg per m^3 at 0°C. At this depth liquid CO_2 will therefore be positively buoyant, and free liquid droplets will rise and evaporate into gas bubbles as pressure drops below the saturation pressure.

Buoyancy

Figure 12.1 shows the densities of CO_2 and seawater versus depth for a range of ocean conditions. Liquid CO_2 is more compressible than water and becomes

Doi:10.1016/B978-1-85617-636-1.00012-2.

neutrally buoyant at depths of 2500–3000 m. Below this depth range, released liquid droplets will sink to the ocean floor.

Hydrate formation

Hydrates are a form of weakly bound molecular complex in which a guest molecule is trapped within a cage of water molecules. The cage of water molecules takes the form of a 12- or 14-faced hedron (dodecahedron or tetrakaidecahedron) having pentagonal or hexagonal faces with an oxygen atom at each vertex, similar to normal ice, as illustrated in Figure 12.2. The stoichiometry of this complex is approximately (CO_2: $6H_2O$), depending on the temperature and pressure of formation.

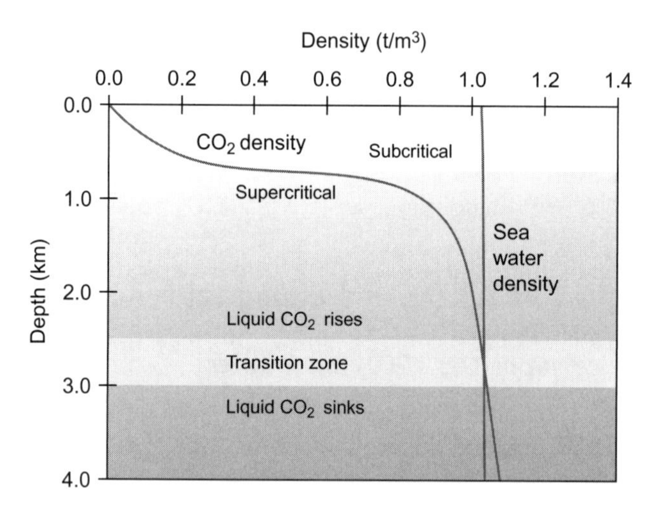

Figure 12.1 CO_2 and seawater density versus depth

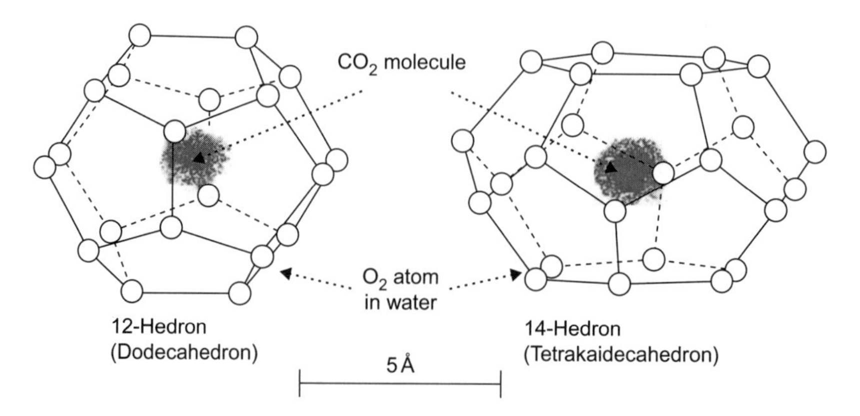

Figure 12.2 CO_2 hydrate structure

CO_2 hydrates can form in all seawater below ~400 m, as long as there is sufficient CO_2 present, and as shallow as 150 m in polar regions where water temperature is close to 0°C. A CO_2 gas bubble or liquid drop at hydrate stable depths will rapidly form a crystalline hydrate skin, which will limit the rate of dissolution of CO_2 into the surrounding seawater. Naturally occurring hydrates have been observed forming around CO_2-rich gas bubbles venting from hydrothermal vents at ~1500 m depth in the mid-Okinawa Trough.

The density of solid CO_2 hydrate is $1.11 \cdot 10^3$ kg per cubic meter, and it will therefore sink in seawater, which has a density of $\sim 1.03 \cdot 10^3$ kg per cubic meter. However, the density of a gas hydrate mass formed from gas bubbles or liquid droplets wrapped in a hydrate skin will depend on the gas or liquid density, as well as the skin thickness and droplet size.

12.2.2 The ocean carbon cycle

As described in Chapter 1, the deep ocean is by far the largest sink in the global carbon inventory, and two interlinked cycles—one physical and one biological—drive the fluxes into and out of that sink.

The solubility pump

The solubility pump, or physical pump, is the term used to describe the physical and chemical process by which CO_2 is transported from the atmosphere to deep ocean waters. The process is illustrated in Figure 12.3, together with the interlinked biological pump described below.

The two key elements of the solubility pump are the dissolution of CO_2 into surface waters and the successive downwelling of surface water and upwelling

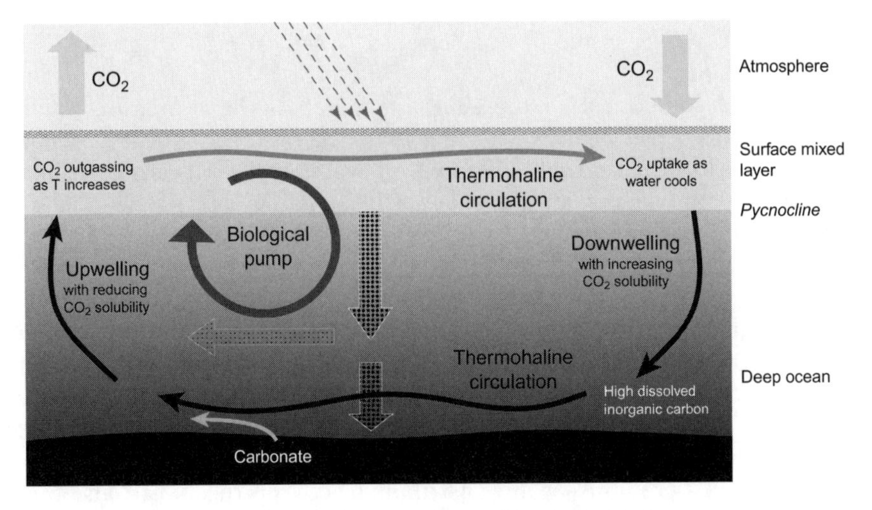

Figure 12.3 The ocean carbon cycle: solubility and biological CO_2 pumps

of deep water via the global thermohaline circulation system. The surface layer of the ocean is well mixed by wind and wave action (hence also known as the mixed layer) and CO_2 in the atmosphere reaches thermodynamic equilibrium with dissolved CO_2 in this layer on a time scale of weeks to months, depending on mixed layer depth and wind conditions. The equilibrium solubility of a gas is directly proportional to its partial pressure (Henry's law):

$$C_{CO_2} = P_{CO_2}/K_{HCO_2} = P_{CO_2}K_{CO_2} \tag{12.1}$$

where C_{CO_2} is the dissolved CO_2 concentration, P_{CO_2} is the partial pressure of CO_2 in the atmosphere, K_{HCO2} is the Henry's law constant, and K_{CO_2} is the solubility of CO_2. The solubility K_{CO_2} is a function of pressure, salinity, and acidity (pH) and, most importantly, of temperature, with solubility increasing from ~0.03 mol-CO_2 per kg at 20°C to ~0.05 mol-CO_2 per kg at 10°C (for typical current atmospheric and oceanic conditions).

The dissolution of CO_2 in water initially forms carbonic acid (H_2CO_3), which will dissociate into bicarbonate and then carbonate ions, the partitioning depending on the water temperature and alkalinity. The total concentration of these species—dissolved free CO_2 ($CO_{2\,(aq)}$), carbonate ions ($CO_3{}^{2-}$), and bicarbonate ions ($HCO_3{}^{-}$)—is known as the dissolved inorganic carbon (DIC). The interactions that establish this balance are as follows:

$$CO_2 + H_2O \leftrightarrow H_2CO_3 \leftrightarrow HCO_3^- + H^+ \leftrightarrow CO_3^{2-} + 2H^+ \tag{12.2}$$

At current average mixed layer conditions, DIC is ~91% bicarbonate and 8% carbonate ions, with the remaining 1% being dissolved free CO_2 plus carbonic acid. The DIC content of surface waters varies from a low of ~1850 μmol-C per kg in the Bay of Bengal to a high of ~2200 μmol-C per kg in the Weddell Sea.

The second component of the solubility pump is the formation of deep waters through meridional overturning that occurs in the North and South Atlantic Oceans (Figure 12.4). A parcel of water moving northward in the Gulf Stream experiences strong evaporative cooling due to wind action, resulting in increasing uptake of CO_2 from the atmosphere as a result of its temperature-dependent solubility, as well as an increase in salinity and density as a consequence of evaporation. By the time a parcel of water reaches the Norwegian Sea, the density becomes high enough to cause it to sink, forming North Atlantic Deep Water (NADW). This water flows back in a southerly direction into the abyssal Atlantic basin. A similar process occurs in the sub-Antarctic South Atlantic, forming Antarctic Bottom Water (AABW). The total rate of downwelling is estimated at 30–40 · 10^6 m^3 per day, and it is roughly evenly distributed between the North and South Atlantic locations.

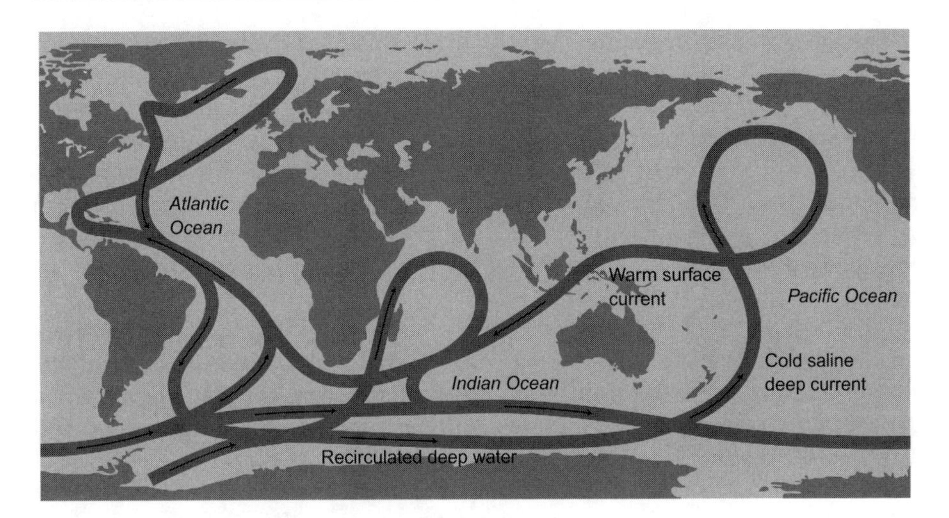

Figure 12.4 Global thermohaline circulation

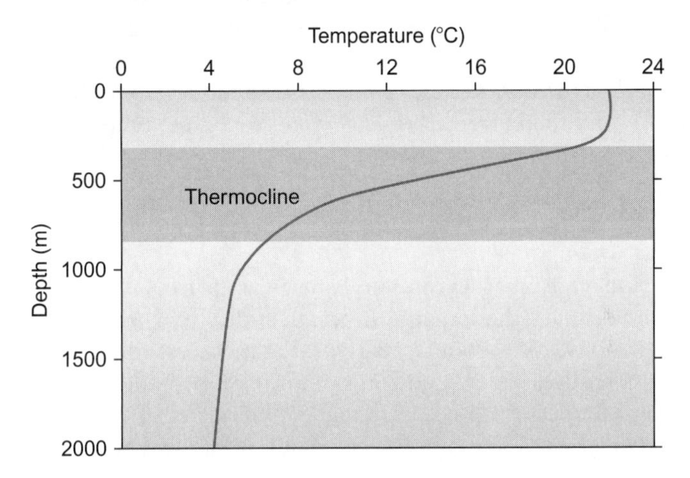

Figure 12.5 Temperature versus depth in a tropical ocean

Other than at these two downwelling locations, a stable layering or stratification occurs in the ocean as a result of temperature and density gradients. This stratification can be seen in the tropical ocean temperature profile shown in Figure 12.5. A warmer, less dense upper layer is separated from the colder, denser deep water by an interval known as the thermocline, where temperature drops rapidly with depth. The thermocline thus marks the limit of the mixed layer, within which fairly rapid thermodynamic equilibrium is reached between the surface water and the atmosphere.

Deep ocean waters, with temperatures down to 2°C and pressures >10 MPa, are highly undersaturated, and it is this excess solubility that enables the ocean to provide a sink for increased atmospheric CO_2 over a centennial time scale.

The deep waters from the Atlantic Basin circulate westward around the Antarctic Ocean basin, branching off and subsequently upwelling in the Indian and Pacific Oceans as well as in the Southern Ocean. The time scale of this circulation system is millennial. Early estimates of the time required for downwelling surface water to fully displace the deep ocean volume based on ^{14}C radioactive tracer measurements put this "ventilation time" of the deep ocean at ~1400 years. Subsequent improved understanding of the processes involved has reduced this time scale to ~250 years for the Atlantic and ~550 years for the Pacific.

When deep waters upwell, CO_2 outgases to the atmosphere at the seawater–air interface since the partial pressure of CO_2 in the seawater is higher than the partial pressure in the atmosphere. In preindustrial times this cycle had reached an equilibrium, with average surface and deep waters containing ~2000 and 2100 μmol-C/kg of DIC respectively. The anthropogenic increase in CO_2 partial pressure in the atmosphere, from 280 ppm in preindustrial times to the current 380 ppm, has resulted in an increase in the average DIC of surface water to ~2080 μmol-C per kg. As a result, the ocean has already sequestered an estimated 500 Gt of anthropogenic CO_2 emissions over the past two centuries.

However, the ability of the surface water to take up additional CO_2 does not increase linearly as atmospheric CO_2 concentration increases. This is because, as dissolved CO_2 and therefore carbonic acid increases, the preferred reaction is:

$$CO_{2(aq)} + H_2O + CO_3^{2-} \rightarrow 2HCO_3^- \quad (12.3)$$

This reaction consumes carbonate ions in a process known as carbonate buffering, and reduces the impact on ocean acidity (pH) that would otherwise arise, since it converts carbonic acid to alkaline bicarbonate ions. If further CO_2 addition reduces carbonate ion concentration to the point that surface waters become undersaturated, carbonate minerals such as $CaCO_3$ will start to dissolve. Under these conditions, calcifying organisms will find it harder to produce shells and coral reefs may dissolve.

For the next 100-ppm increase in atmospheric CO_2 to 480 ppm, the increase in DIC will be only ~70% of the increase that resulted from the historical 100-ppm increase.

The biological pump

The biological pump is the term used to describe the biological mechanisms that control the export of particulate organic carbon (POC) from the surface waters down to the deep ocean. Figure 12.6 illustrates the biological pump and its close interconnection with the solubility pump described earlier.

The starting point for the biological pump is the process of photosynthesis, through which the production of organic carbon takes place in well-lit (euphotic) shallow ocean waters. Photosynthesis starts with the absorption

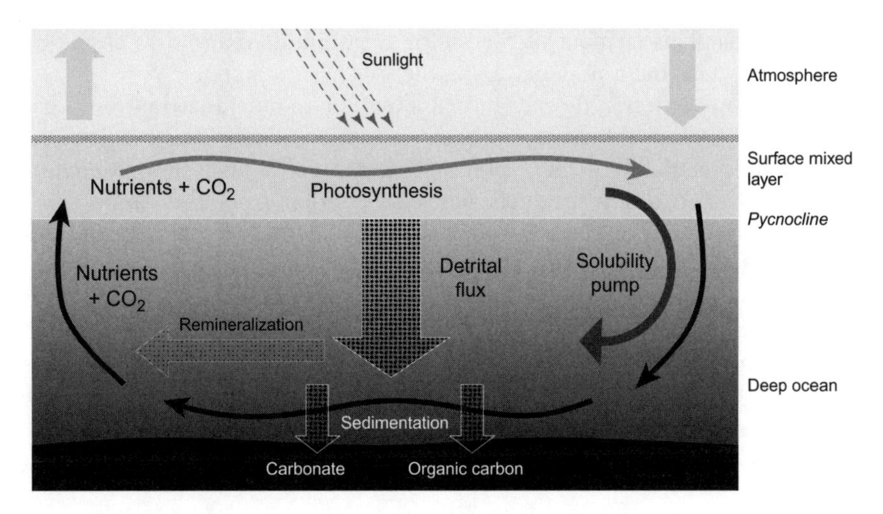

Figure 12.6 The ocean carbon cycle: biological pump

of a photon of light by a chlorophyll molecule, contained in the chloroplast of marine algae or within a photosynthetic bacterium. The energy from the photon is used in a complex sequence of biochemical reactions known as the Calvin cycle (Section 13.2.1), which eventually results in the production of simple sugars that are then used to synthesize other organic compounds and to support the metabolism of the organization.

This set of reactions can be expressed in the highly simplified form:

$$6CO_2 + 6H_2O + \text{photons} \rightarrow C_6H_{12}O_6 + 6O_2 \quad (12.4)$$

which illustrates that CO_2 is consumed and O_2 produced in the overall chain of reactions. CO_2 may be either absorbed directly from the air, as is the case for terrestrial plants, or obtained from CO_2 dissolved in the surface seawater. Phytoplankton also require nutrients such as nitrate, phosphate, and silicic acid, which are delivered to the euphotic zone by upwelling deep ocean waters, and micronutrients, such as iron, which are deposited as windborne (aeolian) dust from arid, upwind land masses. This process of generating organic carbon via photosynthesis is called primary production and is the base of the marine food web.

The biomass of phytoplankton is successively consumed by small, then larger marine animals, which release organic carbon as fecal pellets. The marine food web also results in the production of particulate inorganic carbonate material (PIC) from the shells or tests of calcifying marine organisms such as foraminifera and free-swimming mollusks. These organisms are able to produce carbonate shells from surface seawater because surface waters are supersaturated with carbonate ions. This element of the biological pump is referred to as the hard tissues pump, with the remainder being the soft tissues pump.

Dead phytoplankton and other biomass may be remineralized into inorganic carbon, nitrate, and ammonia by bacterial action, contributing to the total DIC and linking the biological pump with the solubility pump. Remaining dead biomass and decay products will be exported from the euphotic zone as particulate organic carbon, where once again it becomes a nutrient source for deep-water and ocean-floor ecosystems. The falling particulate matter and decay products, both organic and inorganic, are collectively known as "marine snow." Through this process an estimated 0.2 Gt-C per year is deposited onto the ocean floor.

In contrast to surface waters, deep waters are undersaturated in carbonate. Falling PIC therefore dissolves as it reaches these undersaturated waters, further increasing DIC in deep ocean waters.

Long-term atmosphere–ocean CO_2 equilibrium

As noted earlier, deep ocean waters provide a CO_2 solubility sink that is coupled to the atmosphere via the solubility and biological pumps. As a result, changes in atmospheric [CO_2] are damped on a time scale determined by the slow turnover of deep ocean waters. This is illustrated in Figure 12.7, which shows schematically the evolution of atmospheric [CO_2] resulting from a 100-year "pulse" of ~6 Gt-CO_2 per year emissions to the atmosphere. This simple model considers only the partitioning of the emitted volume between atmosphere and ocean and excludes other feedbacks, so the numbers are purely illustrative.

In this simple model 70–80% of the emitted mass is taken up by the ocean over a millennial time scale; using more complete longer-term models, the total uptake of emitted CO_2 into terrestrial and oceanic sinks is estimated to be 85–90%.

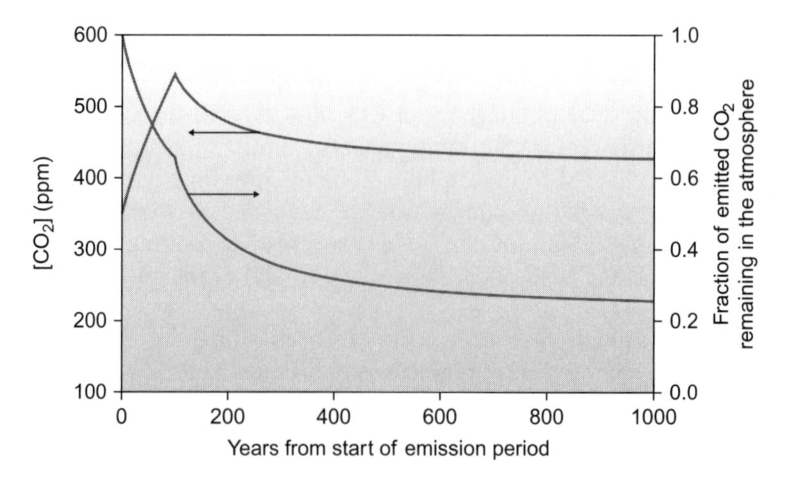

Figure 12.7 Schematic long-term partitioning of an emitted CO_2 pulse

12.3 Direct CO_2 injection

The scale of anthropogenic CO_2 uptake into the oceans has been increasing with the rise in CO_2 partial pressure over the past 200 years and is currently occurring at a rate of ~2.4 Gt-CO_2 per year.

Long-term storage of CO_2 in the ocean relies on its retention in deep waters that have a ventilation time of several centuries, plus the natural tendency of CO_2 to partition into seawater versus the atmosphere. Two basic concepts and a range of technological options have been proposed, as summarized in Table 12.1.

12.3.1 Direct CO_2 dissolution

The potential for deep ocean waters to carry a higher dissolved carbon load is illustrated in Figure 12.8, which shows the approximate DIC profile for

Table 12.1 Options for ocean storage of CO_2

Physical concept	Technological scheme
Direct CO_2 dissolution	Rising plume
	Neutral buoyancy (isopycnal) spreading
	Sinking plume
Liquid CO_2 isolation	Piped feed to seabed lake
	Sinking cooled liquid plus solid CO_2 slurry

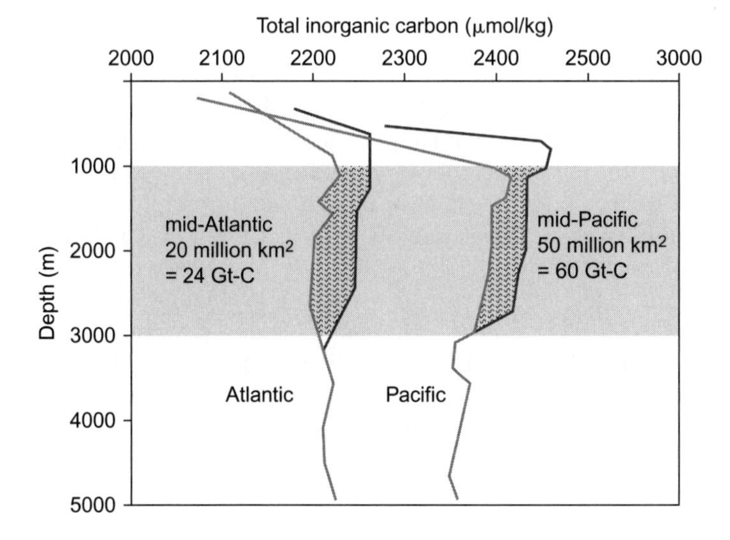

Figure 12.8 Ocean storage capacity for dissolved CO_2

mid-Pacific and mid-Atlantic Ocean locations. An increase in DIC of 50 μmol-C per kg, ~2% of current levels, for the body of water from 1000 m to 3000 m depth, represents an increase in the carbon inventory of >80 Gt-C, equivalent to 14 years of global anthropogenic CO_2 emissions at year 2000 levels.

Direct dissolution below the thermocline is one option to make use of this storage capacity to achieve long-term CO_2 sequestration. Release at depths >500 m would be in liquid form and would be accompanied by the immediate formation of a hydrate shell around individual liquid droplets. At depths shallower than ~2500 m, depending on droplet size and hydrate skin thickness, the resulting plume would be positively buoyant and would rise while individual droplets slowly dissolve. Below ~3000 m release depth, similar behavior would be seen with a sinking plume, while at intermediate depths neutral buoyancy (isopycnal) spreading of the plume would occur.

Results of experiments and field trials of CO_2 dissolution

Laboratory and in situ experiments, mathematical modeling studies, and small-scale field trials have been conducted to investigate the properties, behavior, and ecological impact of CO_2 released into seawater, generally as liquid droplets, larger liquid masses, or liquid–hydrate–water composites.

In situ experiments conducted at the Monterey Bay Aquarium Research Institute (MBARI; see Key References), typically involving the release and observation of liquid droplets by remotely operated vehicles (ROVs), have yielded insights into the dissolution rate of individual hydrate-enclosed droplets and their rising or sinking rate, depending on release depth. Rising droplets are found to have terminal velocities in line with the Stokes law behavior of a rigid sphere, consistent with the presence of a rigid hydrate shell. Dissolution rates of hydrate-enclosed liquid droplets are found to be reduced by a factor of three to four compared to the rate expected for a nonhydrated droplet. Other phenomena such as droplet rafting and formation of a gas phase as a droplet crosses the liquefaction pressure depth have also been observed.

Experimental releases of relatively larger liquid CO_2 volumes have also been performed to investigate the use of sonar as a method of tracking a rising plume. This technique could have importance in monitoring CO_2 storage in subsea aquifers or depleted gas fields, or to monitor CO_2 emitted from hydrothermal vents. Hydrate-enclosed CO_2 droplets were found to be easily detectable using high-frequency sonar (38 and 675 kHz) as a result of the high acoustic velocity difference between liquid CO_2 and seawater. Evolution of a droplet plume formed by the release of 5 liters of CO_2 could be easily tracked over a 150 m ascent, and the overall plume behavior was accurately described by mathematical models. Further experiments using lower acoustic frequencies are envisaged that could monitor seawater density changes resulting from CO_2 dissolution, as well as possible biological responses to the plume.

Other small-scale release experiments have investigated the behavior of CO_2 liquid–hydrate–seawater composite material formed by the injection and vigorous mixing of seawater into a liquid CO_2 stream. Unlike hydrate-enclosed liquid CO_2 droplets, which as noted become negatively buoyant at ~2500–3000 m, this composite material contains a higher proportion of hydrate and becomes negatively buoyant at ~1000 m. This material was extruded in the form of cylindrical pellets 6.5 mm in diameter and ranging in length from 5 mm to 85 mm, with a density that depends on the degree of conversion of the liquid CO_2 to hydrate. More complete conversion to hydrate is desirable since it results in a denser pellet, thereby maximizing the effective depth of disposal.

As for liquid CO_2 droplets, movement and dissolution of the pellets were monitored and estimates of the dissolution rate were obtained. Individual pellets were found to sink between 10 m and 70 m before complete dissolution.

These small-scale experiments have given insight into the fate of individual droplets and small hydrate masses but are unable to address larger-scale effects that would determine the dynamic behavior of a large plume. Mathematical simulations indicate that if a large number of the extruded hydrate pellets were released, the resulting negatively buoyant stream would sink significantly faster and further than individual pellets, as a result of the entrainment of water into a plume and the increasing density of entrained water as CO_2 dissolves into the plume. A plume containing a mass flux of 100 kg-CO_2 per second was predicted to descend ~500 m before the pellets were fully dissolved.

Free ocean CO_2 enrichment experiment

By analogy with the Free Air CO_2 Enrichment (FACE) experiments that have been performed to test the effects of elevated [CO_2] in terrestrial ecosystems (described in Section 13.5.1), a Free Ocean CO_2 Enrichment (FOCE) experiment has been constructed at the MBARI to study the effects of increased [CO_2] in seawater on marine ecosystems.

In the current system, seafloor currents carry a parcel of CO_2-enriched seawater through a series of baffles within a rectangular flume. The CO_2-enriched parcel mixes and reacts with the surrounding seawater and the system is actively controlled to achieve a consistent elevated seafloor [CO_2] level in a 1 meter square test area in the middle of the flume. The FOCE system is designed to be left in place for weeks or months, allowing investigation of the effects of elevated [CO_2] on seafloor animals within the test area.

12.3.2 Liquid CO_2 isolation

The negative buoyancy of unhydrated liquid CO_2 below ~3000 m opens up the possibility of sequestering CO_2 as a liquid lake in an ocean-floor depression or deep trough. Such a lake could be formed either by releasing liquid directly into a depression or by releasing negative-buoyancy droplets or hydrated particles sufficiently close to the bottom that a substantial proportion of the released liquid mass would reach bottom before dissolution was complete.

Naturally occurring ocean-floor CO_2 lakes have been observed in the vicinity of hydrothermal vents, as a result of the liquefaction and pooling of CO_2-rich gases released from these vents. These naturally occurring CO_2 lakes are too shallow to be gravitationally stable and are held in place by overlying sediments and by hydrate crusts or cements in these sediments.

Small-scale ocean-floor experiments have also been performed with liquid CO_2 pools to assess the impact of direct CO_2 injection on deep sea ecosystems. Small pools of liquid CO_2 were created by releasing ~20 liters of liquid into open-topped cylindrical "corrals" on the sea bed at 3600 m depth. At this depth the excess density of the liquid CO_2 keeps the mass constrained within the corral, but slow dissolution results in a dense CO_2-rich, low-pH plume being swept downcurrent across the ocean floor.

The results of these small-scale experiments show that significant reductions in pH will occur in the close vicinity of an ocean-floor CO_2 lake, resulting in high mortality rates in the affected marine ecosystem. Given these results it is not surprising that attempts to perform mesoscale trials have met with considerable environmental opposition.

Numerical studies have also been performed to assess the dissolution rate from a liquid CO_2 lake, both under static conditions and under the influence of ocean-floor currents. These studies show that vertical mixing above such a lake is reduced as a result of the formation of a high-density (~1.5 kg per m^3) boundary layer, and conclude that gravity currents that may result from this high-density water mass will be an important factor to consider in future studies and experiments.

In an environment that is not disturbed by ocean-bottom currents, the dissolution rate from the lake would be reduced as a result of the hydrate layer that would form at the lake surface as well as by the stratification that would result from denser, CO_2-rich water filling the depression above the lake surface. A lifetime in excess of 10,000 years would be possible for a 50 m deep lake with a dissolution rate of $<5 \cdot 10^{-3}$ m per year, although these residence times would be substantially reduced (25–500 times faster) in the presence of moderate to extreme ocean-floor currents.

12.3.3 Prospects for large-scale field trials

The only mesoscale direct injection field trial to reach an advanced planning stage was the so-called CO_2 Ocean Sequestration Field Experiment, that was proposed under the auspices of a project agreement (the International Collaboration on CO_2 Ocean Sequestration) signed by the U.S. Department of Energy, the New Energy and Industrial Technology Development Organization of Japan, and the Norwegian Research Council during the 1997 Third Conference of Parties to the UNFCCC in Kyoto.

The first series of experiments to be designed under this agreement had as its objectives to:

- Investigate CO_2 droplet plume dynamics through qualitative (video) and quantitative methods (pH and velocity measurements)

- Clarify the effects of hydrates on the dissolution of CO_2 droplets through similar measurements
- Trace the evolution of the CO_2-enriched seawater by performing 3D mapping of velocity, pH, and DIC
- Assess the potential biological impact of changes in seawater pH by quantifying changes in bacterial biomass, production, and growth efficiency both in the water column and on the ocean floor.

The final project design envisaged the release of 20–40 t-CO_2 over a period of 1–2 weeks using a 4 cm diameter coiled steel tubing deployed from a ship down to the injection depth at 800 m (Figure 12.9). The buoyant plume of CO_2 droplets, injected from a diffuser assembly on the ocean floor at a rate of ~1 kg per second, was expected to rise ~100 m before dissolving.

Further field experiments under the project agreement were expected to focus on acute and chronic environmental impacts and subsequent recovery, and were planned to be conducted over a time scale of at least a year, comparable to the lifetimes of affected organisms.

The experiment was planned to be hosted by the Natural Energy Laboratory of Hawaii Authority (NELHA) and to take place in the summer of 2001 in the ocean research corridor operated by NELHA at Keahole Point in Kona, on the west coast of Hawaii. However, as a result of strong public opposition the project was unable to secure all the required permits and had to abandon the Hawaii experiment. Subsequently a scaled-down version of the experiment was planned, to release and monitor 5.4 t-CO_2 off the coast of Norway, but failed to secure approval from the Norwegian environment ministry.

Results from small-scale experiments confirm that the dynamic behavior of individual droplets, small hydrate particles, and small-scale masses (in the order of 5 kg) is reasonably well understood and can be modeled with confidence. However, the dynamics and environmental impact of plumes resulting from large-scale CO_2 injection are unknown and will remain so until public

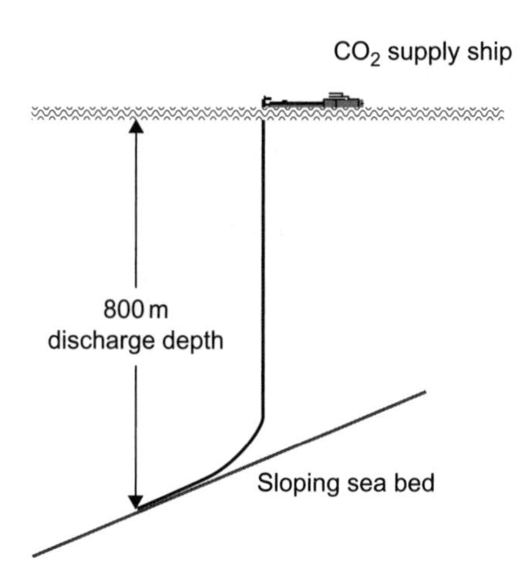

Figure 12.9 Proposed CO_2 Ocean Sequestration Field Experiment configuration

and political perception of the risks of conducting large-scale experiments are overtaken by the perceived risks of not doing so.

12.4 Chemical sequestration

Like the igneous rocks that provide a number of potential feedstocks for mineral carbonation, carbonate rocks, most commonly limestone with a high content of calcium carbonate, are also subject to weathering when exposed on the earth's surface. The weathering or dissolution of limestone by carbonic acid is a key reaction in the geochemical carbon cycle, described in Chapter 1, and proceeds according to the reactions:

$$H_2O + CO_2 \rightarrow H_2CO_3 \quad (12.5)$$

$$CaCO_3 + H_2CO_3 \rightarrow Ca^{2+} + 2HCO_3^- \quad (12.6)$$

Natural limestone weathering removes an estimated 0.7 Gt-CO_2 per year from the atmosphere and results in the transportation of ~2.6 Gt per year of dissolved calcium bicarbonate into the ocean. In the absence of countervailing fluxes, this process would remove all CO_2 from the atmosphere on a time scale of ~4500 years.

The same chemistry can be applied to the capture and sequestration of CO_2 from power plant or cement plant flue gases in a process called the accelerated weathering of limestone (AWL). Flue gases would be bubbled through a reactor containing crushed limestone particles that is continuously wetted by a supply of water. For each tonne of CO_2 captured, the process consumes 2.3 tonnes of $CaCO_3$ plus 0.4 tonnes of water and produces 3.7 tonnes of calcium bicarbonate in solution.

The bicarbonate-rich effluent stream is envisaged to be disposed of by discharge into the ocean and this could have a beneficial environmental effect, since the addition of bicarbonate would counteract ocean acidification. However, if discharged as a 75% saturated bicarbonate solution ($5\ 10^{-3}$ Molar at a CO_2 partial pressure of 150 kPa) this would require a water supply of 10^4 t-water per t-CO_2. This equates to a colossal ~10^8 t-water per day for a 500 MW_e power plant (~1/10 of a cubic km of water) and is some 300 times the volume that would be used for power plant cooling. In addition, some 3.10^4 t-$CaCO_3$ would be required to be transported to the plant, compared to ~4800 t-coal per day fuel requirement.

The prodigious quantities of limestone and particularly water required represent significant hurdles to the large-scale application of the AWL process.

12.5 Biological sequestration

12.5.1 Ocean Iron fertilization

The concept of ocean fertilization as a carbon sequestration strategy arises from the observation that despite the availability of nutrients, primary production of biomass through photosynthesis is limited in some ocean areas, particularly in the Southern Ocean and in the North Pacific. These areas are known as high-nutrient, low-chlorophyll (HNLC).

It was first suggested in the 1930s, by English biologist Joseph Hart, that this might be due to a deficiency of iron in these areas. Iron is required as a micronutrient by the phytoplankton responsible for primary production, and this suggestion was confirmed in the 1980s by oceanographer John Martin, based on experiments demonstrating increased phytoplankton growth in water samples from several HNLC ocean locations after enrichment with iron.

Unlike the primary phytoplankton macronutrients (nitrate, phosphate, and silicic acid) that are delivered into the euphotic zone by the upwelling of nutrient-rich deep ocean waters, iron is delivered to the ocean as a component of windborne (aeolian) dust, carried offshore from arid land masses. The low level of primary production in the Southern Ocean and North Pacific can then be understood as a result of the lack of upwind sources of wind-blown iron-bearing dust.

Results of iron fertilization trials

Fertilization of these HNLC ocean areas by seeding with iron would be expected to increase primary production and increase the flux of organic carbon into the deep ocean where it would be effectively sequestered over a millennial time scale. Since the mid-1990s, 11 iron fertilization trials have been conducted in HNLC areas, six in the Southern Ocean, three in the subarctic North Pacific, and two in the equatorial East Pacific. The typical parameters of these mesoscale experiments are shown in Table 12.2.

Table 12.2 Parameters of mesoscale iron fertilization experiments

Parameter	Range	Description
Iron quantity	300–1000 kg	Commonly known as $FeSO_4$ solution
Patch area	50–100 km^2	
Mixed layer thickness	10–50 m	
Tracers	SF_6, ^{234}Th	Tracing patch movement and particle export

Tracers such as SF_6 and thorium-234 (^{234}Th) are added to the iron enrichment solution for experimental monitoring. SF_6 can be detected in minute quantities and is used to enable subsequent mapping of the location of the iron-enriched patch as it is moved and stirred by ocean currents. The naturally occurring radioactive isotope thorium-234 has a high affinity for particles, and the loss of the isotope from the enriched water indicates the rate at which particles are sinking from the upper ocean.

These experiments have confirmed that iron enrichment does affect the rate of primary production, causing a bloom of phytoplankton. However, they have also demonstrated that iron enrichment results in other changes in the ecology and biogeochemistry of the enriched patch and that many factors then come into play that determine the fate of the bloom and the extent to which export of POC to the deep ocean takes place (Table 12.3).

One example of an adverse impact from iron enrichment is an increased release of nitrous oxide (N_2O) as a result of microbial nitrification of ammonia from the decaying bloom. N_2O is a greenhouse gas with a lifetime in the atmospheric of 120 years and, relative to CO_2, has a Global Warming Potential (GWP) of 296 over a 100-year time scale.

The effectiveness of induced blooms in increasing the export of POC to the deep ocean remains highly uncertain and has been demonstrated in only two of the mesoscale experiments. Estimates of the required rate of iron addition to the rate of carbon export range from 2 to 2000 μmol-Fe per mol-C; the lower end reflects the carbon content of the total biomass produced per mole of added Fe, while the upper end is a more conservative estimate of the net long-term export to the deep water and ocean floor.

Table 12.3 Iron fertilization: impacts and controlling factors

	Impacts and controlling factors
Characteristics of the phytoplankton bloom	Initial chemical and optical conditions Preexisting phytoplankton communities
	Dilution rate of enriching solution
	Iron supply rate and duration
	Macronutrient supply
Local ecological impact of the bloom	Redistribution of phytoplankton grazers in response to increased food availability
	Increased growth and reproduction of phytoplankton grazers
	Changes in species composition and food web structure Increased bacterial activity and biogenic gas generation

Other fertilization options

While iron fertilization aims to supplement a deficient micronutrient, mesoscale experiments have also been conducted in macronutrient-poor waters, an example being the phosphate-deficient eastern Mediterranean Sea. However, as well as unexpected interactions in the marine ecosystem, macronutrient-based fertilization would face an additional challenge due to the substantially greater amount of macronutrient required for phytoplankton growth.

The molar ratio of carbon to the macronutrients nitrogen and phosphorus in phytoplankton is a relatively stable ratio and has been found to be equal to the ratio of these elements in deep ocean water, given by the Redfield ratio (C:N:P = 106:16:1). This congruence demonstrates the closeness of the interaction between the biochemistry of phytoplankton and the chemistry of the deep ocean water body that regulates their environment, and is an example of ecological stoichiometry. This ratio, named after American oceanographer Alfred Redfield (1890–1983), is the reason that the marine photosynthetic reaction (Equation 12.7) is sometimes written as:

$$106CO_2 + 16HNO_3 + H_3PO_4 + 122H_2O \rightarrow C_{106}H_{263}O_{110}N_{16}P + 138O_2 \quad (12.7)$$

where $C_{106}H_{263}O_{110}N_{16}P$ represents the average plankton composition.

In contrast to the relative constancy of the Redfield ratio, the molar ratio of iron to carbon (Fe:C) in phytoplankton is more variable. A value of 2–3 μmol-Fe per mol-C is considered a minimum to sustain cells at zero growth rate, close to the value of 2 μmol-Fe per mol-C measured in HNLC surface water. Increasing iron availability leads to increasing uptake, up to a saturation growth rate at an Fe:C ratio of ~20 μmol-Fe per mol-C. An average value of 6 μmol-Fe per mol-C is commonly used in modeling studies, which compares to 9400 μmol-P per mol-C from the Redfield ratio. If 1% of net biomass production resulting from phosphorus fertilization reaches the deep ocean, sequestration of 1 t-C could require the application of 2.4 t of phosphorus—equivalent to 7.7 t phosphoric acid or 8.9 t ammonium phosphate!

Prospects for implementation

In a 2007 statement of concern, the Scientific Groups of the London Convention (the Convention on the Prevention of Marine Pollution by Dumping of Wastes and Other Matter) stated the view that current knowledge about the effectiveness and potential environmental impacts of iron fertilization was insufficient to justify large-scale operations. They recommended that any proposed operations should be subject to a careful evaluation, including the factors summarized in Table 12.4.

The large-scale application of iron fertilization would have a major impact on ocean biogeochemistry and ecosystems. Understanding these impacts with

Table 12.4 London Convention impact assessment recommendation

	Impact assessment
1	Estimated amounts and potential impacts of iron and other materials to be released with the iron
2	Potential impacts of gases that may be produced by the expected phytoplankton blooms or by bacteria decomposing the dead phytoplankton
3	Estimated extent and potential impacts of bacterial decay of the expected phytoplankton blooms, including reducing oxygen concentrations
4	Types of phytoplankton expected to bloom and the potential impacts of any harmful algal blooms that may develop
5	Nature and extent of potential impacts on the marine ecosystem, including naturally occurring marine species and communities
6	Estimated amounts and time scales of carbon sequestration, taking account of partitioning between sediments and water
7	Estimated carbon mass balance for the operation

the level of confidence that would be required to satisfy the London Protocol requirements, or for verifiable carbon offset accounting, will likely require tens or hundreds of mesoscale experiments.

Given these challenges, the prospect for large-scale implementation of ocean iron fertilization seems remote. However, multidisciplinary studies based on satellite observations have concluded that a drop in primary production in the oceans over recent decades is partly attributable to a decline in the supply of aeolian dust to the oceans. This natural source of iron as a micronutrient for the ocean food web is estimated to have declined by 25% over the two decades ~1980 to 2000 as a result of changes in land use and increasing primary production in arid-zone grasslands.

When the complex interactions between marine biogeochemical and ecological systems are better understood, selective iron fertilization may play a part in a strategy to mitigate some of the consequences of climate change, such as declining primary ocean productivity in traditionally productive regions.

12.5.2 Wave-driven ocean upwelling

An alternative approach to enhancing the rate of CO_2 sequestration in the ocean has been proposed that does not rely on enrichment using external fertilization. This system uses wave-powered pumps to draw cool, nutrient-rich waters to the ocean surface from below the thermocline. Figure 12.10 illustrates the wave-driven ocean upwelling system concept.

Each pumping unit comprises a large-diameter (3–10 m) vertical tube of 100–200 m length, sealed at its base by a flapper valve that only allows water to pass into the tube, and attached to a surface float that lifts the water-filled

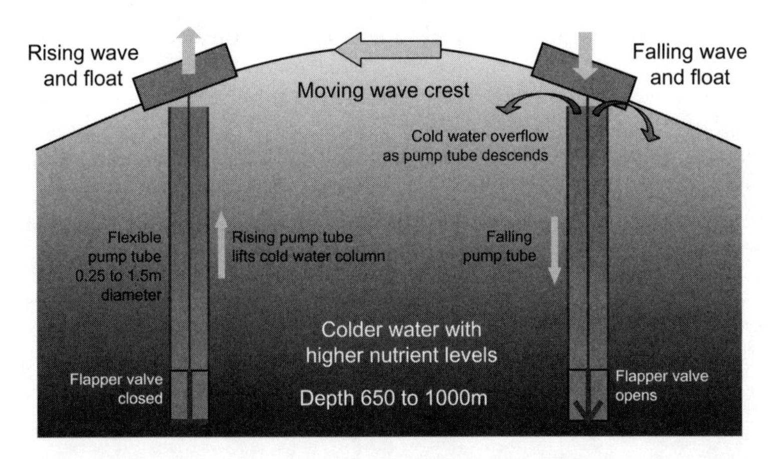

Figure 12.10 Wave-driven ocean upwelling system

tube on each wave crest. As the float and tube start to descend into the next trough, the flapper valve opens, the tube drops, and the water in the upper section of the tube is released onto the ocean surface.

A large-scale application would require an array of such pumping units tethered to nearest neighbors at a separation of 1.5 to 2.5 km, and potentially covering thousands of square kilometers of ocean. However, deep water is also richer in DIC, and bringing this to the surface would result in CO_2 release to the atmosphere. Warming of the lifted cooler water could also result in CO_2 release due to the drop in solubility at higher temperatures. It is therefore questionable whether such an approach would lead to a net export of carbon from the mixed layer to the deep ocean. An initial field trial of the pumping units took place in May 2008.

A megascale application of wave-driven upwelling, $>100{,}000\,km^2$ of the Gulf of Mexico, has also been proposed to reduce the high sea-surface temperatures that contribute to the formation and growth of hurricanes.

12.6 References and Resources

12.6.1 Key References

The following key references, arranged in chronological order, provide a starting point for further study.

Seifritz, W. (1990). CO_2 disposal by means of silicates, *Nature*, **345** (486).

Handa, N. and Ohsumi, T. (eds.). (1995). Direct Ocean Disposal of Carbon Dioxide, *Proceedings of First and Second International Workshops on Interaction between CO_2 and Ocean*, 1991 & 1993, Terrapub, Tokyo. Available at www.terrapub.co.jp/e-library/dod/index.html.

Brewer, P. G., G. Friederich, E. T. Peltzer, and F. M. Orr, Jr. (1999). Direct experiments on the ocean disposal of fossil fuel CO_2, *Science*, **284** (943–945).

IEA GHG. (1999). Ocean Storage of CO_2, IEA Greenhouse Gas R&D Programme. Available at www.ieagreen.org.uk/oceanrep.pdf.

Aya, I., *et al.* (2004). In situ experiments of cold CO_2 release in mid-depth, *Energy*, **29** (1499–1509).
Barry, J. P., *et al.* (2004). Effects of direct ocean CO_2 injection on deep-sea meiofauna, *Journal of Oceanography*, **60** (759–766).
Cicerone, R., *et al.* (2004). The ocean in a high CO_2 world, *Oceanography*, **17** (72–78).
Brewer, P. G., *et al.* (2005). Deep ocean experiments with fossil fuel carbon dioxide: Creation and sensing of a controlled plume at 4 km depth, *Journal of Marine Research*, **63** (9–33).
Rau, G. H., K. G. Knauss, W. H. Langer, and K. Caldeira. (2007). Reducing energy-related CO_2 emissions using accelerated weathering of limestone, *Energy*, **32** (1471–1477).
Karl, D. M. and R. M. Letelier. (2008). Nitrogen fixation-enhanced carbon sequestration in low nitrate, low chlorophyll seascapes, *Marine Ecology Progress Series*, **364** (257–268).

12.6.2 Institutions and Organizations

The institutions and organizations listed below have been active in the research and development work on ocean storage.

Atmocean Inc. (developing wave-driven system to enhance upwelling): www.atmocean.com
Carboocean IP (European Union-funded project aiming to provide an accurate scientific assessment of marine carbon sources and sinks, focusing on the Atlantic and Southern Oceans over the past 200 and next 200 years): www.carboocean.org
Cquestrate ("open source" initiative developing a process to enhance the carbon sink capacity of the oceans and mitigate ocean acidification, by adding calcium oxide to seawater): www.cquestrate.com
Monterey Bay Aquarium Research Institute (MBARI): www.mbari.org
National Maritime Research Institute (NMRI): www.nmri.go.jp/index_e.html
Ocean Carbon & Biogeochemistry: www.us-ocb.org
The Ocean in a High CO_2 world, UNESCO Intergovernmental Oceanographic Commission International Symposium Series: ioc3.unesco.org/oanet/HighCO2World.html
UNESCO Intergovernmental Oceanographic Commission; International Ocean Carbon Coordination Project: http://ioc-unesco.org/, www.ioccp.org
Woods Hole Oceanographic Institution: www.whoi.edu

12.6.3 Web Resources

Ocean Chemistry of the Greenhouse Gases: www.mbari.org/ghgases
The Emerging Science of a High CO_2/Low pH Ocean: www.mbari.org/highCO2
Free Ocean Carbon Dioxide Enhancement Experiment: www.mbari.org/highCO2/foce
Woods Hole Oceanographic Institution, Ocean Iron Fertilization Symposium, September 2007: www.whoi.edu/page.do?pid = 14618
Café Thorium; Radiochemistry Group at WHOI: www.whoi.edu/science/MCG/cafethorium
GLODAP (Global Ocean Data Analysis Project): http://cdiac.esd.ornl.gov/oceans/glodap/Glopintrod.htm
Ocean Carbon and Climate Change (an implementation strategy for U.S. ocean carbon research): www.carboncyclescience.gov/documents/occc_is_2004.pdf

13 Storage in terrestrial ecosystems

13.1 Introduction

Terrestrial ecosystems play an important role in the global carbon cycle, being both the repository of a carbon inventory currently estimated at 2200 Gt-C (600 Gt-C in biomass and 1600 Gt-C in soils) and the source and sink for closely balanced CO_2 fluxes of the order of 120 Gt-C per year taken up from the atmosphere through photosynthesis and emitted through plant and microbial respiration. The combined effect of these fluxes results in a net uptake of carbon into terrestrial ecosystems from the atmosphere currently estimated at 2.8 Gt-C per year.

Changes in this carbon inventory and related fluxes as a result of human activity have also been a major contributor to the atmospheric [CO_2] increase during industrial times, with an estimated 150 Gt-C having been released due to land use changes since 1850. This is a significant quantity when compared to the ~275 Gt-C emitted through fossil fuel combustion and cement production since the mid-1800s, and indicates the potential for the terrestrial ecosystems to have a significant impact in mitigating future increase in atmospheric [CO_2].

Terrestrial ecosystems also respond to local, regional, and global climate variability and change on a variety of time scales and through a multitude of feedback loops. Quantifying the global impact of climate–ecosystem feedback and indeed establishing whether the overall feedback is positive or negative—accelerating or slowing climate change—is not currently possible with any degree of confidence due to the complexity and limited understanding of the processes that cycle carbon between the atmosphere and terrestrial ecosystems, particularly those processes occurring within soils.

Carbon storage in terrestrial ecosystems can be achieved by increasing the flux of CO_2 from the atmospheric into long-lived terrestrial carbon pools, either in or derived from plant biomass, or by reducing the rate of CO_2 emissions from carbon pools in terrestrial ecosystems back into the atmosphere. Examples of long-lived carbon pools in terrestrial ecosystems are:

- Above- and below-ground biomass, such as trees
- Long-lived products derived from biomass (primarily from wood)
- Biochemically recalcitrant (see Glossary) and stabilized organic carbon fractions in soils (inorganic soil carbon)

Doi:10.1016/B978-1-85617-636-1.00004-3

Although there are significant areas of ongoing research directed at gaining a better understanding of soil carbon dynamics under a wide range of biological, physical, and soil conditions, many of the practices that can increase soil carbon pools are based on existing technology and can be applied immediately. Terrestrial carbon storage options therefore have a potential role to buy time while other technologies are being brought to readiness for large-scale deployment.

13.2 Biological and chemical fundamentals

As discussed in Chapter 1, within the global carbon cycle the total inventory of carbon in soils is estimated to be 1600 Gt-C, ~4% of that in the oceans. Vegetation holds a further estimated 600 Gt-C, while annual fluxes of ~120 Gt-C per year occurs between the atmosphere and terrestrial ecosystems as a result of photosynthesis and respiration.

The transfer of CO_2 from the atmosphere into terrestrial vegetation and soils begins with the process of carbon fixation through photosynthesis, in which plants convert CO_2, water, and energy in the form of absorbed photons into organic compounds.

13.2.1 Photosynthesis

C3 photosynthesis

In the plant chloroplast, photosynthesis starts with the so-called light reactions, in which an electron is released from a chlorophyll molecule as a result of the adsorption of a photon. The electron is exchanged between a number of complex organic compounds present in the chloroplast and eventually results in the reduction of nicotinamide adenine dinucleotide phosphate (NADP; $C_{21}H_{29}N_7O_{17}P_3$) to NADPH, and the synthesis of adenosine-5'-triphosphate (ATP; $C_{10}H_{16}N_5O_{13}P_3$) from adenosine diphosphate (ADP; $C_{10}H_{15}N_5O_{10}P_2$).

A sequence of reactions known as the Calvin cycle (or Calvin–Benson–Bassham cycle) then takes place within the chloroplasts, in which the enzyme RuBisCO (ribulose-1,5-bisphosphate carboxylase/oxygenase) catalyzes the fixation of CO_2 to produce glyceraldehyde 3-phosphate (G3P), using the energy stored in NADPH and ATP. The Calvin cycle is shown schematically in Figure 13.1, and the overall equations for the light reaction and the Calvin cycle can be written:

$$2H_2O + 2NADP^+ + 2ADP + 2P_i + \text{light} \rightarrow 2NADPH + 2H^+ + 2ATP + O_2 \tag{13.1}$$

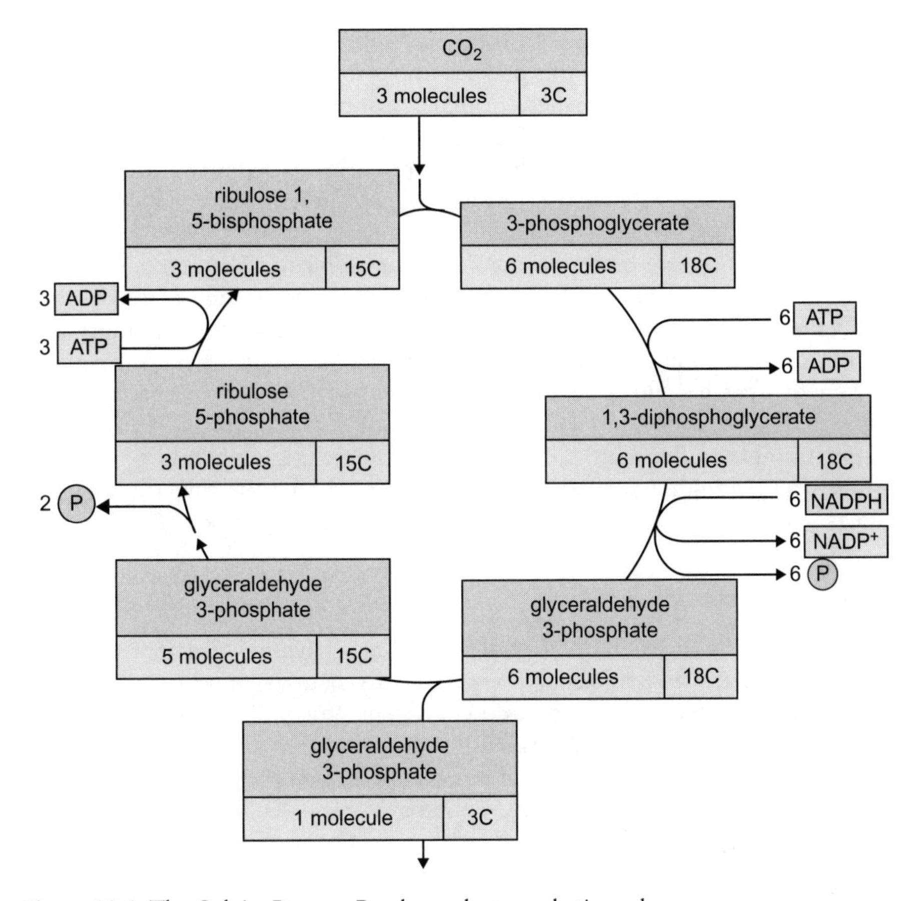

Figure 13.1 The Calvin–Benson–Bassham photosynthetic cycle

and

$$3CO_2 + 6NADPH + 9ATP + 6H^+ \rightarrow G3P + 6NADP^+ + 9ADP + 3H_2O + 8P_i \qquad (13.2)$$

where $NADP^+$ is the oxidized form of NADPH and G3P–$C_3H_7O_6P$ is the end product of the cycle. G3P, the first photosynthetic product, is a three-carbon compound—hence the name C3 photosynthesis. It is the precursor for synthesis of glucose and other organic compounds, such as cellulose, and is thus the starting point for the growth of terrestrial biomass.

Some photosynthetically fixed carbon is lost from the chloroplast by photorespiration, a process that regenerates the CO_2 acceptor ribulose-1,5 bisphosphate that is lost in the oxygenase reaction pathway in which RuBisCO catalyses the fixation of O_2 rather than CO_2. The product of the oxygenation is recycled in the mitochondria to produce serine, an alternative precursor to

G3P, with CO_2 being released. In total ~20–40% of the carbon fixed by C3 photosynthesis is respired by this process of photorespiration.

In contrast to the oceans where primary production is limited by the availability of nutrients rather than CO_2 (Chapter 12), land plants are able to increase primary production with increasing CO_2 availability. Although the response of mature forests is still unclear, vegetation in arid or semi-arid zones is particularly responsive to higher CO_2 concentrations ([CO_2]), since the amount of water loss due to transpiration for a given CO_2 uptake is reduced as [CO_2] increases. This enables a longer growing season and greater biomass production for a given rainfall. However, while this increase in primary production can contribute positively to carbon stocks, studies have shown that the benefit may be mitigated in some ecosystems if the availability of nitrogen is limited—the progressive nitrogen limitation (PNL) theory. In this case, relatively stable soil carbon is decomposed to meet microbial and plant needs for nitrogen, resulting in higher soil respiration and offsetting the effect of increased gross primary production (GPP) on soil carbon stocks.

C4 and CAM photosynthesis

Two variants of the basic photosynthetic process have evolved that minimize water loss and the less efficient photorespiratory pathway, and typically occur in plants adapted for arid or semi-arid conditions. In so-called C4 plants, which include sugar cane, maize, and sorghum, CO_2 is captured in the inner mesophyll layer via the formation of oxaloacetate and malate. These compounds contain four carbon atoms, hence the name C4 photosynthesis. The malate is transported into bundle sheath cells (Figure 13.2), where the CO_2 is released and the standard Calvin cycle proceeds within the chloroplasts. By this mechanism RuBisCO is spatially isolated from oxygen present in the mesophyll, reducing photorespiration. High CO_2 concentrations are achieved at the site of the Calvin cycle without the need for the stomata to be wide open, thus reducing water loss.

Crassulacean acid metabolism (CAM) is another variant, occurring in plants such as pineapple, orchids, and cacti. In CAM plants, the stomata are opened at night, when the temperature is lower and humidity is generally higher, and water loss will therefore be minimized. As in C4 photosynthesis, the oxaloacetate and malate pathway is used to capture and store carbon dioxide. During the day the stomata are closed, reducing oxygen availability for photorespiration, and the CO_2 is released, enabling Calvin cycle (C3) photosynthesis. The benefit that accrues to a C4 or CAM plant to offset the additional cost incurred in the production of malate is a substantial reduction in the loss of water through transpiration.

A feature of photosynthesis that has become an important research tool for tracking plant-derived carbon in the environment is its discrimination in favor of the lighter ^{12}C isotope. Although the heavier ^{13}C isotope has a natural abundance of ~1% in the atmosphere, when CO_2 availability is not the limiting factor for primary production, photosynthesis discriminates against the heavier

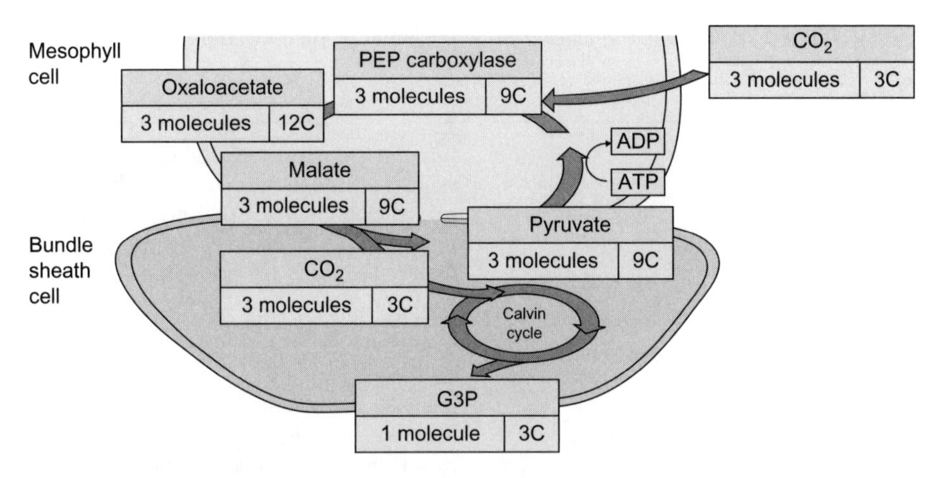

Figure 13.2 Photosynthetic mechanisms in C4 plants

isotope with the result that carbon fixed plant biomass is relatively depleted in ^{13}C. This discrimination is strongest in C3 plants, with C3 biomass being typically 25‰ depleted in ^{13}C (i.e., $\delta^{13}C = -22$ to -30‰ compared to the VPDB carbonate standard; see Glossary), while in C4 and CAM plants the depletion is typically $\delta^{13}C \approx -8$ to -11‰.

Above-ground and below-ground carbon allocation

Up to 50% of the mono- and disaccharides (e.g., glucose and sucrose respectively) produced by plants are delivered to the root system, where they are used to build root biomass, exuded and accessed by soil microbes, particularly the mycorrhizal fungi that form a symbiotic relationship with the plant root, with 5–20% of a plant's total carbon budget being allocated to these fungal associations. The exuded carbohydrates are partly consumed by heterotrophic respiration (see Glossary), releasing some fixed CO_2 back to the atmosphere, and partly used to build microbial biomass. In return the plant gains access to the filamentary hyphae of the fungal mycelium, effectively increasing the surface area and fine scale penetrative capacity of the root system for the uptake of water and nutrients.

Since a plant's primary reproductive task—the production of flowers, fruits, and seeds—takes place mainly above ground, the below-ground allocation of GPP will be optimized to maximize reproductive success under specific ecological and environmental conditions. Some of the factors affecting below-ground allocation are described in Table 13.1.

13.2.2 Biogeochemical processes in soils

The complex symbiotic and competitive processes taking place in the immediate vicinity of root systems, the rhizosphere, have a major impact on the fate of soil organic matter (SOM) and are therefore a key to understanding the

Table 13.1 Factors influencing below-ground carbon allocation in plants

Factor	Description
Nutrient availability	Increased nutrient availability, for example through fertilization, reduces below-ground allocation since the plant is more easily able to meet its nutrient requirements.
Water availability	Similar to nutrients, less GPP needs to be invested below ground when water availability is high.
Soil disturbance	Intensive cultivation may result in a shift in soil microbial community composition toward bacteria rather than fungi, which can result in a compensating increase in below-ground allocation.

potential for carbon storage in soils. Fixed carbon accumulates as above-ground and below-ground biomass and is incorporated into soils through litterfall from leaves and branches, from root growth and mortality, and as a result of root exudates, which are partly consumed in building microbial biomass.

Humification

Carbon typically represents ~57% by weight of organic matter incorporated into soils. This material is progressively broken down and decomposed by detritus-feeding soil animals, as well as by the action of microbes, primarily bacteria and fungi, with ~70% being mineralized to CO_2 within 1 year as a result of root and microbial respiration. The residues from plant biomass and the byproducts from microbial activity (microbial organic matter [MOM]) are a complex mixture of organic compounds including sugars, amino acids, proteins, cellulose, lignin, lipids, tannins, and fragments of these molecules. Humic material is the end product of the process of decomposition (Figure 13.3) and comprises the remnants of plant-, microbe-, and animal-derived compounds, broken down under partially oxidizing conditions into smaller organic molecules that are held together in supramolecular structures by weak (non-chemical) binding forces. Lignin is the least susceptible of biomass residues to decomposition and is therefore a major component of humus and of soil organic carbon stocks.

Complete decomposition of these organic fragments would eventually result in full mineralization to CO_2 and is prevented either by biochemical recalcitrance or physicochemical protection, or by low activity and abundance of decomposer organisms. Biochemical recalcitrance results from the blocking of

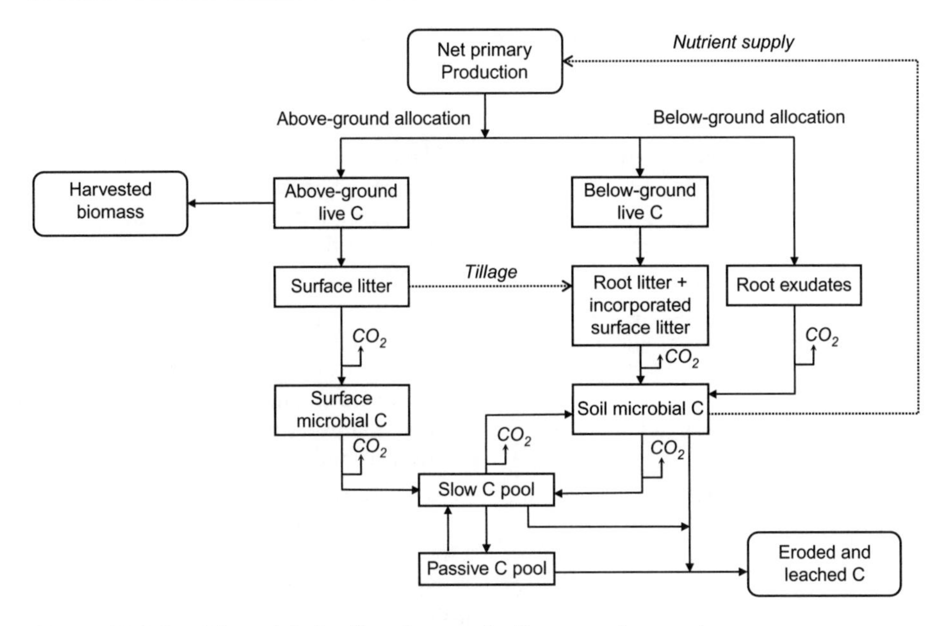

Figure 13.3 Partial model of soil carbon pools, fluxes, and controls

microbial catabolic pathways due to the lack of essential enzymes, while the physicochemical protection of SOM occurs as a result of chemical adsorption onto mineral surfaces, facilitated by the presence of polyvalent cations such as calcium, iron, and magnesium, or the complete envelopment of SOM particles within aggregated soil particles. The ability of soils to physically protect organic material is an important determinant of soil carbon retention and is influenced by a range of factors described in Table 13.2.

The resulting SOM provides a source of nutrients for plant growth, as well as helping to retain water in the soil, further contributing to productivity. Full mineralization of labile and nonlabile SOM will occur if it is exposed to fully oxidative conditions, promoting aerobic microbial activity, for example as a result of disturbance of the aggregated soil structure by tilling. The addition of readily available carbon compounds to soil microorganisms may also promote their decomposer activity on SOM (the so-called primer effect). Conversely, the stability and activity of the enzymes and oxidants needed for microbial decomposition of organic material are reduced under acidic conditions and low temperatures, such as those occurring in northern peatlands, resulting in reduced rates of decomposition.

Soil structure

The structure of soils is an important factor in determining the residence time of soil organic carbon (SOC), playing a role in protecting SOM from microbial access and in the balance between bacterial- and fungal-dominated pathways

Table 13.2 Factors influencing the physical protection of SOM

Factor	Description
Clay content and mineralogy	Clays promote the growth of microbial biomass by providing protection from predation and desiccation, by regulating the soil pH, and by absorbing contaminants detrimental to microbial growth.
Pore-size distribution in aggregate particles	Small pores (<10 μm) in soil micro- and macroaggregates provide refuge to bacteria and fungal hyphae from predation by larger predators such as protozoa and nematodes.
Soil microclimate, including wetting and drying cycles	Repeated wetting and drying leads to soil erosion with SOM export, and to breakdown of soil microaggregates, exposing SOM particles to further decomposition.
Soil aggregate dynamics	Soil disturbance, for example as a result of tillage, results in the breakdown of soil macroaggregates and increased soil aeration, leading to decomposition of SOM.

by which this material is cycled through the food web. Particulate matter in soils can be aggregated at a microaggregate level (aggregate particle diameter <0.25 mm) or at a larger macroaggregate level, as illustrated in Figure 13.4.

The hyphal networks of mycorrhizal fungi play an important role in stabilizing macroaggregates, and loss of these networks can result in a breakdown of macroaggregates and a loss of SOC through increased bacterial activity. As a result of the clay-binding process described earlier, microaggregate particles provide a degree of physical protection to SOM, which becomes less susceptible to microbial action when incorporated into these particles. Microaggregates are thus the essential repository of recalcitrant carbon in soils, alongside unprotected black (pyrogenic) carbon.

When perturbed by changes in land-use, land-management practices or environmental conditions, soil carbon stocks will tend towards a new quasi-equilibrium. Carbon stocks will tend to be increased by practices or conditions that increase primary production and below-ground allocation or result in an increase in the quantity of above-ground biomass or other organic matter returned to the soil. Similarly, practices or conditions that reduce the rate of decomposition in the soil or reduce soil loss through erosion will also lead to increased carbon stocks. Table 13.3 summarizes some of these factors.

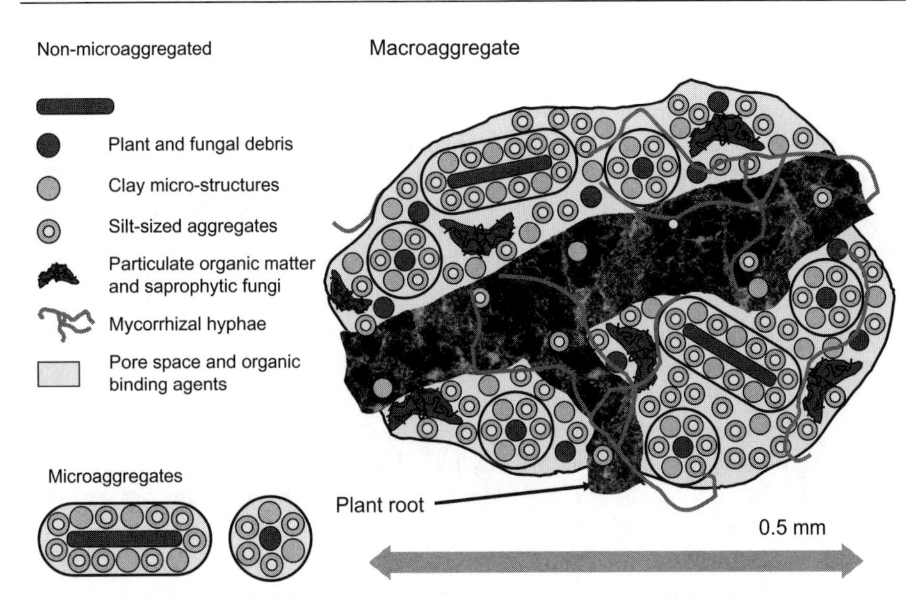

Figure 13.4 Macro- and microaggregate particles in soil (after Jastrow and Miller, 1988)

Table 13.3 Factors increasing soil carbon stocks and related practices

Factor	Related practices
Increased return of biomass and retention of below-ground inputs to the soil	Conservation tillage
Addition of exogenous organic matter	Composts and manures (grazing)
Addition of mineral nutrients	Fertilization
Reduced soil disturbance	Tillage, grazing, logging
Reduced soil aeration	Tillage

13.2.3 Modeling climate–ecosystem interactions

Most current global climate models that include interactions with the terrestrial carbon cycle rely on two simple feedback mechanisms:

- Increasing [CO_2] and rising temperature stimulate photosynthetic production, increasing the flux of CO_2 from the atmosphere into the terrestrial ecosystem and providing a negative feedback to rising [CO_2].
- Rising temperatures stimulate respiratory activity, resulting in increased respiratory flux of CO_2 from terrestrial ecosystems to the atmosphere and a positive feedback on mean temperatures.

While both of these mechanisms are still the subject of study, the net observed effect is currently a global uptake of carbon into terrestrial ecosystems, with rising [CO_2] and mean global temperatures, which may be gradually overtaken by the increasing respiratory release of CO_2, the pace of this reversal depending on the assumed parameters in the models. Figure 13.5 shows schematically the range of predictions of net carbon uptake from a number of such climate models under the IPCC SRES-A2 carbon emissions scenario in which global CO_2 emission from energy production rises to ~16 Gt-C per year in 2050, without significant CCS deployment.

The range of model predictions differ by about 1/3 of the emissions volume under this scenario up to 2030, and diverge more strongly thereafter. Clearly this difference represents a significant uncertainty for global climate change prediction, necessitating an improvement in the fundamental understanding of the processes and interactions involved.

Figure 13.3 showed a simple high-level model of soil carbon pools and flows at the next level of detail beyond the simple two-factor ([CO_2], T) approach described earlier. The model shows the pathway from net primary production (NPP) to the long-lived passive pool of soil carbon, and indicates a number of points at which respiratory activity contributes to the flux of CO_2 back to the atmosphere. For simplicity, only two process linkages are shown in the figure: tillage, incorporating surface litter into the subsurface litter pool, and the supply of nutrient from subsurface microbial activity into NPP.

This model provides a framework that allows the incorporation of more detailed models of specific pools and processes. For example, Figures 13.6 and 13.7 show further levels of detail of the soil carbon pools and the interactions between soil microbial carbon and the progressively longer-lived slow carbon

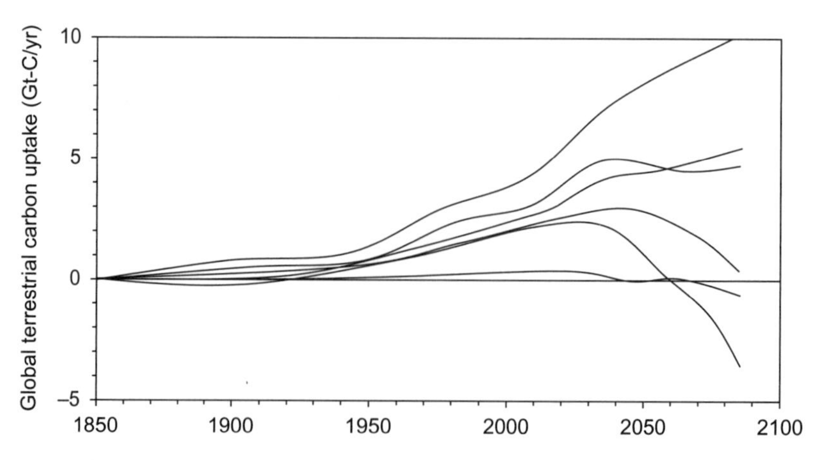

Figure 13.5 Range of predicted terrestrial carbon uptake from climate–ecosystem models

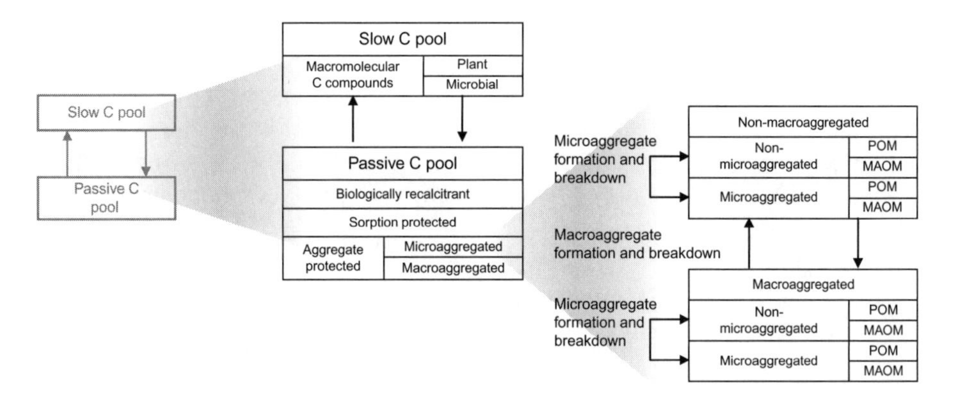

Figure 13.6 A simple conceptual model of soil carbon pools

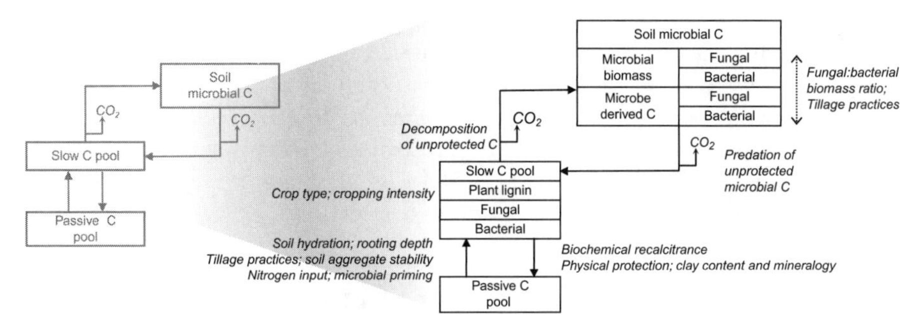

Figure 13.7 Second-level model of soil microbial, slow and passive carbon pools

and passive carbon pools. The figures indicate some of the factors that influence these pools and processes, which are further discussed later.

Many of the interactions occurring within soils are highly complex and in many cases counterintuitive, as is well illustrated by considering microbial priming—one of the factors indicated in Figure 13.7 as controlling the degree to which "passive" carbon may become available for further biological processing. Some soil carbon may be stabilized due to the absence of microbial activity, for example in deeper soils where a lack of readily available substrates limits microbial activity. A change to a deeper rooting pattern may be induced by a reduction in local precipitation or by increasing [CO_2], leading to higher NPP and increased below-ground allocation. Cultivating plants with higher below-ground carbon inputs and deeper roots may also cause increased below-ground allocation. Deeper rooting in turn will cause the addition of labile carbon in the form of roots and root exudates into deeper soils, stimulating microbial activity, which may then result in the decomposition of previously stable soil carbon. The addition of labile carbon to deep soils may therefore result in an overall drop in the soil carbon pool.

A realistic coupling of such a soil carbon model to climatic conditions will require many more factors than simply [CO_2] and mean temperature to adequately reflect the complex interactions. For example, a longer but still far from complete list of factors affecting soil carbon processes would include:

- Sensitivity of CO_2 fertilization effect to local factors such as water and nitrogen availability
- Impact of higher temperature on water availability and consequently on NPP for different soil and vegetation types
- Changes in the mean precipitation and in the timing and frequency of rainfall within that average
- Impact of climate extremes (temperature, rainfall, wind) on soil hydrological conditions, aggregation, erosion, and leaching of soil carbon

Models that incorporate the many complex processes occurring within the terrestrial ecosystems and their interaction with climate on multiple spatial and temporal scales will be needed to accurately assess the nature of climate–ecosystem coupling and the carbon storage potential of the options described in the following section. Long-term ecosystem-scale observations and experiments will also be required in order to improve understanding and validate the simulation models. Some of the areas of focus of research are covered in the discussion of specific options below, and in Section 13.5.

13.3 Terrestrial carbon storage options

Table 13.4 summarizes the range of approaches that may be taken with the aim of increasing carbon storage in terrestrial ecosystems, including products derived from them.

13.3.1 Agricultural carbon storage

Agricultural practices can make a significant contribution to carbon storage, at low cost, by increasing the soil carbon pool and reducing the emission of greenhouse gases (GHGs) resulting from soil biochemical processes (CO_2, N_2O, CH_4), as well as by producing biomass as an energy feedstock, offsetting fossil fuel use.

Agricultural carbon storage (ACS) refers to the management of vegetation and soils in agroecosystems to increase carbon stocks in the organic matter of these ecosystems or to lower the existing net rate of release of CO_2 or other GHGs to the atmosphere. The IPCC (2000) report on land use, land-use change, and forestry estimated that ~1 Gt-C per yr could be stored in the short term as a result of the regrowth of perennial vegetation and improvements in land-management practices in croplands, grasslands, and forests.

Table 13.4 Approaches to increasing soil carbon stocks

Terrestrial carbon storage strategy	Tactical approaches
Changing land use to ecosystems that sustain higher soil carbon stocks	Reclaiming degraded land to reestablish soil C pools
	Afforestation and reforestation; reducing forest degradation and deforestation
	Reinstatement of wetlands, particularly coastal systems
Increasing net primary production for any land use	Use of high-yield crops Perennial planting, undercrops Managed fertilization
Minimize and delay the return of carbon in plant products to the atmosphere	Extend lifetime of wood products Recycling of wood products Use of biomass in zero-emission power generation Carbonization of plant residues as stable biochar (see Glossary)
Increasing the fraction of NPP that finds its way into soil	Genetic manipulation of plants to increase below-ground carbon allocation and recalcitrant carbon production
	Modifying soil chemistry to maximize below-ground allocation
Manipulating processes in soils to maximize carbon retention in stable SOC	Use of soil amendments to maximize SOC physicochemical protection
	Manipulation of microbial genetics to maximize biological protection of SOC and production of biochemically recalcitrant microbial compounds

Increasing the amount of crop residue available to be converted into humus can move long-term soil carbon stocks to a higher equilibrium level over a decadal time scale. Intensified cropping and conservation tillage are examples of agricultural practices that increase this availability. Biologically altered organic inputs, such as manure, can result in greater sustainable increases in SOC due to the inclusion of material that decomposes less rapidly than unaltered crop residues. These practices—reduced tillage, crop residue retention, increased carbon inputs—also result in increased levels of microbial biomass and retention in soils of MOM, further increasing soil carbon content.

When an agricultural practice such as conservative tilling is put in place, the soil carbon content will increase until the adsorptive capacity of the soil is saturated and a new equilibrium is reached. The additional carbon inventory will remain in the soil while the practice is continued, but will be released within a few years and the previous equilibrium reestablished if the practice is discontinued.

Conservation tillage

Tilling breaks down the aggregated structure of the soil, both at the macroaggregate level by physical action and at the microaggregate level by subsequent dehydration of binding clays. This allows previously protected labile SOM to be exposed to microbial decay and to loss by wind and water erosion. The net effect of this loss of SOM may be mitigated if the eroded material is deposited in downstream sediments, particularly if these are protected from decomposition. The interlinked processes of erosion and deposition and their impact on net carbon stocks are poorly understood and are the subject of ongoing research.

Soil disturbance through tillage has a significant impact on the activity and make-up of the microbial community, with a reduction in tillage intensity resulting in increases in;

- Fungal to bacterial biomass ratio
- Mycorrhizal colonization
- Total soil carbon and nitrogen content
- Macroaggregated soil structure

Natural ecosystems have fungal-dominated microbial communities and are able to sustain the decomposition of organic material and recycling of nutrients without the supplementary inputs required in intensively managed agroecosystems. By avoiding the interruption of nutrient-recycling processes and maintaining the water-holding capacity of soils, practices that reduce soil disturbance therefore minimize the need for increased fertilization and irrigation to sustain soil productivity and soil carbon stocks.

Conservation tillage (CT) has the primary aim of reducing soil erosion and is defined by the IPCC as any tillage and planting method in which 30% or more of the crop residue remains on the soil after harvesting. These methods, which include no till, ridge till, minimum till, and mulch till as well as drill planting, minimize the disturbance of the soil and may allow SOC to increase to a higher equilibrium level. Crop residues either remain on the surface in the case of no tillage, or are partially incorporated into the soil surface layer for other CT practices. The change to CT can increase SOC stocks by 0.1–0.5 t-C per ha-yr depending on the soil type, and typically requires a 10- to 50-year period to reach a new equilibrium. Reversion to previous intensive tillage practice would result in a return to the previous SOC equilibrium levels, but over a shorter period.

Current research in tilling practice includes:

- Impact of tilling practices on microbial biomass and soil organic carbon and nitrogen under various crops
- Relative importance of the degree of soil disturbance, soil moisture content, and crop residue placement on fungal to bacterial biomass ratios under reduced or no tillage

Crop selection, rotation, and intensified cropping

The amount of crop residue available to be converted into humus can be increased by a number of practices that intensify cropping, such as:

- Elimination of fallow periods in seasonally cold or dry environments
- Selection of crop varieties for high yield of above-ground and below-ground biomass
- Selection, breeding, or genetic manipulation of crops for increased production of recalcitrant material, primarily lignin
- Application of fertilizers and other additives to increase crop biomass

Intensified cropping increases the amount of above-ground biomass available to contribute to SOC, and also increases below-ground input to SOC through root and mycorrhizal fungal biomass. The resulting increased availability of above-ground organic inputs will have maximum impact on SOC when applied in conjunction with conservation tillage.

Roots tend to decompose more slowly than leaf litter, possibly as a result of the higher proportion of complex organic compounds in roots that are more slowly transformed and therefore contribute to a longer SOC residence time. Selection, breeding, or genetic manipulation of crop varieties that result in higher below-ground NPP allocation can therefore contribute to increasing soil carbon stocks.

The establishment of perennial vegetation as a biomass crop, on previously tilled croplands, combines the advantages of intensified cropping and no tillage with the added advantage that fossil fuel use can be offset through biomass conversion. Assessment of the impact of such crops on carbon storage in soils is an ongoing research area and is important in order to fully evaluate the carbon impact of bioenergy crops.

Current research in crop selection, rotation, and intensified cropping includes:

- Understanding the genetic controls of above-ground versus below-ground biomass production
- Understanding the impact of cover cropping and other organic farming practices on soil microbial communities and MOM
- Understanding how the plant–microbe interactions in the rhizosphere influence SOC inventory and longevity
- Controlling plant–microbe interactions in the rhizosphere to maximize carbon storage, through plant breeding or transgenic plant development
- Understanding the influence of various crops, including bioenergy crops, on soil carbon stocks for varying climatic conditions and soil types

Managing soil biogeochemistry

The SOC pool is susceptible to the chemical conditions in the soil, and modification of soil chemistry can be used to influence the rate of accumulation of humic material, as well as the redistribution of organic carbon into deeper subsurface layers.

Sorption of organic molecules onto mineral surfaces protects these humic fragments from further mineralization and is aided by the presence in the soil of polyvalent cations. Addition of lime (CaO) or minerals containing iron and manganese oxides provides a source of these cations and can result in an increase in the sorption and protection of organic carbon. However, the shift in pH due to lime addition may also promote microbial decomposition of SOM.

Organic carbon remains susceptible to continued mineralization in the upper soil layers, particularly in the rhizosphere, where respiratory activity is high. Soil carbon can therefore be increased by moving organic carbon from the upper soil layers into deeper, less oxidative layers. Adsorbed organic carbon can be released into pore water by the addition of anions such as phosphate or sulfate that will compete for sorption sites on mineral surfaces. Under suitable hydrological conditions, vertical pore water flows can then carry dissolved organic carbon to deeper soil layers, where it can be adsorbed onto unsaturated mineral surfaces. As well as the primary aim of plant nutrition, fertilizer addition can thus be managed to achieve the additional objective of transporting carbon deeper into soils, increasing total soil carbon stocks.

Current research in soil biogeochemistry for carbon storage includes:

- Understanding the processes that control the accumulation, movement, distribution, and residence of carbon in the vertical soil profile (e.g., impact of soil mineralogy, soil chemistry, and plant carbon input quality on SOC storage and residence time)
- Understanding the spatial and temporal variability of these processes
- Development of methods to assess the storage potential of soils
- Understanding how soil aggregation, pore network properties, and aggregate stability affect SOC capacity and microbial activity
- Impact of climate change on soil respiration

Manipulation of microbial communities

Soil microbial communities, comprising predominantly fungal and bacterial species, facilitate some of the key biochemical processes that determine the fate of SOC as well as the emissions of GHGs such as N_2O and CH_4. These processes include decomposition of animal and microbial biomass and resulting supply of metabolites to soil organisms; mobilization of plant nutrients such as iron and phosphorus; nitrogen cycling, including fixation and denitrification; and metabolic removal of soil contaminants.

While the biochemistry of these processes is increasingly known, their sensitivity to microbial community composition and dynamics is less well understood. Due to their different physiologies, the two main microbial groups interact differently with soil properties and processes so that community composition and activity may have an impact on soil carbon cycling and storage.

Bacterial respiration mineralizes ~60% of the fixed organic carbon metabolized from decaying plant material, while fungi respire only 30–40%, which translates to a higher microbial growth efficiency (MGE), expressed as the weight fraction of metabolized carbon that is incorporated into microbial biomass and byproducts. A shift in the microbial community toward fungal dominance would therefore increase SOM by reducing the loss through respiration and increasing MOM. This shift toward a higher fungal to bacterial biomass ratio is one factor that could increase SOM under no-till cultivation, since tillage interrupts fungal growth by breaking up the network of fungal mycellia.

The stability of macroaggregates in soils is also enhanced when the soil microbial community is fungal-dominated. Since the incorporation of SOM into macroaggregates provides physical protection, fungal dominance reduces carbon mineralization by reducing bioavailability, leading to increased soil carbon accumulation.

A better understanding of these processes would open up the possibility of modification by direct or indirect means in order to maximize carbon storage and minimize GHG emissions. Direct methods could potentially include genetic engineering of microbial genes that control or modify specific functions, while indirect means would include altering soil chemistry and pH in order to enhance existing functions.

The complexity of microbial communities involved in biochemical soil processes provides the opportunity to use the population dynamics of these diverse communities as indicators of the varying status of soil—for example, as indicators of nutrient availability or SOC content. Research studies using DNA sequencing techniques have confirmed correlations between soil status indicators, such as SOC, and the relative abundance of different microbial species, which respond differently to changing soil conditions.

Current research areas in microbial processes in soils include:

- Understanding fungal and bacterial contributions to the decay pathways from above- and below-ground plant residues to stable humified SOC
- Identification of key microbial species or groups that maximize SOC
- Understanding how microbial activity can be shifted toward increased SOC longevity for varying soil types and chemistry, and for different plant residue types
- Understanding the relative importance of soil disturbance, soil moisture, and residue placement on microbial community structure and microbial biomass
- Understanding the relative influence of biochemical recalcitrance, physical protection in aggregates, and chemical protection by clay association in controlling the fate of fungal and bacterial MOM and their relative contribution to soil carbon stocks
- Understanding the role of soil animals in carbon decomposition and transport

13.3.2 Changes in land use

Strategies to increase the terrestrial organic carbon stocks through changes in land use typically involve the conversion of lands to uses that reduce soil disturbance and maximize the content and longevity of organic carbon.

Table 13.5 Relative contributions to carbon stocks of different land uses

Contribution to terrestrial carbon stocks	Land type	Commentary
High	Forest	High productivity, longevity, unfavorable decomposition conditions (boreal forests), and minimal soil disturbance
	Wetlands	No soil disturbance
	Grasslands	No soil disturbance
	Pasture	Reduced soil disturbance
	Croplands	Frequent soil disturbance
Low	Degraded land	Low existing carbon stock

Land uses may be ordered in terms of their effective contribution to terrestrial carbon stocks, as shown in Table 13.5.

The terrestrial carbon stock can therefore be increased by any change in land use that moves up the scale, for example the restoration of degraded land to any other use, the conversion of croplands to pasture or wetlands, or afforestation of any land type.

Wetland management and restoration

Wetlands, including estuaries, mangrove forests, marshes, peatlands, and northern tundra, cover 7% of the earth's surface and, aside from those in cold climates, are some of the most productive of terrestrial ecosystems, contributing an estimated 10% of primary production. These areas accumulate large stocks of below-ground organic carbon, and therefore present an opportunity to increase terrestrial carbon storage both by protecting and enhancing existing wetlands as well as by restoring those that have been degraded. Northern peatlands, for example, are estimated to have accumulated 300–450 Gt-C since the end of the last glacial period—equal to 50–70% of the carbon currently present in the atmosphere. The restoration of tidal salt marshes is estimated to achieve a SOC accumulation rate in the order of 2 t-C per ha-yr, similar to low-grade forest growth, while algal productivity in these ecosystems can add a similar amount. Depending on the type of plant material and the degree of inundation, carbon stored in these types of wetland can have a residence time of decades to millennia due to the very slow rate of decomposition under anoxic conditions and, in northern peatlands for example, also under low temperatures.

Draining of wetlands for uses such as agriculture, forestry, or urban and industrial development results in a loss of organic material due to an increase in aerobic microbial decomposition, with the consequent release of CO_2 to the atmosphere. Reestablishment of wetlands requires a restoration of previous hydrological flows by removing or plugging drainage systems or by artificially diverting water flows. Attempts to restore wetlands have shown that the loss of SOC is not easily reversed, with created wetland SOC being slow to accumulate, pointing to complexities in the overall ecosystem that are difficult to recreate. In freshwater environments the increased carbon stock will be offset by an increase in the emission of methane released through decomposition if the new water table is close to the surface. This is not an issue for coastal wetlands such as tidal salt marshes, since methane emissions are not significant in a saline environment.

The development and eventual deployment of approaches to increase terrestrial carbon storage follows a similar methodology to the RD&D process in other technology areas. Identified approaches are initially subject to site testing to understand the effects on carbon stocks and fluxes at a site scale. The evaluation of these trials also needs to consider other actual or potential environmental impacts in order to achieve full accounting for stocks and fluxes of carbon, as well as other GHGs. Mathematical models are developed and validated by field trial results, allowing sensitivity analyses to be performed over a wide range of site types and conditions.

Current RD&D efforts in wetland management and restoration are summarized in Table 13.6.

Forestry management, afforestation, and reforestation

The carbon stocks in forests can be enhanced by reducing logging and deforestation, for example by the protection particularly of old-growth forests, by the regeneration of secondary and degraded forests with standing biomass and SOC below their potential values, and by the application of specific silvicultural practices. Afforestation of croplands can increase carbon stocks by 1–10 t-C per ha-yr over periods of 50–100 years, while improvements in management practices in existing forests can contribute 1–2 t-C per ha-yr over a period of 10–20 years.

Forestry management practices

Carbon stocks in standing forest biomass and in forest soils can also be maximized through specific management practices, including the optimization of harvesting schedules, forest fertilization, and the application of low-impact logging methods.

The timing and quantity of timber extracted from commercial forests during thinning and harvesting, whether through selective- or clear-felling, are traditionally driven by the end product requirements, with the aim of maximizing commercial returns. If carbon storage is also included as an objective, the optimal harvesting strategy shifts toward selective harvesting with a minimum cutting diameter as well as longer rotation cycles, although the latter also increases the risk of carbon loss through major disturbances such as fires, windthrow, and insect outbreak.

Table 13.6 RD&D efforts in wetland management and restoration

RD&D area	Description
Quantifying existing carbon stocks and fluxes	Quantifying stocks and fluxes by wetland type to enable estimates of carbon storage potential
	Dynamics of carbon cycling in coastal wetlands and response to sea-level changes
	Full accounting for climate forcing of wetlands (CH_4, N_2O, albedo, water vapor)
Wetland management practices	Evaluation of current wetland management and restoration practices to assess their impact on carbon storage
	Identification and development of new wetland-restoration approaches to ensure that created wetlands mimic natural wetlands
	Impact of nutrients on primary production and decomposition rates in wetland systems
Wetland vulnerability	Assessing the degree to which existing wetland carbon stocks are vulnerable to human activity and climate change
	Identifying and developing wetland management practices and technologies to reduce loss of carbon from northern latitude wetlands
Wetland ecosystem process modeling	Construction and verification of integrated ecosystem–climate–hydrological models to gain quantitative insights into the relationships between wetland environmental factors and soil carbon storage potential
Monitoring and verification techniques	Methods to assess net carbon uptake and monitor methane and CO_2 fluxes, particularly if the degree of inundation varies due to changing hydrological fluxes or climatic conditions

Forest fertilization by the addition of nitrogen, typically as ammonium nitrate (NH_4NO_3), is practiced in many commercial forestry operations worldwide with the aim of increasing the biomass in stem and branches. Fertilization is most beneficial on low- to medium-grade sites, where nutrition is the limiting factor for growth. The incremental carbon storage rate depends on many factors including tree species, climate, and fertilization dosage and timing, with various investigations indicating carbon storage rates in the range 0.2–0.8 t-C per ha-yr can be achieved for periods of 10–20 years.

Studies have shown that nitrogen addition can also aid in retaining SOC by reducing microbial activity and the attendant respiration of CO_2 from the soil.

Clearly a tree will allocate NPP to root production and root exudates only if this is necessary to build the root networks and sustain the microbial activity needed to deliver nutrients and water to the plant. If nutrients are more readily available, for example as a result of fertilization, below-ground allocation will be reduced and with it the attendant root and microbial respiration. While higher nitrogen levels have been shown to suppress SOC mineralization, whether this aids in carbon storage will depend on the balance between the reduced decomposition of existing SOC and the reduced input from new root and microbial biomass.

Low-impact logging aims to minimize the disturbance of forest soil during logging while ensuring that the remaining trees, new trees, and other vegetation retain maximum potential for growth and carbon storage. Particularly in tropical rainforests, clear-felling results in substantial carbon emissions as a result of soil erosion and subsequent decomposition of SOC.

An evaluation of the net carbon storage resulting from forest management, including forest fertilization, must also account for the carbon fluxes resulting from other practices such as fire management. While fires have an important role in rejuvenating forest and grassland ecosystems, in semi-arid areas or under drought conditions fires pose a significant risk to carbon storage in standing forest biomass. Appropriate fire-management practices may include mechanical removal of undergrowth and controlled fires, depending on the specific conditions. The carbon flux resulting from fire-management practices must be accounted for when evaluating the net carbon storage in existing or reestablished forests.

Afforestation and reforestation

Afforestation of agricultural or other nonforested lands, as well as increasing the peripheral tree cover on crop or pasture lands or in urban areas, is currently considered to be the largest potential contributor to the increased storage of carbon in the terrestrial ecosystem.

Within wetland ecosystems, restoration of hardwood forests in bottomland (floodplains and marshes) is a storage approach that combines the high primary productivity of wetlands with the slow release of SOC under flooded conditions. Other practices, such as the planting of trees as riparian buffer zones along stream and the planting of cover crops below trees, can also sustain the SOC stocks by preventing soil erosion.

The longevity of the carbon stocks established through afforestation can be extended by increasing the demand for longer-lasting wood products, and also by maximizing the lifetime of wood products, for example through recycling.

13.4 Full GHG accounting for terrestrial storage

While the various land-use and -management approaches described above have the potential to enhance carbon storage in terrestrial ecosystems, a wide range of indirect effects need to be considered in arriving at a comprehensive assessment of the net impact on the fluxes of CO_2 and other GHGs. Agriculture is a major contributor to the flux of both nitrous oxide (N_2O) and CH_4 to the atmosphere, both of which have significantly higher global warming potential over a 100-year time scale than CO_2 (GWP_{100yr} = 23 and 296 respectively).

Microbial activity is responsible for the major loss of nitrogen from the soil, as N_2 and N_2O, and changes in land use and practices that affect microbial communities, whether directly or indirectly, will affect these emissions. Practices such as tillage, quantity and timing of application of nitrogen fertilizers, and drainage conditions will all have an impact on the rate of nitrogen loss from soils.

Methane is emitted from soils under highly reducing conditions such as occur in fresh-water wetlands and other flooded soils. Wetland restoration or irrigation practices that dramatically change the oxidation status of soils, such as periodic flooding, will therefore tend to increase CH_4 emissions. Coastal wetlands are an exception, as noted above, since CH_4 emissions are negligible from saline environments.

Forests also have indirect climate impacts as a result of their low albedo and high water vapor input to the atmosphere. Typically the albedo of forests is 30–70% of the albedo of grasslands, for coniferous and deciduous forests respectively, while the evaporative loss from a forest with a high leaf area index can exceed that from a water surface.

Other more remote GHG impacts from changes in land-use or -management practices can arise from fossil fuel consumption for the production, transport, and use of agricultural machinery and other inputs such as fertilizers, mineral additives, and biocides.

13.5 Current R&D focus in terrestrial storage

Carbon storage in terrestrial ecosystems is a relatively new R&D area, although a number of programs have been established in the last decade. The initial aim of these programs is to improve our understanding of the fundamental physical, chemical, and biological processes that control the accumulation and fate of carbon stocks, to develop measurement and manipulation techniques that allow those processes to be quantified, and to develop verifiable models of those processes to allow extrapolation of experimental results to other environmental conditions.

The development and demonstration of specific strategies to enhance carbon storage in terrestrial ecosystems awaits the outcome of these fundamental investigations, of which two examples are described in the following sections.

13.5.1 *Free air CO_2 enhancement experiments*

Free air CO_2 enrichment (FACE) experiments are an important class of mid-scale experiments that investigate the carbon cycle in a variety of terrestrial ecosystems, with emphasis on the impact of elevated [CO_2] and in some cases also elevated ozone concentration ([O_3]). A typical FACE facility consists of a number of plots within which the trace gas concentration is maintained at an elevated level during the daylight hours of photosynthetic activity.

The Aspen FACE, operated by the Brookhaven National Laboratory and the Michigan Technological University, is a typical forest ecosystem example. The layout of this facility, shown in Figure 13.8, consists of 12 circular experimental plots, each 30 m in diameter, within which [CO_2] and [O_3] can be controlled. The trace gases are introduced from a set of vertical vent pipes positioned on the circumference of each plot, with gas being vented only from those pipes upwind of the plot. The effect of elevated [CO_2] is usually investigated by introducing an excess [CO_2] of ~200 ppm or ~50% above ambient concentration. Sensors mounted within the canopy provide feedback to a computer control system that maintains the trace gas concentration at the desired level, and these systems are typically able to maintain the annual average [CO_2] within ±10% of the target level for >90% of the canopy volume within each experimental plot, although short-term excursions from the target level are larger.

Since the plots are unconfined, other aspects of the natural environment remain unchanged. The design allows assessment of the effects of these gases, either alone or in combination, on many attributes of the ecosystem, including above- and below-ground growth and soil carbon.

FACE experiments have been constructed in a range of ecosystems, from forests to desert scrub, and the key features of a number of these experiments are summarized in Tables 13.7 and 13.8. Forest FACE experiments to date have focused on young trees in temperate environments (including some species from boreal forests), and experiments in mature boreal forests or in any tropical forests are notably absent from the list. Despite the importance of understanding climate feedback effects in these ecosystems and their potential significance for terrestrial carbon storage, many of these experiments struggle for funding.

Figure 13.8 Aspen FACE experimental configuration (Courtesy; Michigan Technological University. Photo Credit David F. Karnosky.)

Table 13.7 Free air CO_2 enhancement (FACE) experiments in forest ecosystems

FACE facility	Organization	Location	Ecosystem	Species	Key parameters
Aspen FACE	Brookhaven National Laboratory, Michigan Technological University	Harshaw Experimental Forest, Rhinelander, Wisconsin, USA	Temperate deciduous forest	Aspen (*Populus tremuloides*) Birch (*Betula pendula*) Sugar Maple (*Acer Saccharum*)	9 FACE + 3 ambient plots $[CO_2] = 560$ ppm $[O_3] = 1.5 \times$ ambient
Bangor FACE	University of Wales	Henfaes Experimental Farm, Bangor, UK	Temperate deciduous forest	Birch (*Betula pendula*) Alder (*Alnus glutinosa*) Beech (*Fagus sylvatica*)	4 FACE + 4 ambient plots 8 m diameter rings $[CO_2] =$ ambient + 200 ppm
Duke FACE	Brookhaven National Laboratory	Duke Forest, Orange County, North Carolina, USA	Temperate pine forest	Loblolly pine (*Pinua taeda*)	4 FACE + 8 ambient plots 30 m diameter rings $[CO_2] =$ ambient + 200 ppm
Euro FACE	University of Tuscia	Tuscania, Viterbo, Italy	Temperate deciduous forest	White Poplar (*Populus alba*)	3 FACE + 3 ambient plots
				Black Poplar (*Populus nigra*)	22 m diameter rings
				Eastern Cottonwood (*Populus deltoides*), and hybrids.	$[CO_2] = 380, 200$ ppm
ORNL-FACE	Oak Ridge National Laboratory	Oak Ridge National Environmental Research Park, Oak Ridge, Tennessee, USA	Temperate deciduous forest	American Sweetgum (*Liquidambar styraciflua*)	2 FACE + 3 ambient plots 25 m diameter rings $[CO_2] = 565$ ppm

Table 13.8 Free air CO_2 enhancement (FACE) experiments in nonforest ecosystems

FACE facility	Organization	Location	Ecosystem	Species	Key parameters
AG FACE	Grains R&D Corporation and the University of Melbourne	First site; Horsham, Victoria, Australia Second site; Walpeup, Victoria, Australia	Agroecosystem	Wheat (*Triticum* spp.)	8 FACE + 8 ambient plots $[CO_2]$ = 350, 550 ppm Site 1, 12 m diameter rings; site 2, 4 m diameter rings
Arizona FACE	University of Arizona	Maricopa Agriculture Centre, Maricopa, Arizona, USA	Agroecosystem	Cotton (*Gossypium hirsutum*) Wheat (*Triticum* spp.) Sorghum (*Sorghum* spp.)	4 FACE + 4 ambient plots $[CO_2]$ = ambient + 200 ppm
BioCON FACE	University of Minnesota	Cedar Creek Ecoscience Reserve, Minneapolis, Minnesota, USA	Grassland	16 species in 4 groups; warm-season grasses (C4), cool-season grasses (C3), forbs, and nitrogen-fixing legumes	3 FACE + 3 ambient plots

(*Continued*)

Table 13.8 (Continued)

FACE facility	Organization	Location	Ecosystem	Species	Key parameters
					20 m diameter rings, each with ~60 2 m × 2 m plots, $[CO_2]$ = 368, 56 ppm
Nevada Desert FACE	Brookhaven National Laboratory, University of Nevada	Mojave Desert, Mercury, Nevada, USA	Desert scrub	Creosote Bush (*Larrea tridentate*), Bur-sage (*Ambrosia dumosa*)	3 FACE + 6 ambient plots 23 m diameter rings $[CO_2]$ = ambient + 50%
Oz FACE	CSIRO and James Cook University	Yabulu Refinery, Townsville, Queensland, Australia	Coastal tropical savannah	Kangaroo Grass (*Themeda triandra*), Eucalyptus, and Acacia	4 FACE + 2 ambient plots 15 m diameter rings $[CO_2]$ = 370, 460, 550 ppm
Soy FACE	University of Illinois	Urbana-Champaign, Illinois, USA	Agroecosystem	Soy (*Glycine max*): 16 rings under elevated $[CO_2]$ and $[O_3]$; Maize (*Zea mays*): 8 rings under elevated $[CO_2]$	16 FACE + 8 ambient plots 15 m diameter rings $[CO_2]$ = 550 ppm $[O_3]$ = 1.2 × ambient

These experiments allow a wide range of questions relating to carbon storage in terrestrial ecosystems to be investigated in a controlled environment. Some of these key research topics are as follows:

- Evaluating the sinks and sources of CO_2 in terrestrial ecosystems to determine the limits of natural carbon storage potential
- Understanding the processes that control CO_2 fluxes between the atmosphere and terrestrial ecosystems from molecular to landscape scales
- Developing quantitative methods to measure CO_2 fluxes
- Understanding the biogeochemical processes, conditions, and interactions (on cellular, individual plant, community, site, and landscape scales) that control the rate of storage and the longevity of SOC (e.g., quantifying the impact of below-ground allocation on carbon storage; genetic and environmental controls on NPP allocation to root production; processes that affect the longevity of NPP allocated to below-ground biomass)
- Understanding the influence of climatic and other feedback mechanisms, including increasing [CO_2], on biogeochemical cycles, microbial communities, etc.
- Understanding the extent to which an experimental [CO_2] step-change can be considered representative of the gradual atmospheric increase
- Designing technical and management approaches that have the potential to enhance processes that contribute to, and inhibit those that oppose, carbon storage
- Developing conceptual and mathematical models of these exchange processes to allow extrapolation over longer time scales and differing environmental conditions
- Development of techniques for in-situ soil status measurements, including the use of microbial population responses to changes induced by land-management practices or environmental conditions
- Assessment of the carbon storage consequences of land-management strategies through site measurements and modeling (forest harvesting strategies, pastureland vegetation type, fertilization, and grazing)
- Transferring experimental results to natural landscapes, which exhibit a wide variety of age classes, ecosystem compositions, soil types, nutrient availabilities, and disturbance regimes

13.5.2 CSiTE terrestrial sequestration R&D program

The U.S. Department of Energy-sponsored Consortium for Research on Enhancing Carbon Sequestration in Terrestrial Ecosystems (CSiTE) was established in 1999 as a multi-institutional research effort involving the Argonne, Oak Ridge, and Pacific Northwest National Laboratories and a number of partnering universities. CSiTE aims to addresses a number of research questions that are critical to protect stored carbon and enhance carbon accrual into terrestrial environments, namely:

- What are the physical, chemical, and biological processes controlling the input, distribution, and longevity of carbon in soils?
- How can these processes be exploited to enhance terrestrial carbon uptake?
- How do terrestrial carbon-storage strategies relate to and influence other approaches to climate change mitigation?

- What is the long-term potential for terrestrial carbon storage to materially affect climate change?

In the period up to 2006, the CSiTE collaboration undertook laboratory, field, and modeling studies in cropland, forest, and grassland ecosystems, which resulted in progress in a number of research areas, including:

- Understanding of factors controlling the mechanisms and rates of accumulation of soil organic matter
- Development of new methods to investigate the role of microbial communities in soil carbon dynamics
- Identification of new manipulation concepts for enhancing soil carbon storage
- Development of improved modeling tools for soil processes

In 2007 a 5-year research program was started, focusing on the bioenergy crop switchgrass (*Panicum virgatum*), and with the overall objective of investigating sustainable biofuel production with accompanying enhancement of soil carbon stocks. The key objectives of the five experimental themes that make up this program are summarized in Table 13.9.

Mechanistic models of soil carbon processes will also be further developed, building on the outcomes of the five experimental themes, in order to improve forecasting of soil carbon dynamics and enable an evaluation of the tradeoffs and synergies between bioenergy crop production and enhancing soil carbon storage.

Table 13.9 CSiTE R&D themes to enhance carbon storage in terrestrial ecosystems

R&D theme	Key experimental objectives
Soil carbon inputs	Investigation of the differences and intra-annual variation in root production, mortality and decomposition, and root and microbial respiration for different switchgrass varieties and fertilization treatments
Soil structural controls	Improved understanding of the ways in which soil structure controls the transformation of organic carbon inputs, through both biotic and abiotic humification processes, and subsequent stabilization as SOM
Microbial community function and dynamics	Understand the influence of switchgrass varieties and crop-management practices on soil microbial community function and structure, and the impact of changes in these communities on soil carbon accrual and storage

(*Continued*)

Table 13.9 (Continued)

R&D theme	Key experimental objectives
Humification chemistry	Understanding humification chemistry, including identification and optimization of the chemical factors that can be manipulated to enhance storage, and development of measurement techniques to rapidly assess whether carbon stocks are increasing or declining in a soil
Carbon transport within soils	Evaluation of the processes that control the transport of solid- and solution-phase carbon through the soil profile and its accumulation in deep soils, including the influence of soil type, carbon inputs, and chemical effects resulting from fertilizers and additives to enhance soil surface humification

Finally an integrated evaluation is planned, drawing on these experimental and modeling results to estimate the potential for enhancing soil carbon storage under bioenergy crops across the full range of soils, climatic conditions crops, and management practices at a national scale. This will allow an assessment of the economic competitiveness of dedicated bioenergy crops integrated with enhanced soil carbon storage as a GHG mitigation strategy. Final results from the program are expected in 2011.

13.6 References and Resources

13.6.1 Key References

The following key references, arranged in chronological order, provide a starting point for further study on storage in terrestrial ecosystems.

Benemann, J. R. (1993). Utilization of carbon dioxide from fossil fuel burning power plant with biological systems, *Energy Conversion and Management*, **34** (999–1004).

Jastrow J. D. and R. M. Miller. (1998). Soil aggregate stabilization and carbon sequestration: Feedbacks through organomineral associations. In Lal R., J. M. Kimble, R. F. Follett, and B. A. Stewart (eds.). *Soil Processes and the Carbon Cycle*, CRC Press, Boca Raton, FL.

Rosenberg, N. J., R. C. Izaurralde, and E. L. Malone (eds.). (1999). *Carbon Sequestration in Soils: Science, Monitoring and Beyond*, Battelle Press, Columbus, OH.

IPCC. (2000). *Land Use, Land-Use Change and Forestry*, Cambridge University Press, Cambridge, UK. (see also Web Resources below).

Kimble, J. M., R. Lal, and R. F. Follett (eds.). (2002). *Agricultural Practices and Policies for Carbon Sequestration in Soil*, CRC Press, Boca Raton, FL.

Post, W. M., *et al.* (2004). Enhancement of carbon sequestration in US soils, *BioScience*, **54** (895–908).
Lorenz, K. and R. Lal. (2006). Subsoil organic carbon pool, *Encyclopedia of Soil Science*, Taylor & Francis, Boca Raton, FL.
CSiTE, DoE Consortium for Research on Enhancing Carbon Sequestration in Terrestrial Ecosystems (2006). *Five Year Science Plan 2007–2011*. Available at http://csite.esd.ornl.gov/presentations/CSITE_Master_Sep_06_FINAL.pdf.
Heimann, M. and M. Reichstein. (2008). Terrestrial ecosystem carbon dynamics and climate feedbacks, *Nature*, **45** (289–292).

13.6.2 Institutions and Organizations

The institutions that have been most active in the research and development work on carbon storage in terrestrial ecosystems are listed below, with relevant web links.

CSiTE (Carbon Sequestration in Terrestrial Ecosystems): http://csite.esd.ornl.gov/
Free air CO_2 enrichment (FACE) experiments:

Arizona FACE: http://130.199.4.11/FACE/Locations/Arizona.htm
Aspen FACE: http://aspenface.mtu.edu/resquest.htm
Bangor FACE: www.bangorface.org.uk/
BioCON Face: www.biocon.umn.edu/index.html
BNL FACE research: www.bnl.gov/face/faceProgram.asp
Brookhaven (Duke Forest) FACE: http://face.ornl.gov/index.html
Euro FACE: www.unitus.it/euroface/
Mohave Desert FACE: www.unlv.edu/Climate_Change_Research/NDFF
Oak Ridge FACE: http://face.ornl.gov/index.html
Oz FACE: www.cse.csiro.au/research/ras/ozface/
Soy FACE: http://soyface.illinois.edu/index.htm

International Centre for Research in Agroforestry (ICRAF): www.worldagroforestry.org
U.S. Department of Energy Terrestrial sequestration research: www.fe.doe.gov/programs/sequestration/terrestrial
U.S. Geological Survey, National Wetlands Research Center: www.nwrc.usgs.gov

13.6.3 Web Resources

Eliasch Review (an independent report to the UK government on financing global forests): www.occ.gov.uk/activities/eliasch.htm
IPCC; Special Report on Land Use, Land-Use Change and Forestry: www.grida.no/publications/other/ipcc_sr/?src=/climate/ipcc/land_use/241.htm
Kansas State University Soil Carbon Center (information portal on soil carbon storage): http://soilcarboncenter.k-state.edu
Ohio State University, Carbon Management and Sequestration Center: http://senr.osu.edu/cmasc
U.S. Environmental Protection Agency (carbon storage in agriculture and forestry): www.epa.gov/sequestration/index.html
U.S. Department of Agriculture, Natural Resources Conservation Service (soil taxonomy): http://soils.usda.gov/technical/classification/taxonomy

14 Other sequestration and use options

While CO_2 is used in a wide range of industrial processes, from the carbonation of soft drinks to the production of fertilizers, very few applications result in a reduction of CO_2 emissions since products such as urea and methanol have a very short lifetime, typically in the order of 1 year, before the CO_2 is released to the atmosphere. Of those applications that do reduce emissions, few have the potential to grow to a scale that would materially affect global emissions.

Two industrial applications are considered in the following section—the production of precipitated calcium carbonate by the carbonation of various alkaline wastes and the enhanced use of CO_2 in the cement industry—that have the potential for growth to a material scale and also result in reduced CO_2 emissions to the atmosphere.

The use of CO_2 emitted from large stationary sources as a feedstock for algal biofuel production also has the potential to grow to a significant global scale. Although this is not a CCS option, since combustion of the biofuel will release the CO_2 into the atmosphere (unless captured), the substitution of fossil fuels with biofuels produced in this way could reduce the emission of fossil fuel CO_2 to the atmosphere in the short term, although any actual effect on long-term fossil fuel usage is not yet clear. Biofuel production from microalgal biomass generated using CO_2 captured from large stationary sources is discussed in Section 14.2.

14.1 Enhanced industrial usage

14.1.1 Precipitated calcium carbonate production

As described in Chapter 10, alkaline wastes from a number of industrial processes, such as ash from coal combustion and municipal waste incinerator (MWI) or slag from steelmaking, are potential feedstocks for mineral carbonation. The potential for generating high-value products such as precipitated calcium carbonate (PCC) from these wastes, while also capturing CO_2, is an additional economic incentive. A number of options for CO_2 capture and storage projects that produce PCC have been proposed or are under active commercial development, including:

- Mineral carbonation of alkaline wastes by direct air or flue gas capture
- Calcium and magnesium carbonate precipitation from seawater by flue gas capture

Doi:10.1016/B978-1-85617-636-1.00014-6.

Whether these uses can be considered to constitute storage of CO_2 will depend on the longevity of the end product and on whether CO_2 is also captured from the eventual recycling of the product.

Steel mill with mineral carbonation of slag by direct air capture

As a precursor to the fully integrated steel mill option described in Section 10.3.1, the use of steel slag (or other alkaline industrial wastes) in a direct air capture carbonation scheme is a possible early-stage demonstrator project that is being pursued by several companies.

This simple, low-tech option, illustrated in Figure 14.1, consists of a number of large basins (each ~10,000 m^2) into which the prepared waste material is loaded. Water is sprayed onto the basin as a fine mist, absorbing CO_2 as the drops fall through the air. The solution dissolves calcium hydroxide ($Ca(OH)_2$) as it trickles through the bed and is recycled, dissolving more CO_2 and $Ca(OH)_2$, and precipitating $CaCO_3$ on each pass through the system.

Water consumption, as a result of evaporation and droplet loss through entrainment in the wind, would be a significant proportion of the overall operating cost of such a system. Stolaroff *et al.* have estimated a sequestration cost of \$8 per t-CO_2 sequestered for a scheme that would sequester 32 kt-CO_2 per year in 140 kt of steel slag or 680 kt of concrete waste. (See Key References.)

Other PCC production options

Similar integration opportunities exist in other industries, such as the paper industry, where the capture of CO_2 emissions from a paper plant could be achieved producing an otherwise energy-intensive feedstock (PCC) while also consuming waste from other industries.

Global consumption of PCC as a feedstock for the paper, pharmaceutical, and plastics industries is a relatively modest 10 Mt per year but this could be expanded by one or two orders of magnitude by increasing the use of PCC in the ~2 Gt per year global cement industry.

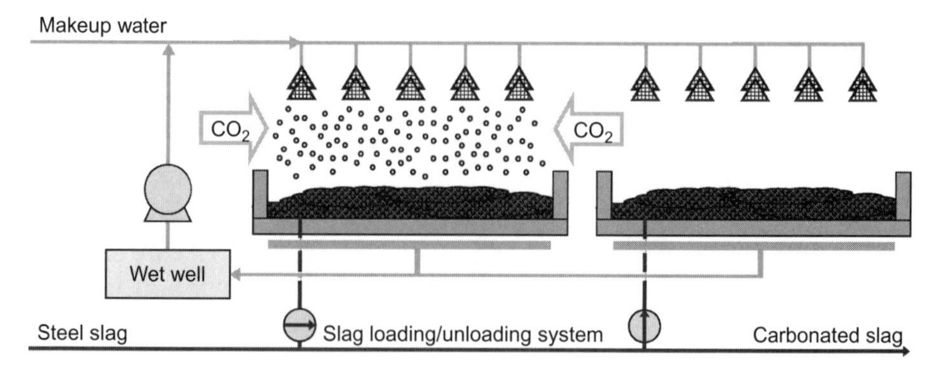

Figure 14.1 Direct air capture of CO_2 by steel slag carbonation

Capture of CO_2 from power plant flue gas by absorption into seawater and precipitation of calcium and magnesium carbonates for use as cement additives has also been proposed. Seawater contains 0.01 mol-Ca per kg and 0.05 mol-Mg per kg, and precipitation of the carbonates can therefore capture 1 t-CO_2 in 370 t-seawater, producing 2.1 tonnes of carbonates, comprising 81% $MgCO_3$ and 19% $CaCO_3$. Precipitation requires a pH of ~10, so that addition of an alkali such as NaOH is necessary for a seawater feed with a pH of ~8.

Calera Corp. opened a demonstration site in August 2008, adjacent to the Dynegy Inc. power plant at Moss Landing, Monterey County, CA. The pilot initially uses a synthetic flue gas to produce up to 10 t per day of carbonates and, with a 0.2 Mt per day seawater supply system on site, could eventually capture ~600 t-CO_2 per day from a flue gas slip stream from the adjacent power plant.

14.1.2 Enhanced use in the cement industry

In addition to the opportunities for capture at cement plants that have been discussed in previous chapters, an interesting and potentially material storage opportunity also exists where cement finds its end use in the curing of concrete products. The use of MgO-based rather than CaO-based cements also offers the opportunity for reduced emissions on production and additional capture during product curing.

Accelerated CO_2 curing of concrete products

When concrete is produced by mixing cement, water, and sand or other aggregate material, the hardening process occurs as a result of the hydration of the silicate and aluminate compounds in the cement, such as allite (Ca_3OSiO_4) and bellite (Ca_2SiO_4), according to reactions:

$$2Ca_3OSiO_4 + 6H_2O \rightarrow 3CaO.2SiO_2.3H_2O + 3Ca(OH)_2 \tag{14.1}$$

$$2Ca_2SiO_4 + 4H_2O \rightarrow 3CaO.2SiO_2.3H_2O + Ca(OH)_2 \tag{14.2}$$

These hydration processes initially form an open microcrystalline structure, which is progressively filled in and strengthened by further hydration products as the process continues over a time scale of hours to weeks. Hydration will continue as long as water is present, and poured concrete is typically cured under a layer of water or covered by an impermeable membrane for 1–2 weeks to ensure maximum strength is achieved.

Precast concrete products can also be cured for 12–24 hours in a steam kiln, at carefully controlled temperatures in the range 55–75°C, in order to accelerate early strength gain, although steam production adds an additional energy and emissions cost to the production process. Curing in a CO_2-rich atmosphere for ~1 hr at 20°C has been demonstrated to be a viable and faster alternative to steam curing, which eliminates this energy cost and offers a route for the cement industry to significantly reduce its net emissions.

In the curing process, CO_2 reacts with the cement hydration products, principally calcium hydroxide ($Ca(OH)_2$), to produce calcium carbonate. This carbonation reaction, which causes concrete to absorb CO_2 from the atmosphere, has traditionally been considered to be undesirable. Although limited carbonation increases strength and makes concrete less permeable and more resistant to shrinkage and cracking, excessive carbonation reduces the pH of water in the cement pores, potentially leading to corrosion of steel reinforcing bars. This is not a concern for many precast concrete products, such as concrete masonry blocks.

In principle, 1 t-cement is capable of absorbing 1 t-CO_2 during the hydration process if all hydration products are carbonated, while 0.5 t-CO_2 per t-cement is considered a more reasonable uptake limit. In practice, uptake is limited by the formation of a carbonated crust in the first 2–10 mm of the product, in which rapid carbonate infilling of pores reduces permeability by three to five orders of magnitude, preventing CO_2 ingress deeper into the product. RD&D work in the concrete industry is ongoing, focusing on increasing CO_2 uptake and penetration into the product.

Magnesium silicate cement

Compared to conventional cement produced using a limestone feedstock, the use of magnesium oxide (MgO) as the intermediate cement product, based on a magnesium silicate (talc) feedstock, results in a significant reduction in the CO_2 footprint. Emissions due to production are reduced from ~1.0 t-CO_2 to 0.2–0.4 t-CO_2 per t-cement produced, while absorption during hydration and curing is increased from ~0.3 t-CO_2 to 1.0 t-CO_2 per t-cement, giving an overall negative emission of ~0.6 t-CO_2 per t-cement produced and used.

When cured in a high-CO_2 atmosphere, the strength of pure MgO-based concrete blocks can exceed that of equivalent Portland cement (PC) blocks. Alternatively, the use of MgO as an additive in PC products increases the permeability of the concrete product, leading to increased CO_2 uptake during curing.

Commercial trials of MgO cements are being progressed by a number of companies spun off from recent RD&D projects.

14.2 Algal biofuel production

Having started the discussion of capture technologies at the point of combustion of fossil fuels, this final section on enhanced industrial usage considers the options to regenerate fuels from CO_2 emissions.

Here the focus is on microalgal biomass generation with capture of CO_2 from stationary sources such as power plants, cement plants, and municipal

Table 14.1 Advantages of aqueous microalgae cultivation for biomass production

System feature	Aqueous cultivation advantage
Land usage	With efficient access to water, CO_2, and nutrients, microalgae can lead to higher productivity per unit of land usage compared to terrestrial crops
Process simplicity	Simpler system with fewer process variables than terrestrial agroecosystem crops, making process engineering for higher yield simpler and more generic
Continuous production and harvesting	Avoids the unproductive period during which new terrestrial crops are becoming established, maximizing photosynthetic production period
Nutrient control	Enables continuous optimization of levels of CO_2 and nutrients (nitrogen and phosphorus)
Specific productivity	Photosynthetic NPP is not "wasted" in constructing roots, stems, etc., which have low or zero yield of end product (e.g., compared to oilseed crops)
Land and water quality	Aqueous cultivation of algae can use poor-quality land and water that is unsuitable for conventional crop production

waste plants. The aqueous cultivation of microalgae has a number of advantages that can lead to higher productivity per unit of land usage when compared to cultivation of biomass crops in terrestrial agroecosystems, as summarized in Table 14.1.

These advantages are balanced by the cost of systems to supply CO_2 and for continuous cultivation and harvesting, which are a major cost factor due to low biomass intensity resulting from the low concentration and microscopic nature of algal cells.

14.2.1 Algal biomass production systems

Algal cultivation in shallow open constructed ponds has been considered as a CO_2 capture method for power plant flue gases, and such open systems are typically able to convert 1–2% of total incident solar energy into stored chemical energy under the high intensity levels typical of full sunlight. This compares to a typical 0.1–0.2% of sunlight converted into biomass for terrestrial crops grown under similar outdoor conditions, although some crops, such as irrigated sugar

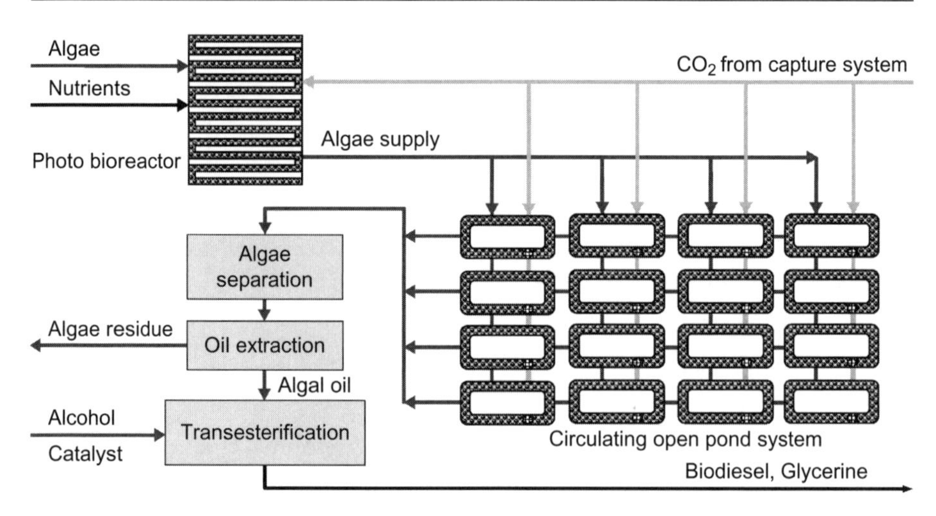

Figure 14.2 Algae farm using bioreactor algae supply and open circulating ponds

cane, can achieve similar productivity to algae cultures. To capture the CO_2 from a given power plant, the algae cultivation ponds would therefore have to be large enough to capture solar energy equal to at least 20 and possibly up to 100 times the output of the power plant.

Algae farms based on open ponds, as illustrated schematically in Figure 14.2, are used commercially to produce algal strains such as spirulina for use as "nutraceuticals." Systems of this type are operated by algae producers in the United States, including Earthrise Farms, CA, and Cyanotech Corp, HI, and in several other countries, principally China and India.

For use in a large-scale capture project, individual ponds would range up to 10 hectare (ha) in area, compared to typically 0.5 ha at present, with a total area of >10,000 ha being required to capture a maximum 30% of the CO_2 released from a 500 MW coal-fired power plant.

This type of open pond system was demonstrated in the late 1980s on a 0.1 ha scale at the U.S. Department of Energy site in Roswell, NM, achieving a 3.5% visible light conversion efficiency (roughly half of this level based on total solar input) and producing an extrapolated 70 t-biomass per ha-yr during the summer months.

A further example of a microalgal capture project being proposed by Debacsa, a subsidiary of the Spanish group Aurantia, will treat flue gas from a cement plant in Jerez, Spain. If this pilot phase performs as expected, a demonstration-scale project employing closed photobioreactors will start in 2012 and will use microalgal-derived biofuel to power a 30 MW cogeneration plant.

The limitations of open pond systems, and photobioreactors described in the following section, include:

- Poor light conversion efficiency
- Incomplete capture due to the inability to sustain a high [CO_2]

- Contamination of the culture by unwanted algal species and algal predators
- Heat losses or overheating, requiring heating or cooling and/or limiting site applicability
- High cost of harvesting and processing

Contamination may be particularly problematic in open systems if an unwanted local species outcompetes a species that has been specifically selected for some characteristic such as high lipid production. However, contamination is also an issue in closed systems, which may be more difficult to clean once contaminated.

Light conversion efficiency and saturation

Although algae are able to photosynthesize under very low light levels, their ability to convert incident light saturates at a light intensity roughly equivalent to a bright overcast day (about one tenth of full direct sunlight). At higher intensities overall light-conversion efficiency is reduced and incident light above this saturation intensity (I_S) is wasted as heat. For the roughly 10-fold increase in light intensity from I_S up to a full sunlit day, only a ~2- to 3-fold increase in photosynthetic conversion will occur in a static aqueous culture, so that 70–80% of solar energy is wasted as a result of saturation.

Incident light above the saturation intensity can still be efficiently used if the duration of illumination is very short (<1 millisecond) and the dark period is about five times longer. Although of scientific interest, such short "flashing light" effects are not yet of any practical significance in algal mass cultures.

Several solutions that have been proposed and tested to overcome this saturation limitation and increase light conversion efficiency are summarized in Table 14.2.

One aspect of the genetic engineering of algal strains is the possibility of reducing the concentration in microalgae of so-called antenna pigments, often primarily chlorophyll, to improve the utilization of photons captured for photosynthesis. This would reduce the amount of nonproductive photon capture that occurs at the surface while increasing irradiance deeper in the culture, thus increasing overall productivity.

Development and demonstration of closed algae production systems

A number of companies are developing closed photobioreactors, typically using one or more of these approaches to increase light conversion efficiency beyond that achievable in open ponds. Temperature, oxygen, and nutrient levels can also be more easily and continuously controlled in closed reactor systems, further improving productivity. The design details of these reactors are not yet in the public domain, although the general concepts have been known for over 50 years: the use of vertical orientations and light-concentrating systems with optical fibers, as well as flashing light effects achieved through very thin, high-density reactors.

Table 14.2 Advantages of aqueous microalgae cultivation for biomass production

Approach	Description
Turbulent mixing	Increasing the turbulence of the culture under illumination results in increased light-conversion efficiency since underilluminated culture is frequently brought up to the saturation intensity level
Light redistribution	Use of optical systems to conduct high-intensity incident sunlight and distribute more uniformly throughout a deep culture
Reactor geometry	Vertical reactor geometry to increase the surface to volume ratio and achieve more uniform illumination close to the saturation intensity
Selection and genetic engineering of algal species	Variations in light sensitivity of different species of algae provide the opportunity to optimize light conversion efficiency through species selection and engineering (e.g., reducing the "antenna" pigment)

One example is GreenFuel's 3D Matrix System (3DMS), which increases area to volume ratio in excess of 500-fold compared to open ponds, and is reported to have achieved productivity of 1.0–1.7 t-biomass per ha-d (ash-free, dry biomass weight) in field trials capturing flue gas CO_2 from power plants, including a trial at the APS Redhawk 1.06 GW gas-fired combined-cycle plant near Phoenix, AZ.

The photobioreactor approach typically involves bubbling CO_2 through the reactor, with carryover gas containing any excess CO_2 that is not absorbed by the culture being emitted to the atmosphere. The flue gas can also be bubbled through the culture liquid external to the reactor, in a separate "bioscrubber." CO_2 absorption and capture has to be integrated with O_2 desorption to avoid O_2 toxicity in the cultures, a major problem and limiting factor in any closed photobioreactor.

A two-stage algae production system that combines closed photobioreactors with open ponds has been developed by HR BioPetroleum in Hawaii. The closed reactor is used to produce a pure feed culture, which is then transferred into the open ponds for batch cultivation (see Huntley and Redalje, 2007, in the Key References). This approach achieves the main advantage of the closed photobioreactor in eliminating contamination, since residence time in the open ponds is reduced, while also minimizing cost through use of the open pond system.

A small pilot system, consisting of several closed reactors and open ponds with a capacity of 600 m^3, has been in operation at Kona in Hawaii since 2004 and is being expanded to 2.5 hectares in a joint venture with Royal Dutch

Shell. If this expanded pilot is successful, a 1,000 hectare demonstration-scale facility is planned to come into operation in 2010, with future expansion envisaged to a commercial-scale 20,000 hectare plant that would produce 1.2 Mt per year of biofuel.

14.2.2 Fuel production from algal biomass

Although algal biomass can be used without further processing as a feedstock for fish farming (aquaculture) or as a solid fuel for power generation, the main organic components can also be processed into high-value products: proteins can be used as animal feed, carbohydrates can be fermented to produce bioethanol, and lipids (oils and fats produced for energy storage) can be used for biodiesel production.

Algae are potentially well suited as a feedstock for biodiesel production, since some strains can produce lipid levels up to 60% of the dry biomass weight when certain nutrients are limited, although it has yet to be demonstrated that algae with a high lipid content (>30% dry weight) can also be productive in mass cultures.

Lipids can be extracted from algal biomass by a number of processes, which may include drying or the use of solvents, depending on the algal species, strain, and growing conditions. Two potential processes are as follows:

- Cell disruption followed by emulsification with recycled oil and centrifuging
- Cold press extraction of 75–80% of the oil, followed by solvent extraction using hexane (C_6H_{14}) or diethyl ether ($C_2H_5OC_2H_5$), the solvents then being separated and recovered by distillation. Extraction can also be achieved using a supercritical fluid such as CO_2 or methanol.

Extracted lipids are mostly in the form of triglycerides, in which a long-chain fatty acid is attached to each of the three hydroxyl (OH) groups of a glycerol molecule ($C_3H_5(OH)_3$). Conversion to biofuels (diesel or jet fuel) involves the process of transesterification, shown in Figure 14.3, in which the triglyceride is catalytically reacted with a simple alcohol, such as methanol or ethanol, transforming the triglyceride into three methyl or ethyl esters of the fatty acids and releasing the glycerol molecule.

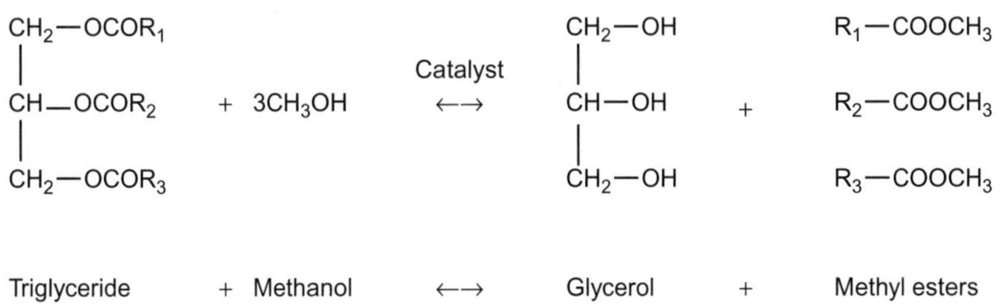

$$\begin{array}{lclclcl}
CH_2{-}OCOR_1 & & & & CH_2{-}OH & & R_1{-}COOCH_3 \\
| & & & \text{Catalyst} & | & & \\
CH{-}OCOR_2 & + & 3CH_3OH & \longleftrightarrow & CH{-}OH & + & R_2{-}COOCH_3 \\
| & & & & | & & \\
CH_2{-}OCOR_3 & & & & CH_2{-}OH & & R_3{-}COOCH_3 \\
 & & & & & & \\
\text{Triglyceride} & + & \text{Methanol} & \longleftrightarrow & \text{Glycerol} & + & \text{Methyl esters}
\end{array}$$

Figure 14.3 Transesterification of triglyceride to produce biodiesel

The catalyst for the reaction is a strong base such as sodium or potassium hydroxide (NaOH, KOH). Separation of the reaction products is by settling due to density differences, and the light biofuel fraction is then finished by washing and drying.

14.2.3 Research focus in algae and biodiesel production

Current R&D work aims to increase algal productivity towards a target level of 100 t-biomass per ha-yr with an oil content target of 40% triglycerides, in order to minimize land usage and reduce the cost of production, harvesting, and processing to the point where algal biofuels become cost competitive with traditional fuels.

The following areas are being addressed by ongoing R&D programs, which are increasingly being conducted in the private sector in view of the anticipated commercial opportunities:

- Design of photobioreactors to overcome light saturation effects and achieve other productivity benefits
- Protection of the open pond cultures against invasion by other algal strains
- Genetic engineering to reduce antenna pigment concentration and increase oil yield
- Optimization and control of culture medium conditions (nutrients, acidity, etc.) to maximize product yield
- Development of low-cost harvesting systems for specific algal strains (e.g., bioflocculation)
- Processing of harvested algal biomass to extract oil and yield high-value residues
- In-situ transesterification of lipids without prior extraction from algal biomass

14.3 References and Resources

14.3.1 Key References

The following key references, arranged in chronological order, provide a starting point for further study.

Burlew, J. S. (ed.) (1953). *Algal Culture From Laboratory to Pilot Plant*, Carnegie Institution for Science, Washington, DC. Available at www.ciw.edu/publications_online/algal_culture.

Benemann, J. R. and W. J. Oswald (1996). Systems and economic analysis of microalgae ponds for conversion of CO_2 to biomass, Final Report to U.S. Department of Energy National Energy Technology Laboratory. Available at www.osti.gov/bridge/servlets/purl/493389-FXQyZ2/webviewable/493389.pdf.

Benemann, J. R. (1997). CO_2 mitigation with microalgae systems, *Energy Conversion and Management*, **38** (Supplement) (S475–S479).

Sheehan, J., T. Dunahay, J. R. Benemann, and P. Roessler (1998). A look back at the US Department of Energy's aquatic species program: Biodiesel from algae, NETL report NETL/TP-580-24190. Available at www.nrel.gov/docs/legosti/fy98/24190.pdf.

Stolaroff, J. K., G. V. Lowry, and D. W. Keith. (2003). Using CaO- and MgO-rich industrial waste streams for carbon sequestration, *Energy Conversion and Management*, **46** (687–699).

Huntley, M. and D. G. Redalje. (2007). CO_2 mitigation and renewable oil from photosynthetic microbes: A new appraisal, *Mitigation and Adaptation Strategies for Global Change*, **12** (573–608).

Hoenig, V., H. Hoppe, and B. Emberger (2007). Carbon capture technology: Options and potentials for the cement industry, European Cement Research Academy report TR 044/2007. Available at www.ecra-online.org/downloads/Rep_CCS.pdf.

14.3.2 Institutions and Organizations

The following institutions and companies that have been active in the research and development work on other sequestration and usage options.

AlgaeLink (manufacturer of commercial-scale algae cultivation equipment and algae-to-fuel technology): www.algaelink.com

Calera Corp. (CO_2 capture for PCC production as a cement additive): www.calera.biz

Carbon Sciences Inc. (CO_2 capture for PCC and biofuel production): www.carbon-sciences.com/01/technology_co2fuel.html

Carbon Sense Solutions Inc. (CO_2 accelerated concrete curing): www.carbonsense solutions.com/pages/projects.html

GreenFuel Technologies Corp.: www.greenfuelonline.com

HR BioPetroleum Inc. (biofuel production from algae): www.hrbp.com

Novacem Ltd.; Magnesium silicate cement (www.novacem.com).

RWE (algae project): www.rwe.com/generator.aspx/konzern/fue/stromerzeugung/co2-minimiertes-kraftwerk/algenprojekt/language = en/id = 687990/algenprojekt-page.html

14.3.3 Web Resources

Biodiesel Magazine: www.biodieselmagazine.com

Oilgae (information portal relating to algal-derived biofuel): www.oilgae.com

Part IV

Carbon dioxide transportation

15 Carbon dioxide transportation

The large-scale deployment of CCS will require the development of a transportation infrastructure to interconnect capture and storage sites, in all but the few cases in which these two are fortuitously co-located. With a single 500-MW_e power plant delivering ~4–5 Mt-CO_2 per year for storage regional solutions such as pipeline networks with capacities in the range of 50–100 Mt-CO_2 per year are likely to be developed to reduce costs and provide flexibility. While road and rail transport have been used at a smaller scale, pipeline and marine transport are the only realistic options for transportation on this scale, and only these options are considered here. Key components of these two transportation options are shown in Figure 15.1.

The technology for pipeline and marine transport is already well developed, with many CO_2 pipelines in operation and with marine transport of liquefied petroleum gas (LPG) being straightforwardly scaleable for large-scale CO_2 handling. This chapter therefore focuses on the current technologies and applications, and the further development and optimization needed to bring these to a regional scale.

A future large-scale CO_2 transport system is likely to include both pipelines and ships, as is the case for the current global market in natural gas. Ships provide a flexible and easily scaleable option, which will be economically preferred for large transport distances and for smaller volumes, while pipelines will provide regional trunk lines and may also be constructed on an intercontinental scale (>3000 km) if quantities in excess of ~20 Mt-CO_2 per year need to be transported over such distances.

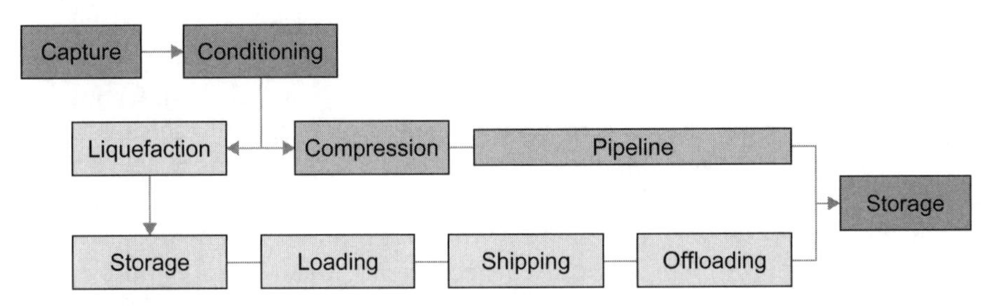

Figure 15.1 CCS transportation system elements

Doi:10.1016/B978-1-85617-636-1.00015-8.

15.1 Pipeline transportation

Although CO_2 transportation in pipelines is possible in the vapor phase, all existing pipelines are designed to carry a dense phase and therefore to operate above the CO_2 critical pressure ($P_c = 7.38\,MPa$). This has the advantage of smaller pipelines and lower pressure drops for a given mass flow rate.

15.1.1 Pipeline engineering fundamentals

Fluid flow

The pressure drop due to friction per unit length of pipeline, $\Delta p/\Delta L$ (Pa/m), depends on the diameter and internal roughness of the pipeline, the fluid flow rate, and the density and viscosity of the transported fluid, according to the Darcy–Weisbach equation:

$$\Delta p/\Delta L = f_D\, \rho\, u^2/2D \quad (15.1)$$

where u is the fluid velocity (m^3/s), D is the pipeline diameter (m), ρ is the fluid density (~877 kg/m^3 for CO_2 at 11 MPa and 25°C), and f_D is the dimensionless Darcy friction factor. The Darcy friction factor depends on the Reynolds number (Re) and the surface roughness of the pipe (ε). The Reynolds number is defined as:

$$Re = 4\dot{m}/\mu\pi D \quad (15.2)$$

where $\dot{m}$ is the mass flow rate (kg/s) and μ is the fluid viscosity ($7.73 \cdot 10^{-5}$ Pa.s for CO_2 at 11 MPa and 25°C). For example, the flow in a 0.4 m diameter pipeline transporting 3.5 Mt-CO_2 per year will have a Reynolds number of $4.5 \cdot 10^6$. The Darcy friction factor can be determined iteratively using the Colebrook–White equation:

$$f_D^{-1/2} = -2\log_{10}\{\varepsilon/3.7D + 2.51/Re\, f_D^{1/2}\} \quad (15.3)$$

where the surface roughness ε is typically $4.6 \cdot 10^{-5}$ m for commercial steel pipe. For the example used above, with $Re = 4.5 \cdot 10^6$, f_D will be ~0.014, giving an estimate of the pipeline pressure drop of ~15.2 Pa per meter, or 1.5 MPa per 100 km.

This calculation gives an estimate of the pressure drop over a pipeline but does not consider the change in fluid properties with temperature and pressure, particularly close to the critical point. This is illustrated in Figure 15.2, which shows the density and compressibility of CO_2 at typical pipeline operating temperatures and pressures.

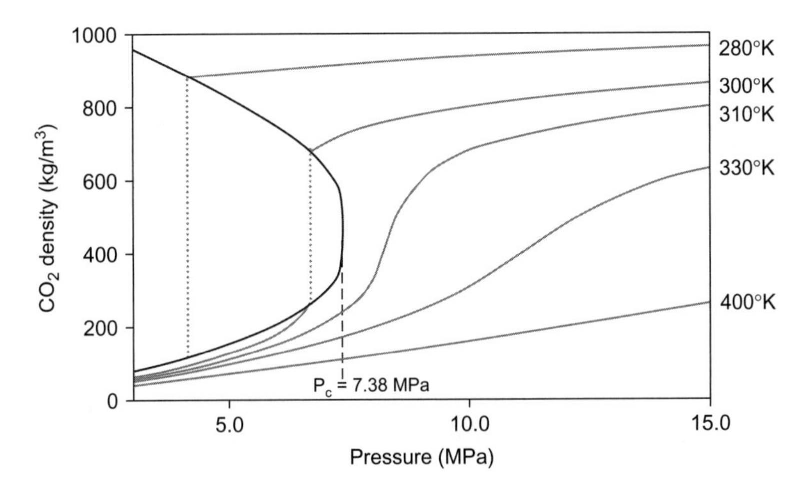

Figure 15.2 Density and compressibility of CO_2

At pressures <9 MPa and temperatures >25°C (>298°K), both density and compressibility vary considerably with P and T. At the extreme, density drops by 50% for a 10°C temperature increase from 25–35°C (ca. 300−310°K) at pressures ~8 MPa.

The transported dense phase may be either a supercritical fluid or a subcooled liquid, depending on whether the temperature is above or below the critical temperature ($T_c = 31.1$°C). Subcooled liquid transportation has the advantage of higher density (Figure 15.2) and lower compressibility, leading to smaller pipe diameters and lower pressure drops compared to a supercritical fluid. Cooling to <15°C would typically be required, depending on distance, ambient conditions, and pipeline insulation, to ensure that temperature remains below T_c along the full length of a pipeline segment. The cost trade-off between these two options will be strongly influenced by the ambient conditions, with subcooled liquid being more likely to be preferred in cooler climates or for subsea pipelines.

The flow-determining properties of density and compressibility are also significantly affected by the presence of impurities such as water, hydrogen sulfide (H_2S), and methane. Under these circumstances, the accurate prediction of flow behavior requires complex flow modeling, particularly in any pipeline segments where the operating pressure drops below 10 MPa. Operating experience of CO_2 pipelines to date has been mostly with natural CO_2 transported for EOR use. For CCS transportation, the impact on dense-phase fluid properties of trace quantities of SO_x, NO_x, and, in the case of precombustion capture, H_2 is an area of current research.

Pipeline and equipment materials issues

The economic imperative to minimize capital cost dictates the use of carbon steel for pipeline construction, and the properties of this material, particularly its strength and corrosion resistance, determine the basis for design of the system.

Table 15.1 EU Dynamis recommended CO_2 specification for transportation and storage

Component	Concentration limit	Consideration
H_2O	500 ppm	Prevention of free water
H_2S	200 ppm	Health and safety
CO	2000 ppm	Health and safety
SO_x	100 ppm	Health and safety
NO_x	100 ppm	Health and safety
O_2	<4 vol%	For aquifer storage
	<1000 ppm	Technical limit, for EOR
CH_4	<4 vol%	For aquifer storage
	<2 vol%	For EOR
$N_2 + Ar + H_2$	<4 vol% total	Lower for H_2 in view of economic value of its energy content

Pipeline corrosion

If the water concentration in the CO_2 stream exceeds the solubility limit at the pipeline operating pressure and temperature, free water will be present and will result in the formation of carbonic acid (H_2CO_3), which will corrode carbon steel pipelines. This type of corrosion is slow and results in pitting of the steel surface, eventually leading to pinhole leaks. The corrosion risk can be mitigated by dehydration of the CO_2 stream to below the solubility limit and inclusion of a corrosion allowance in the design wall thickness of the pipeline. Similarly, the concentration of hydrogen sulfide in the fluid stream must be kept below the solubility limit to prevent H_2S corrosion, while the concentration of H_2 from precombustion capture must be limited to avoid hydrogen embrittlement.

A recommended quality specification for pipeline transportation of CO_2 has been published by the European Union's Dynamis project, as summarized in Table 15.1.

Carbon steel pipelines also require external corrosion protection, which may be achieved by coatings, cathodic protection, or both.

Pipeline fracture failure

Two types of fracture failure can occur in pipelines in CO_2 service: brittle fractures from stress corrosion cracking (SCC) and ductile fractures. Brittle fractures occur where stresses are concentrated as a result of corrosion, causing small cracks that can then grow catastrophically. The ability of a material to resist brittle fracturing once a small crack has developed is known as the toughness, or fracture toughness. It is recognized that CO_2 pipelines are at greater risk of failure from long-running axial brittle fractures as a result of the severe Joule–Thomson cooling and embrittlement that occurs around any leak, for example as a result of carbonic acid corrosion.

Ductile fractures occur when stresses exceed the normal tensile strength of the material, resulting in failure after plastic deformation. Large pressure transients that can occur as a result of phase changes during depressurization are a potential cause of ductile failure specific to CO_2 pipelines, and management of this hazard requires careful operational control.

In the CO_2 pipeline systems operating in the United States it is common practice to install fracture or crack arrestors to impede the explosive propagation of axial fractures. These typically consist of glass fiber wrapped around the pipe and embedded in an epoxy resin, although steel hoops are also used. Arrestors are placed every 100–3000 m, depending on the class of location in terms of the population density of the surrounding area, and dual arrestors are often installed side by side to ensure that any crack is fully arrested.

A further important element of the pipeline design is the inclusion of isolation or block valves at intervals along the pipeline in order to mitigate the risk of a major release of CO_2 in the event of a pipeline rupture. These would be closed in by the pipeline control system in the event that a leak is detected, minimizing the pipeline inventory that would be released. For a given block valve separation, a safe distance can be calculated at which CO_2 released in a major rupture would be dissipated to a safe level. The optimal block valve separation is then a balance between leak risk, including the additional risk of leakage from the valves, the safe distance in relation to local population density, and cost—both capital cost and additional operating cost for valve maintenance. Typical distances between block valves vary from 5 km to 40 km, depending once again on the class of location.

Other materials issues

Other materials issues relevant to pipelines and related equipment in CO_2 service are summarized in Table 15.2.

Flow assurance: hydrate formation

Clathrates are a class of chemical substance in which a molecule of a certain type is trapped within a cage or lattice formed from molecules of a second type; if the cage is formed from water molecules, the structure is known as a hydrate.

Hydrates of CO_2 ($CO_2 \cdot 6H_2O$) can be formed in a pipeline if free water is present and the operating temperature drops below ~15°C. The resulting ice-like solid can plug the pipeline and can be extremely difficult to remove once in place.

The simplest way to avoid this flow assurance problem in CO_2 pipelines where the operating temperature may fall into this range (particularly subsea pipelines) is to dehydrate the gas stream to ensure that free water never condenses. This is also required in order to avoid carbonic acid corrosion, and

Table 15.2 Materials considerations for CO_2 service

Materials issue	Description and mitigation
Elastomeric seals	Diffusion of CO_2 into elastomeric materials, used for example in pump and valve seals, can result in explosive decompression if the operating pressure drops suddenly. Selection and testing of low permeable and harder elastomers is required to minimize CO_2 diffusion and decompression risk.
Lubrication	The poor lubricating properties of dry CO_2 affect pump design and internal pipeline cleaning and scraping (known as "pigging"). Petroleum-based and some synthetic lubricants harden and become ineffective in CO_2 service; special inorganic greases are required.

results in a maximum permissible water content for CO_2 transportation of $3\text{–}5 \cdot 10^{-4}$ kg per cubic meter (300–500 ppm).

Hydrate formation is a well-known risk in natural gas wells and pipelines and is managed by the injection of hydrate inhibitors, most commonly methanol, which lower the hydrate formation temperature to below the minimum expected operating temperature. Injection of inhibitors may also be applicable in CO_2 transportation operations when transient low temperatures are expected, for example as a result of Joule–Thomson cooling during depressurization for maintenance.

15.1.2 Pipeline operating considerations

The operating characteristics of CO_2 pipelines are significantly affected by the pressure and temperature variability of density and compressibility, and operating procedures specific to CO_2 service are needed to cater for this (Table 15.3).

15.1.3 Pipeline transportation systems: current and planned

Pipelines for CO_2 transportation have been in operation, primarily to deliver CO_2 to EOR projects, since the 1970s. Table 15.4 summaries the key features of a number of these systems.

The Cortez, Canyon Reef, and Sheep Mountain pipelines noted in the table are part of a network of pipelines that stretches from Colorado to Texas, connecting a number of natural and anthropogenic CO_2 sources to supply gas for EOR in the Permian Basin oilfields in Texas. The network includes the world's largest CO_2 hub located in Denver City, TX, which currently handles up to 80 Mt-CO_2 per year.

Table 15.3 Operational considerations for CO_2 pipeline systems

Operating issue	Description and mitigation
Depressurization	Rapid depressurization will result in extreme cooling. Careful control of depressurization is required, for example during maintenance operations, to avoid thermal stresses on the pipeline, valves, and other equipment that may cause embrittlement and fracture failure. Rapid depressurization in the early stage of blowdown can be slowed or interrupted to prevent very low pipeline and equipment metal temperatures and solid CO_2 formation. Similar problems need to be addressed on repressurization.
Pressure transients	Unlike natural gas or oil pipelines, which operate at pressures (oil) or temperatures (gas) far from the critical point, CO_2 pipelines operate with both P and T relatively close to the critical point. Pressure transients can cause significant density and compressibility changes, and in extreme cases a phase transition. CO_2 pipelines experience the so-called slinky effect under pressure transients, the equivalent of a water hammer for an incompressible fluid.
Leak detection	Detection of leaks is more difficult than with hydrocarbon pipelines due to the high temperature dependence of compressibility of the fluid. Due to the extreme cooling experienced during a leak, aerial surveying using thermal imaging is the most effective leak-detection method for CO_2 pipelines.
Linepack	Linepack occurs when a pump feeding a pipeline segment continues to operate after the pipeline is closed in downstream. This results in a pressure increase and buildup of pipeline inventory that may be difficult to release safely. This can be mitigated with appropriate control system design and hardware redundancy.

15.1.4 Pipeline transportation case studies

Texas Gulf Coast CO_2 network

The current network of CO_2 pipelines in operation in the United States extends to ~6200 km, and studies indicate that this will need to be extended by an additional 17,500–37,000 km between 2010 and 2050, depending on the CO_2 stabilization scenario adopted and the consequent scope of CCS implementation. The west Texas network has been progressively developed, and continues to be extended, in response to the demand for CO_2 to drive EOR projects in the Permian Basin.

Table 15.4 Currently operating and planned long-distance CO_2 pipelines

Pipeline, start-up year Origin and destination	**Length (km)**	**Diameter (m)**	**Operating pressure (MPa)**	**Capacity ($Mt\text{-}CO_2$ per year)**
Canyon Reef Carriers, 1972 McCamey gasification plant to SACROC oilfield, TX	224	0.4		5.2
Cortez, 1984 McElmo Dome CO_2 field, CO, to Denver City Hub, TX	803	0.6	9.5–13.6	19.3
Sheep Mountain, Bravo Dome CO_2 field, CO, to Denver City Hub and Texas oilfields	653	0.5 and 0.6	8.4–13.9	9.5
Weyburn, 2000 Gasification plant NF, USA, to Weyburn oilfield, Canada	328			5.0
Bati Raman, 1983 Donan field, Turkey	90			1.1
Budafa, Hungary	33	0.15–0.3	14.1	
Snøhvit, Norway	153	0.2		0.7
Planned				
Alberta, 2011 Gasification upgrader to EOR fields	240	0.3–0.4	Subject to design	5.5
Green Pipeline, 2010 Jackson Dome to Texas oilfields extension	500	0.6		

The additional economic incentive resulting from an EOR demand would also be the driver for a potential CO_2 supply network in the Texas Gulf Coast region, where geological studies indicates that 1.5 billion barrels of oil could be recovered from existing fields by CO_2 flooding. A pipeline network has been envisaged to supply 31 of the largest candidate fields with 40–50 $Mt\text{-}CO_2$ per year captured from 11 gas- and coal-fired power stations in the region. The key parameters of the network are summarized in Table 15.5.

COOTs North Sea CO_2 network

The potential for EOR is also the driver behind a CO_2 network that could eventually extend across the U.K., Norwegian, and Dutch sectors of the North

Table 15.5 Key parameters of a proposed Texas Gulf Coast CO_2 pipeline network

Parameter	Description
Network dimensions	1600–1820 km, 0.2- to 1.0-m-diameter carbon steel pipelines
CO_2 supply	
Gas-fired plant	6 stations, ranging from 460 to 1420 MW_e, total 4.6 GW_e Supplying 27 Mt-CO_2 per year
Coal-fired plant	5 stations, ranging from 180 to 892 MW_e, total 2.6 GW_e Supplying 25 Mt-CO_2 per year
Operating pressure	9.0–14.0 MPa, with two booster compressor stations
Total CO_2 sequestration	0.8–1.3 Gt-CO_2 over a 20- to 25-year economic life

Sea. The first link of this network is already in place with StatoilHydro's 153-km Snøhvit pipeline, and the possible future phases of development are illustrated in Figure 15.3.

A preliminary front-end engineering design (FEED) study has been undertaken for a possible first U.K. phase of such a network (shown in the figure as segment A), which envisages a 500 km, 0.7 m diameter trunk line, initially delivering 5 Mt-CO_2 per year from a number of U.K. east coast capture sites to North Sea EOR projects. The preliminary design basis specifies operation at 10.0–19.0 MPa, a 40-year design life, and flexibility for capacity increase up to 15 Mt-CO_2 per year, with a possible start-up in 2014. The system would include a number of inline Ts to allow connection of future spur lines to connect to various oil fields along the route.

The 800 MW_e Eston Grange coal-fired IGCC plant being developed by Coastal Power Ltd. (a joint venture between Centrica plc and Progressive Energy Ltd.) is being considered as a possible initial CO_2 source for the pipeline, and the pipeline transport and disposal infrastructure would be owned and operated by COOTs Ltd., another Centrica and Progressive joint venture.

15.1.5 RD&D for pipeline transportation in CCS projects

Although CO_2 pipelines have been in use since 1972, and the Permian Basin EOR network alone currently transports close to 30 Mt-CO_2 per year, significant knowledge gaps remain regarding the transport of CO_2 in CCS applications. These issues are summarized in Table 15.6.

The Pipeline Research Council International (PRCI), a global collaborative RD&D organization for the pipeline industry, has also commenced a project to review the state of the art in CO_2 transport and storage in order to establish RD&D priorities in the areas of pipeline design, materials, operations, and

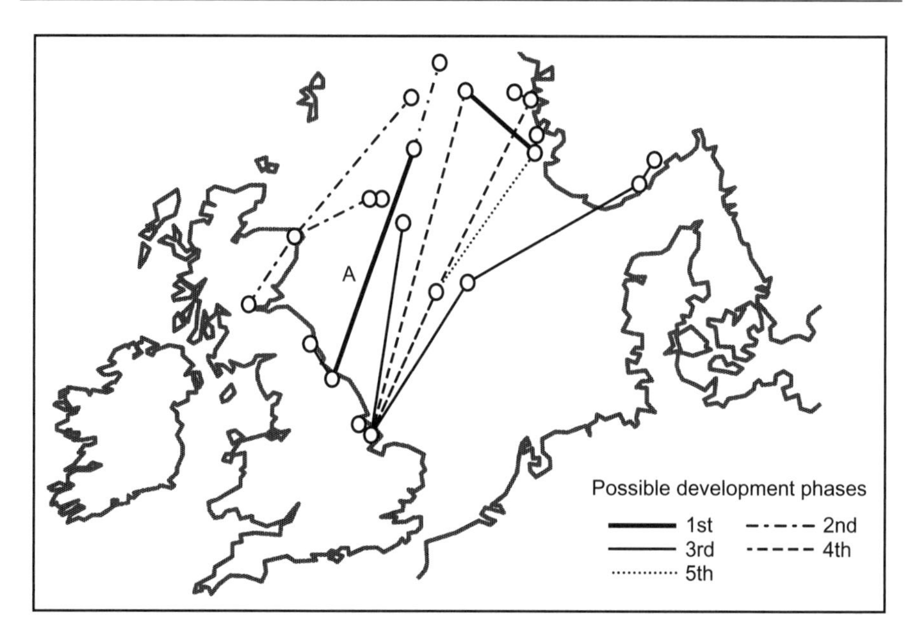

Figure 15.3 Possible development phases of a North Sea CO_2 transport network

integrity assessments and to develop a research plan to address these needs. A similar study is being undertaken by Det Norske Veritas (DNV) as a European Joint Industry Project with the aim of reviewing available knowledge and identifying gaps in order to develop best-practice guidelines for CO_2 transport in both onshore and subsea pipelines.

Compared to other aspects of CCS technologies, ongoing RD&D projects addressing the challenges of pipeline transportation are relatively few.

In association with the Norwegian research institute SINTEF, StatoilHydro has established a CO_2 pipeline test rig at its Research Centre in Trondheim to investigate and validate models for depressurization, two-phase flow transients, and heat transfer in CO_2 pipelines, as well as assessing the effect of impurities. The facility is also used for the training of staff operating StatoilHydro's Snøhvit pipeline.

Enel/Eni have also established a dense-phase CO_2 transport pilot loop at the Enel Brindisi power station, as a precursor to the transport of CO_2 from the plant for injection into Eni's depleted Stogit gas field at Cortemaggiore, Piacenza, ~850 km to the north.

15.2 Marine transportation

Transportation of liquefied hydrocarbon gas in very large LGP and liquefied natural gas (LNG) carriers is a major contributor to global energy trade, with LNG trade alone reaching 174 Mt in 2007, and provides a body of proven technology

Table 15.6 Current RD&D topics for CCS pipeline systems

RD&D topic	Description
Physical properties	Impact of impurities on water solubility and other properties. Viscosity measurement of CO_2 mixtures. Optimization of equation of state representations of properties of CO_2, including the impact of impurities, to allow accurate pipeline modeling.
Ductile fracture propagation	Burst testing of line pipe under CO_2 service to validate fracture propagation models developed for hydrocarbon service. Fracture propagation behavior of subsea CO_2 pipelines. Applicability of current crack arrestors and assessment methods.
Corrosion prediction	Experimental investigations of corrosion rates in CO_2 with free water present, including the impact of impurities in the gas stream. Development of corrosion prediction models for different steel specifications. Identification of suitable corrosion inhibitors.
Material properties and compatibility	Testing to establish material compatibility for CO_2 service, including impact of impurities in the CO_2 stream.
Flow modeling and validation	Experimental investigations to validate flow modeling and establish pipeline control algorithms. Investigation of two-phase flow regime, including impact of impurities, and consequences for operating strategies. Development of models for heat transfer, thermodynamic, and transport properties of CO_2 mixtures during depressurization.
CO_2 hydrate management	Investigation of conditions for hydrate formation and dissociation, and potential hydrate inhibitors. Determine water content limits to avoid hydrate formation.

and operating experience as a basis for the development of marine options for CO_2 transportation. Current large LNG carriers can reach 270,000 m^3 and could carry 0.3 Mt-CO_2 at the anticipated transport conditions for CO_2.

15.2.1 Optimal physical conditions for marine transport

Since LNG (methane) and LPG (propane and butane) can be liquefied by cooling at atmospheric pressure, large carriers transport LNG and LPG as a liquid in refrigerated tanks at or slightly above atmospheric pressure (up to ~30 kPa). This type of fully refrigerated carrier is not suitable for the transport of CO_2, which has a triple point pressure of 520 kPa, since CO_2 is either a solid or gas at atmospheric pressure.

Transport efficiency is maximized if the density of liquid CO_2 is as high as possible and the density increases rapidly with decreasing pressure, reaching

1200 kg per cubic meter at the triple point. The optimal condition for CO_2 transport is thus at a pressure and temperature slightly above but as close to the triple point as can be operationally managed; the risk is the formation of solid CO_2 (dry ice) if the pressure drops below the triple point pressure during loading and unloading. The likely operating conditions for a large CO_2 carrier would therefore be a temperature of between −50°C and −54°C and a pressure of 0.6–0.7 MPa, which is close to the operating condition of existing semi-pressurized LPG carriers.

15.2.2 Design, development, and future deployment of CO_2 carriers

The existing global fleet of operating CO_2 carriers is limited to a few small vessels with capacities of 1000–1500 m^3 transporting liquefied food-grade CO_2 at a temperature of −30°C and a pressure of 1.8–2.0 MPa. Current shipped volume is ~1 Mt-CO_2 per year. In anticipation of a growth in volume as a result of CCS projects, the Norwegian shipping company I.M. Skaugen AS specifically designed six LPG carriers of 10,000 m^3 capacity to be CO_2-capable, and these have been operating since 2003, although not yet in CO_2 service.

The next generation of midrange CO_2 carriers has been the subject of a number of research and engineering projects, including one led by StatoilHydro together with SINTEF, Vigor/Fabricom, and Navion/Teekay. The project designed a 177 m long by 31 m wide vessel capable of transporting 20,000 m^3 of liquefied CO_2 (~24 kt-CO_2) at 0.7 MPa and approximately −50°C in four to six tanks. The design, illustrated in Figure 15.4, included an offloading turret to enable direct CO_2 discharge at offshore installations.

The use of marine transport is being considered in two potential CCS projects currently under development. As part of the shortlisted project in the U.K. Government's CCS Demonstration Competition, led by RWE nPower and Peel Energy, I.M. Skaugen AS and Teekay Corp. will develop a transport infrastructure solution that may include marine transport for storage in depleted gas reservoirs in Tullow Oil's Hewett Field in the U.K. sector of southern North Sea (see Section 11.5.1). This field, although currently producing, is expected to be depleted and would otherwise be ready for abandonment by 2014, the target date for the CCS demonstration project start-up.

15.2.3 Operational aspects of marine transportation

The operational aspects of marine transportation of CO_2 are largely understood as a result of existing experience with small-scale CO_2 shipping, as well as the transfer of experience from large-scale LPG and LNG operations.

Integration of the continuous capture process with batch transportation in ships requires liquefaction and storage facilities at the loading terminal. In the initial loading operation, the cargo tanks would first be pressurized with gaseous CO_2 to avoid the Joule–Thomson cooling and dry ice formation that would occur if liquid CO_2 was admitted to an unpressurized tank.

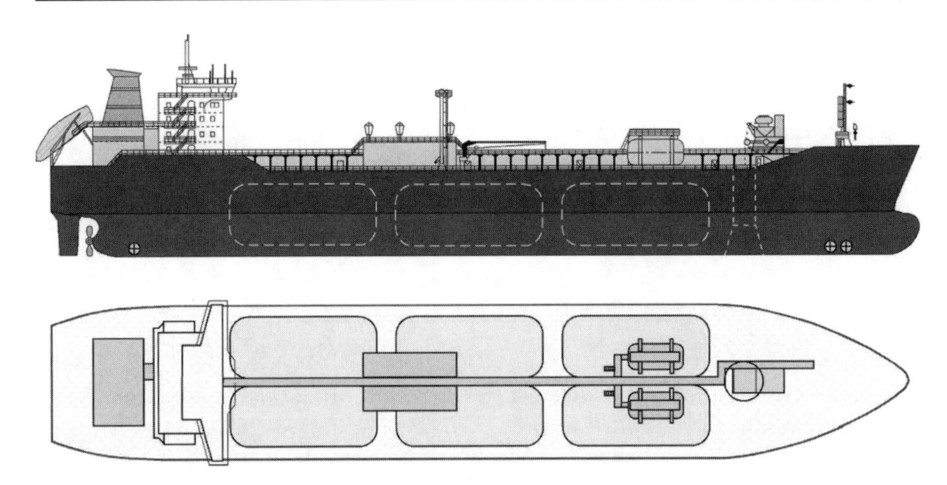

Figure 15.4 Conceptual arrangement for CO_2 ship
(Courtesy; Statoil/SINTEF/Teekay/Vigor, used with permission from StatoilHydro)

In LNG or LPG transport, boil-off gas evolved during the voyage can be used to fuel the ship, whereas in the case of CO_2 transport, boil-off gas would probably be released to the atmosphere. Net emission could be reduced by recovering cold for use in onboard refrigeration or air conditioning, or onboard reliquefaction may also be possible.

Offloading may be into a similar terminal facility, with onward transportation by pipeline to storage sites. Alternatively, offloading may occur at offshore locations for direct injection into the storage site, by connecting the ship to a submerged turret loading (STL) system. In either case the cargo tanks are left charged with dry gaseous CO_2 in preparation for the next loading cycle.

While offshore unloading may seem like a simple solution, it would be operationally complicated because, without storage facilities, the offloading rate would be limited by the achievable injection rate of the wells. Perhaps more importantly, managing the reheating and pressurization of the liquid back to a supercritical state, avoiding phase transitions and operational problems due to the cryogenic temperatures, would require additional offshore facilities and heat input. There would be little opportunity to recover the work expended to achieve cryogenic conditions: essentially the cold is lost!

15.3 References and Resources

15.3.1 Key References

The following key references, arranged in chronological order, provide a starting point for further study.

Skovholt, O. (1993). CO_2 transportation systems, *Energy Conversion and Management*, **34** (1095–1103).

IEA GHG. (2004). *Ship transport of CO_2*, Report PH4/30, IEA Greenhouse Gas R&D Programme, Cheltenham, UK.

Gale, J. and J. Davison. (2004). Transmission of CO_2—safety and economic considerations, *Energy*, **29** (1319–1328).

Svensson, R., *et al.* (2004). Transportation systems for CO_2—application to carbon capture and storage, *Energy Conversion and Management*, **45** (2343–2353).

Aspelund, A., *et al.* (2006). Ship transport of CO_2, *Chemical Engineering Research and Design*, **84** (A9) (847–855).

U.K. Department for Business Enterprise and Regulatory Reform. (2007). *Development of a CO_2 Transport and Storage Network in the North Sea*. Available at www.berr.gov.uk/files/file42476.pdf

Dooley, J., *et al.* (2008). Comparing existing pipeline networks with the potential scale of future US CO_2 pipeline networks, *Proceedings of the Ninth International Conference on Greenhouse Gas Control Technologies*, Elsevier, Oxford, UK.

Visser, E., *et al.* (2008). Dynamic CO_2 quality recommendations, *International Journal of Greenhouse Gas Control*, **2** (478–484).

15.3.2 Institutions and Organizations

The institutions that have been most active in the research and development work on CO_2 transport are listed below, with relevant web links.

DNV (developing a CO_2 pipeline and transportation standard under the CO2PIPETRANS Joint Industry Project, due early 2010): www.dnv.com

EC/EU Dynamis project (developing CO_2 quality recommendation for pipeline transportation): www.dynamis-hypogen.com

I.M. Skaugen SE (marine transport of CO_2): www.skaugen.com/index.php/new-projects/large-scale-seaborne-co2

Kinder Morgan CO_2 Company (KMCO2; CO_2 pipeline infrastructure and EOR oilfield operator): www.kindermorgan.com/business/co2

SINTEF (Norwegian R&D organization active in CO_2 transportation and process technology): www.sintef.no/Home/Petroleum-and-Energy/SINTEF-Energy-Research/Energy-Processes/Process-Technology/

StatoilHydro (CO_2 pipeline infrastructure in the Norwegian North Sea): www.statoilhydro.com/en/TechnologyInnovation

Part V

Carbon capture and storage information resources

16 Further sources of information

Organizations, initiatives, projects, and other predominantly online resources relating to CCS technologies

16.1 National and international organizations and projects

16.1.1 International Organizations and Projects

CDM	UNFCCC Clean Development Mechanism	www.cdm.infccc.int
CSLF	Carbon Sequestration Leadership Forum	www.cslforum.org
	Environmental Research Web	www.environmentalresearchweb.org
GCP	Global Carbon Project	www.globalcarbonproject.org
IEA CCC	International Energy Agency Clean Coal Centre	www.iea-coal.org.uk
IEA CTI	International Energy Agency Climate Technology Initiative	www.climatech.net
IEA GHG	International Energy Agency GHG Research Programme	www.ieagreen.org.uk www.co2captureandstorage.info
IEA GREENTIE	International Energy Agency Greenhouse Gas Technologies Information Exchange	www.greentie.org
IETA	International Emissions Trading Association	www.ieta.org
IPCC	Intergovernmental Panel on Climate Change	www.ipcc.ch
UNEP GRID-Arendal	United Nations Environmental Programme, Information Services	www.grida.no www.climatewire.org

Doi:10.1016/B978-1-85617-636-1.00016-X.

16.1.2 National organizations and projects

	Australasia	
CO2CRC	Co-operative Research Centre for Greenhouse Gas Technologies	www.co2crc.com.au
	Carbon Forum Asia	www.carbonforumasia.com
	Canada	
CCCSTN	Canadian Carbon Capture and Storage Technology Network	www.co2network.gc.ca
CCPC	Canadian Clean Power Coalition	www.canadiancleanpowercoalition.com
ICO2N	Canadian CO_2 capture and storage initiative	www.ico2n.com
PTRC	Petroleum Technology Research Centre	www.ptrc.ca
	Japan	
CREPI	Central Research Institute of Electric Power Industry	http://criepi.denken.or.jp/en/
NEDO	New Energy and Industrial Technology Development Organization	www.nedo.go.jp/english/index.html
RITE	Research Institute of Innovative Technology for the Earth	www.rite.or.jp/index_e.html
	Europe	
ACCSEPT	European project addressing CCS acceptance	www.accsept.org
BERR	UK Department for Business Enterprise and Regulatory Reform, CCS	www.berr.gov.uk/energy/sources/sustainable/ccs/page42320.html
	CO_2 reduction plan, Netherlands	www.co2-reductie.nl
CACHET	EU-funded precombustion project	www.cachetco2.eu
CCP	CO_2 capture project	www.co2captureproject.org
CCSA	Carbon Capture and Storage Association	www.ccsassociation.org.uk
CLIMIT	Norwegian R&D programme	www.climit.no
CO2NET	EU CO2NET knowledge-sharing network	www.co2net.eu/public
CO2STORE	EU-funded DSF joint industry research project	www.co2store.org

ENCAP	European power generation consortium project for ENhanced CAPture of CO_2	www.encapco2.org
ETP ZEP	European Technology Platform on Zero-Emission Fossil Fuel Power Plants	www.zero-emissionplatform.eu
EU CCS	European Commission CCS site	http://ec.europa.eu/environment/climat/ccs
Gassnova SF	Norwegian State enterprise for CCS	www.gassnova.no
IC	Imperial College Carbon Capture and Storage Research	www.imperial.ac.uk/carboncaptureandstorage
IFP	Institute Française du Pétrole	www.ifp.fr
SCCS	Scottish Centre for Carbon Storage	www.geos.ed.ac.uk/sccs
	U.K. Office of Climate Change	www.occ.gov.uk/
	United States	
Berkeley Labs	Lawrence Berkeley National Laboratory Climate Change and Carbon Management	http://esd.lbl.gov/CLIMATE http://esd.lbl.gov/GCS
Big Sky	Big Sky Carbon Sequestration Partnership	www.bigskyco2.org
CCSP	Carbon Cycle Science Program	www.carboncyclescience.gov
	U.S. Climate Change Technology Program	www.climatetechnology.gov
CDMC	Climate Decision Making Center, Carnegie Mellon University	www.cdmc.epp.cmu.edu
CMI	Princeton Environmental Institute Carbon Mitigation Initiative	www.princeton.edu/~cmi/
CWC	Ohio State University, The Climate, Water, Carbon Program	http://cwc.osu.edu
DOE	Department of Energy (Fossil energy)	www.fossil.energy.gov
	DoE Sequestration	www.carbonsequestration.us
DOE NETL	DoE National Energy Technology Laboratory	www.netl.doe.gov
GCEP	Stanford University Global Climate and Energy Project	www.gcep.stanford.edu
MIT	MIT Carbon Capture and Sequestration Technologies	www.sequestration.mit.edu
PCO2R	Plains CO_2 Reduction Partnership	www.undeerc.org/pcor/
PointCarbon	Carbon market information portal	www.pointcarbon.com

(Continued)

UC3	University of Utah; Utah Clean Coal Program	www.uc3.utah.edu
West Carb	West Coast Regional Carbon Sequestration Partnership	www.westcarb.org

16.2 Resources by technology area

16.2.1 Clean coal-fired generation

Air Products and Ceramatec	Oxygen separation using ion transport membranes	www.airproducts.com/markets/gasification, www.ceramatec.com
FutureGen	U.S. DoE FutureGen Initiative	www.fossil.energy.gov/programs/powersystems/futuregen/
	FutureGen Alliance	www.futuregenalliance.org/
GTC	Gasification Technologies Council	www.gasification.org
IEA	IEA Clean Coal Centre	www.iea-coal.org.uk
IEA GHG	IEA GHG oxyfuel combustion network	www.co2captureandstorage.info/networks/oxyfuel.htm

16.2.2 Adsorption

PSA Plants	Pressure-swing adsorption trading portal	www.psaplants.com
Adsorption.org	Knowledge-sharing site for adsorption-based systems and technologies	www.adsorption.org

16.2.3 Membranes and molecular sieves

LANL	Los Alamos National Lab	www.lanl.gov
NanoGLOWA	Nanostructured Membranes Against Global Warming; EU-funded academic and industrial consortium	www.nanoglowa.com
NATCO	Cynara membrane separation systems	www.natcogroup.com
Media and Process Technology Inc.	Technology developer specializing in advanced ceramic membranes	www.mediaandprocess.com
Membrana	Industrial membrane producer	www.membrana.com

Membrane Guide	Industry portal for membrane suppliers	www.membrane-guide.com
Praxair	Molecular sieve air separation unit	www.praxair.com
UOP	Separex membrane separation systems	www.uop.com

16.2.4 Cryogenic and distillation systems

Air Liquide	Cryogenic air separation systems	www.airliquide.com
Air Products	Cryogenic air separation systems	www.airproducts.co.uk
Linde AG	Cryogenic air separation systems	www.linde.de
Praxair	Cryogenic air separation systems	www.praxair.com
Sulzer	Distillation column equipment	www.sulzerchemtech.com

16.2.5 Mineral carbonation

CCT	Carbon Trap Technologies, L.P.	www.carbontrap.net
LANL	Los Alamos National Lab	www.lanl.gov
UBC	University of British Colombia, Earth and Ocean Sciences Department	www2.ocgy.ubc.ca/research/dipple/UBC_Carbonation

16.2.6 Geological storage

AERI	Alberta Energy Research Institute	www.aeri.ab.ca
APCRC	Australian Petroleum Energy Co-operative Research Council	www.apcrc.au
ARC	Alberta Research Council	www.arc.ab.ca
Battelle Columbus Lab	Carbon Management Strategies for Industry	www.battelle.org
CO2GeoNet	European Network of Excellence on Geological Storage of CO_2	www.co2geonet.com
CO2ReMoVe	Research into Monitoring and Verification technologies	www.co2remove.eu
CO2SINK	EU Geological Storage Research Project	www.co2sink.org
CSIRO	Commonwealth Scientific and Industrial Research Organisation	www.csiro.au
Lawrence Berkley Lab	Lawrence Berkeley National Laboratory Geological Storage Research	http://esd.lbl.gov/GCS

(*Continued*)

MGSC	Midwest Geological Sequestration Consortium	www.sequestration.org

16.2.7 Ocean storage

Atmocean	Developing proprietary wave-driven technology to enhance upwelling	www.atmocean.com
Climos	Developing carbon offsets through ocean iron fertilization	www.climos.com
Cquestrate	Open source initiative to develop ocean acid reduction	www.cquestrate.com
MBARI	Monterey Bay Aquarium Research Institute	www.mbari.org
OANET	Ocean Acidification Network; scientific information network	www.ocean-acidification.net
OCN	Ocean Nourishment Corporation	www.oceannourishment.com
PICHTR	Pacific International Center for High Technology Research, Hawaii	www.pichtr. orgco2experiment.org
	The Ocean in a High CO_2 World: International Science Symposium Series	http://ioc3.unesco.org/oanet/HighCO2World.html
TOS	The Oceanography Society	www.tos.org

16.2.8 Terrestrial ecosystem storage

CSiTE	Carbon Sequestration in Terrestrial Ecosystems	http://csite.esd.ornl.gov/
CASMGS	Consortium for Agricultural Soils Mitigation of Greenhouse Gases	http://agecon2.tamu.edu/people/faculty/mccarl-bruce/acs/

16.3 CCS-related online journals and newsletters

Biodiesel Magazine: www.biodieselmagazine.com

Carbon Capture Journal: www.carboncapturejournal.com

Clean Coal Today (National Energy Technology Laboratory quarterly newsletter): www.netl.doe.gov/technologies/coalpower/cctc/newsletter/newsletter.html

International Journal of Greenhouse Gas Control (published by Elsevier); abstracts available online at www.sciencedirect.com/science/journal/17505836

NETL Carbon Sequestration Newsletter: http://listserv.netl.doe.gov/mailman/listinfo/sequestration

17 Units, acronyms, and glossary

A compendium of units, conversion factors, and acronyms and a glossary of common terms related to CCS technologies

17.1 CCS units and conversion factors

Prefixes		
G	= giga	$= 10^9$
M	= mega	$= 10^6$
k	= kilo	$= 10^3$
m	= milli	$= 10^{-3}$
μ	= micro	$= 10^{-6}$
n	= nano	$= 10^{-9}$

Quantity	SI unit
Permeability	$mol\ m^{-1}\ s^{-1}\ Pa^{-1}$
Permeance	$mol\ m^{-2}\ s^{-1}\ Pa^{-1}$

Conversion factors		
1 Å	$= 10^{-10}$ m	= 0.1 nm
1 bar	= 100 kPa	
1 barrer	$= 3.4\ 10^{-16}\ mol\ m^{-1}\ s^{-1}\ Pa^{-1}$	
	$= 10^{-10}\ cm^3(STP)\ cm^{-1}\ s^{-1}\ cmHg^{-1}$	
1 GJ	= 277.8 kWh	= 0.278 MWh
1 tonne (t)	= 1000 kg	= 1 Mg
1 t-C	$= 3.66$ t-CO_2	
1 t-CO_2	= 0.273 t-C	

Doi:10.1016/B978-1-85617-636-1.00017-1.

17.2 CCS-related acronyms

A	
AABW	Antarctic Bottom Water
ACFC	Activated carbon fiber cloth
ACS	Agricultural carbon storage (or sequestration)
AER	Adsorption-enhanced reforming
AGR	Acid gas removal
ALW	Accelerated limestone weathering
AR4	Fourth Assessment Report (IPCC 2007)
ARD	Afforestation, reforestation, and deforestation
ASCM	Adsorption-selective carbon membranes
ASU	Air separation unit
A_T	Total alkalinity
AZEP	Advanced Zero-Emission Power Plant
B	
BCA	Below-ground carbon allocation
BECS	Biomass energy with carbon storage (or sequestration)
BFB	Bubbling fluidized bed
BFLM	Bulk flow liquid membrane
BIGCC	Biomass integrated gasification combined cycle
BOS	Basic oxygen steelmaking
C	
CAAA	Clean Air Act Amendments
CAM	Crassulacean acid metabolism
CAP	Chilled ammonia process
CAT	Carbon abatement technologies
CBM	Coal bed methane
CC	Combined cycle
CCGT	Combined-cycle gas turbine
CCS	Carbon capture and storage (or sequestration)
CCSD	Centre for Coal in Sustainable Development
CCSTN	Canadian Carbon Capture and Storage Technology Network
CCT	Clean coal technology
CDCL	Coal direct chemical looping

(*Continued*)

CDM	Clean Development Mechanism
CDMC	Climate Decision Making Center
CERs	Certified emission reductions
CFBC	Circulating fluidized bed combustion
CFCs	Chloroflurocarbons
CHP	Combined heat and power
CLC	Chemical looping combustion
CLM	Contained liquid membrane
CLR	Chemical looping reforming
[CO_2]	CO_2 concentration
CMSM	Carbon molecular sieve membranes
CSiTE	Consortium for Research on Enhancing Carbon Sequestration in Terrestrial Ecosystems
CSLF	Carbon Sequestration Leadership Forum
CT	Conservation tilling
CTI	Climate Technology Initiative
CTD	Conductivity temperature depth
CVI	Chemical vapor infiltration

D

DEA	Diethanolamine
DIC	Dissolved inorganic carbon
DOE	U.S. Department of Energy
DMEPG	Dimethyl ethers of polyethylene glycol
DNV	Det Norske Vertia
DSF	Deep saline formation

E

ECBM	Enhanced coal bed methane recovery
ECCP	European Climate Change Programme
EGR	Enhanced gas recovery
EIA	Environmental impact assessment
	Energy Information Agency (U.S. Department of Energy)
EMC	Electrochemically modulated complexation
EOR	Enhanced oil recovery
EOS	Equation of state
EPRI	Electric Power Research Institute
ERT	Electrical resistance tomography

(Continued)

ESA	Electric-swing adsorption
ESP	Electrostatic precipitator
ETP ZEP	European Technology Platform for Zero Emission Fossil Fuel Power Plants
ETS	European Trading System
EU	European Union
F	
FACE	Free air carbon dioxide enrichment
FAR	First Assessment Report (IPCC 1990)
FBC	Fluidized bed combustion
FCCC	Framework Convention on Climate Change
FEED	Front-end engineering design
FGD	Flue gas desulfurization
FOCE	Free ocean carbon dioxide enrichment
G	
GFBCC	Gasification fluidized-bed combined cycle
GHG	Greenhouse gas
GIIP	Gas initial in place
GPP	Gross primary production
GTL	Gas to liquids
GW	Gigawatt
GWP	Global warming potential
H	
H	Enthalpy
HAT	Humid air turbine
HDS	Hydrodesulfurization
HETP	Height equivalent to theoretical plate
HFCLM	Hollow-fiber contained liquid membrane
HFMC	Hollow-fiber membrane contactor
HHV	Higher heating value
HNLC	High nutrient, low chlorophyll
HP	High pressure
HR	Heavy product reflux
HRSG	Heat recovery steam generator
HTCs	Hydrotalcites

(*Continued*)

I	
IEA	International Energy Authority
IGCC	Integrated gasification combined cycle
IGFC	Integrated gasification fuel cell
IGHAT	Integrated gasification humid air turbine
IL	Ionic liquid
ILM	Immobilized liquid membrane
INSAR	Interferometry for synthetic aperture radar
IP	Intermediate pressure
IPCC	Intergovernmental Panel on Climate Change
IRCC	Integrated reforming combined cycle
ITM	Ion transport membrane
IUPAC	International Union of Pure and Applied Chemistry
K	
kPa	Kilopascal
kW	Kilowatt
L	
LEERT	Long electrode electrical resistance tomography
LHV	Lower heating value
LIDAR	Light detection and ranging
LNG	Liquefied natural gas
LM	Liquid membrane
LP	Low pressure
LPG	Liquefied petroleum gas
LPP	Light product pressurization
LR	Light product reflux
LULUCF	Land use, land use change and forestry
M	
MAOM	Mineral associated organic matter
MBARI	Monterey Bay Aquarium Research Institute
MCFC	Molten carbonate fuel cell
MCM	Mixed conducting membrane (or medium)
MDEA	Methyldiethanolamine
MEA	Monoethanolamine

(*Continued*)

MFA	Model free adaptive
MGA	Membrane gas absorption
MGE	Microbial growth efficiency
MIEC	Mixed ionic electronic conductors
MMM	Mixed matrix membranes
MMV	Monitoring, measurement, and verification
MOM	Microbial organic matter
MPa	Megapascal
MSC	Molecular sieve carbon
MSW	Municipal solid waste
MW	Molecular weight
MW_e	Megawatts (electric power)
MW_{th}	Megawatts (thermal power)
MWI	Municipal waste incineration
N	
NADW	North Atlantic Deep Water
NELHA	Natural Energy Laboratory of Hawaii Authority
NGCC	Natural gas combined cycle
NGL	Natural gas liquids
NOAA	National Oceanic & Atmospheric Administration (U.S. Dept. of Commerce)
NO_x	Mono-nitrogen oxides (NO, NO_2)
NPP	Net Primary Production
NSPS	New Source Performance Standards
O	
OIF	Ocean iron fertilization
ONS	Ordered nanoporous silicas
OOIP	Original oil in place
OTM	Oxygen transport membrane
P	
P	Pressure
P_c	Critical pressure
PC	Pulverized coal
	Portland cement
PCC	Pulverized coal combustion
	Precipitated calcium carbonate

(*Continued*)

PCFBC	Pressurized circulating fluidized-bed combustion
PCOR	Plains CO_2 Reduction Partnership
PEG	Polyethylene glycol
PEI	Polyethyleneimine
PF	Pulverized fuel
PFBC	Pressurized fluidized-bed combustion
PIC	Particulate inorganic carbon
PNL	Progressive nitrogen limitation
PNNL	Pacific Northwest National Laboratory
POC	Particulate organic carbon
POM	Partial oxidation of methane
	Particulate organic matter
POX	Partial oxidation
ppb	Parts per billion
PPCC	Pressurized pulverized coal combustion
ppm	Parts per million
PRCI	Pipeline Research Council International
PSA	Pressure-swing adsorption
PTSA	Pressure and temperature swing
P–V	Pressure–volume
PVT	Pressure, volume, temperature
PWHT	Postweld heat treatment
Q	
QRA	Quantitative risk assessment
R	
RD3	Research, development, demonstration, and deployment
RDF	Refuse-derived fuel
ROV	Remotely operated vehicle
RTIL	Room-temperature ionic liquid
S	
S	Entropy
SACS	Saline aquifer CO_2 storage
SAPO	Silicoaluminophosphate
SAR	Second Assessment Report (IPCC 1996)
SC	Supercritical

(Continued)

SCC	Stress corrosion cracking
SCPCC	Supercritical pulverized coal combustion
SCR	Selective catalytic reduction
SEM	Scanning electronic microscopy
SER	Sorption-enhanced reaction
SEWGS	Sorption-enhanced water–gas shift
SIC	Soil inorganic carbon
SLM	Supported liquid membrane
SMB	Simulated moving bed
SMBC	Soil microbial biomass carbon
SMR	Steam methane reforming
SNCR	Selective noncatalytic reduction
SOC	Soil organic carbon
	Soil organic compound
SOCCR	State of the Carbon Cycle Report
SOFC	Solid oxide fuel cell
SOM	Soil organic matter
SO_x	Oxides of sulfur (SO, SO_2, SO_3)
SRES	Special Report on Emissions Scenarios (IPCC 2000)
STIG	Steam injected gas turbine
STL	Submerged turret loading
STP	Standard temperature and pressure (IUPAC; 0°C, 100 kPa)
T	
T	Temperature
T_{ALK}	Total alkalinity
TAR	Third Assessment Report (IPCC 2001)
TBCA	Total below-ground carbon allocation
T_c	Critical temperature
T_C	Total carbon content
T_{CO_2}	Total CO_2 content
TCP	Terrestrial carbon program
THC	Thermohaline circulation
TIC	Total inorganic carbon
TOC	Total organic carbon
TRL	Technology readiness level
TSA	Temperature-swing adsorption
TSIL	Task-specific ionic liquid

(Continued)

U	
UCG	Underground coal gasification
UNEP	United Nations Environment Programme
UNFCCC	United Nations Framework Convention on Climate Change
USC	Ultrasupercritical

V	
V	Volume
V&V	Validation and verification
VOC	Volatile organic compounds
VPDB	Vienna Peedee Belemnite
VPSA	Vacuum- and pressure-swing adsorption
VSA	Vacuum-swing adsorption
VSP	Vertical seismic profile

W	
WAG	Water-alternate-gas
WGSR	Water–gas shift reaction

X	
XRD	X-ray diffraction

Z	
ZEC	Zero-emission coal
ZECA	Zero-Emission Coal Alliance
ZEP	Zero-emission power
ZET	Zero-emissions technologies
ZEIGCC	Zero-emissions integrated gasification combined cycle

17.3 CCS technology glossary

A

Absorption

Physical or chemical absorption is a process in which atoms or molecules enter the bulk phase of a gas, liquid, or solid material and are taken up within the volume. See *Adsorption.*

Adiabatic

An adiabatic process is a thermodynamic process in which no heat is transferred to or released from the working fluid. An adiabatic process that is also reversible is isentropic.

Adsorption

A physical or chemical process in which molecules of a gas or liquid adhere to the surface of a solid adsorbent material. See also *Pressure-swing adsorption, Temperature-swing adsorption, Electrical-swing adsorption.*

Aquifer

A porous and permeable geological formation containing water. Saline aquifers contain water that is nonpotable without desalination.

Austenite

Austenite is a face-centered cubic crystallization phase of iron that, for pure iron, is stable between 912°C and 1394°C. With the addition of alloying elements such as Cr, Ni, and Mn, this so-called γ phase can be made to remain stable down to room temperature, as a result of the precipitation of carbides (e.g., Fe_3C, Fe_7C_3) as the steel solidifies. Austenitic steels are Fe-Cr-Ni alloys with 16–25% Cr, 1–37% Ni, and <0.24% C. The original austenitic manganese steel was invented in 1882 by Sir Robert Hadfield.

Azeotrope

An azeotrope is a mixture of two or more chemicals (e.g., CO_2 and methane) such that the ratio of constituents in the vapor phase is the same as in the liquid phase. As a result of this constant ratio, the components in the azeotrope cannot be separated by distillation unless an additive is introduced to break the azeotrope.

B

Barrer

The Barrer, named after Richard Barrer, a New Zealand chemist and pioneer in the measurement of gas permeation, is a CGS unit of gas permeability commonly used to measure the permeance of oxygen through contact lens materials. One Barrer equals 10^{-10} cm^2 s^{-1} cmHg^{-1} or, in SI units, $7.50 \cdot 10^{-18}$ m^2 s^{-1} Pa^{-1}. The unit is used to express membrane permeability in gas separation applications.

Biochar

Biochar is charcoal resulting from the pyrolysis of biomass – either as a result of natural fires or as a soil amendment.

Brayton cycle

A thermodynamic cycle that describes the operation of internal combustion engines, including gas turbines. The cycle comprises compression of the working fluid, heating by combustion with an injected fuel, expansion against a piston or turbine, and cooling to return the working fluid to the initial state.

C

Carbonate buffering

A geochemical effect that results in the acidity (pH) of ocean water being stabilized. If ocean pH increases, the ionization of H_2CO_3 and subsequently of the bicarbonate ion HCO_3^- results in the release of H^+, resulting in a stabilizing drop in pH. The resulting carbonate ion may then be precipitated as calcium carbonate ($CaCO_3$). If ocean pH drops (acidity increasing), carbonate ions recombine with H^+, reducing H^+ availability and increasing pH. While the carbonate ion is available, CO_2 dissolution in the ocean does not affect acidity as a result of this buffering, but once the carbonate ion concentration becomes depleted, further CO_2 addition results in acidification.

Carnot cycle

The Carnot cycle is an idealized reversible thermodynamic cycle that consists of a succession of four processes operating on the working fluid: an isothermal expansion at a temperature T_1, a reversible adiabatic (q.v.) expansion to temperature T_2, an isothermal compression at temperature T_2, and finally a reversible adiabatic compression to the original state of the working fluid to complete the cycle. The French physicist Sadi Carnot first proposed and studied this idealized cycle in the 1820s, in establishing the first theoretical framework to describe the steam engine.

Chemical absorption

An absorption process in which the sorbate chemically combines with the sorbent

Chemical looping combustion

Chemical looping combustion (CLC) is a circulating combustion process in which the oxygen required for combustion is delivered to the combustion reactor as a metal oxide rather than as free oxygen. Reduction of the oxide releases oxygen for combustion of the fuel, and the reduced metal oxide is conveyed to a second reactor where it is reoxidized, completing the loop.

Chemical vapor infiltration

A membrane fabrication process in which the pores of a silica membrane are infiltrated with a gaseous or liquid precursor that is then oxidized. This process plugs the pores to form a dense membrane in which separation occurs by solution–diffusion through the pore-filling material.

Claus process

An industrial process for the recovery of sulfur from a gas stream containing >25% H_2S, invented by the chemist Carl Claus. In this process one-third of the H_2S in the feed gas stream is first combusted to produce SO_2 ($2H_2S + 3O_2 \rightarrow 2SO_2 + 2H_2O$), which is then catalytically reacted with the remaining H_2S to form elemental sulfur plus water in the Claus reaction: $2H_2S + SO_2 \rightarrow 3S + 2H_2O$. The process is commonly applied to the H_2S-containing product streams from acid gas treatment processes.

Clean Development Mechanism

The Clean Development Mechanism (CDM) is a project-based mechanism under the Kyoto Protocol to the UN Framework Convention on Climate Change (UNFCCC) that enables the generation and issue of Certified Emission Reductions (CERs) from eligible projects.

Combined cycle

When heat recovered from the working fluid of one thermodynamic cycle is used to heat the working fluid of another cycle, the cycles are described as combined. Most commonly applied to the heat recovery from Brayton cycle (gas turbine) exhaust gas to drive Rankine (steam) cycle power generation. Either cycle may be a general industrial process, and the add-on cycle is generically known as a topping cycle if at higher temperature or a bottoming cycle if at lower temperature.

D

Dissolved inorganic carbon

Dissolved inorganic carbon (DIC) is the sum of all inorganic carbon species in a solution, including CO_2, carbonic acid, and bicarbonate and carbonate anions. The average DIC in the world's oceans is $\sim 2 \cdot 10^{-3}$ mol-C per kg.

E

Electric-swing adsorption

Electric swing adsorption (ESA) refers to an adsorption–desorption cycle in which sorbent regeneration is achieved using an applied electric field to heat the sorbent and release the sorbate.

Endothermic

An endothermic reaction requires an input of energy in the form of heat to enable the reaction to proceed. The enthalpy change of reaction ΔH is positive in an endothermic reaction ($\Delta H > 0$). See also *Exothermic.*

Enthalpy

Enthalpy is a thermodynamic property of a system, measured in kJ per mol, and is equal to the internal energy of the system (U) plus the product of pressure (p) and volume (V) of the system ($H = U + pV$). Enthalpy change (ΔH) is related to the change in entropy (ΔS) by $\Delta H = T\Delta S + V\Delta p$.

Entropy

Entropy is a thermodynamic quantity, measured in kJ per °K-mole. It is a measure of the degree of disorder within a system, with higher entropy corresponding to a greater degree of disorder.

Exothermic

An exothermic reaction releases energy, typically in the form of heat as the reaction proceeds. The enthalpy change of reaction ΔH is negative in an exothermic reaction ($\Delta H < 0$). See also *Endothermic*.

F

Facilitated transport membrane

In a facilitated transport membrane, the permeate is transported across the membrane attached to a carrier.

Ferrite

Ferrite is a crystallization phase of iron with a body-centered cubic crystal structure. Ferritic steels are iron–chromium alloy steels with 10–25% Cr, 2–4% Mo, <1% Ni, and C content <0.75%.

Fischer–Tropsch process

The Fischer–Tropsch process is a chemical reaction, most commonly catalyzed by iron or cobalt, in which a mixture of carbon monoxide and hydrogen is converted into a synthetic liquid hydrocarbon. Feedstock for the process is commonly natural gas (gas to liquids [GTL]) or syngas from coal or biomass gasification.

Fluidized bed reactor

Fluidization of a bed of solid particles is achieved by the controlled injection of gas into the bed, causing the particles to be suspended in the flowing gas by viscous forces. The efficient heat and mass transfer that occurs in such a bed makes this an effective medium for various reactions such as solid fuel combustion and gasification.

Functional membrane

Functionalization of a membrane is the process of incorporating an active chemical group (the functional group) into the matrix or onto the pore surface of a membrane in order to enhance separation of a specific species through surface interactions with the functional group. An example of a functional membrane is the amine-modified silica membrane for CO_2 separation. Membrane separation properties are strongly influenced by the density and the surface chemistry of the functional group.

G

Gas turbine cycle

See *Brayton cycle*

Gasification

Gasification is the high-temperature conversion of carbonaceous fuels, such as municipal wastes, biomass, petroleum residues, or coal, into a mixture of carbon monoxide and hydrogen by partial oxidation ($C \rightarrow CO$) or steam reforming ($C \rightarrow H_2 + CO$). The resulting gas (syngas) may be used directly as a fuel or as an intermediate product for the production of liquid fuel or hydrogen.

Glass transition temperature

The glass transition temperature is the temperature (T_g) below which an amorphous material exhibits a brittle, glassy structure. Above T_g, bonds within the material weaken and it becomes rubbery (soft and deformable without fracturing). The glassy versus rubbery distinction is important for polymeric membranes.

H

Heterotroph

In contrast to autotrophs, which produce organic compounds or substrates directly from inorganic carbon, heterotrophs obtain organic carbon by feeding on autotrophs or other heterotrophs. Plants and some bacteria use photosynthesis to produce organic substrates and are thus autotrophs, while other bacteria, fungi, and animals that feed on plant matter are heterotrophs.

Higher heating value (HHV)

The higher heating value, or gross calorific value, of a quantity of fuel is the amount of heat released when the fuel, initially at 25°C, is combusted and the combustion products are returned to 25°C. Cooling results in release of the latent heat of vaporization of the water content of the combustion products, which is therefore included in the HHV (cf. *LHV*).

Hydrocracking

Hydrocracking is a chemical process in which a catalyst, in the presence of an elevated partial pressure of hydrogen, is used to break and rearrange long chain hydrocarbons.

Hydrodesulfurization (Hydrotreating)

Hydrodesulfurization is a process used to remove sulfur from unfinished oil products such as gasoline and jet fuel. Removal of sulfur is achieved by contacting the unfinished products with hydrogen at high pressure in the presence of a catalyst. Desulfurization improves the performance of later catalytic processes, used to upgrade fuel products, and is necessary to meet environmental standards – by reducing sulfur dioxide (SO_2) emissions in final combustion of the fuel.

Hypersorption

Hypersorption refers to an adsorption process using a moving or simulated moving bed of sorbent. The moving bed concept was initially patented in 1922, as a method to separate syngas components from coal gasification, although the hypersorption name was only applied from the 1940s.

I

Intergovernmental Panel on Climate Change

Established by the World Meteorological Organization (WMO) and the United Nations Environment Programme (UNEP) in 1988, the aim of the Intergovernmental Panel on Climate Change (IPCC) is to assess the state of knowledge of all aspects of the climate system and of climate change, whether due to natural variability or resulting from human activities.

Isosteric heat (enthalpy) of adsorption

The quantity of heat (enthalpy) generated when a differential quantity of a sorbate is adsorbed onto a solid surface at constant pressure.

J

Joule Thomson cooling

When a gas expands, work must be done to separate molecules against the weak attractive van der Waal's forces between molecules. If the expansion occurs without heat being exchanged with the surrounding environment, the thermal energy of the gas is expended to do this work, and the gas is cooled as a result. The effect was discovered in 1852 by physicists J.P. Joule and W. Thomson (Lord Kelvin), and is used in many industrial processes requiring cooling, including the liquefaction of gases.

L

Lower heating value (LHV)

The lower heating value, or net calorific value, of a quantity of fuel is the amount of heat released when the fuel, initially at 25°C, is combusted and the combustion products are returned to 150°C. Since the end point is above

100°C the latent heat of vaporization of the water content of the combustion products is not released and is therefore not included in the LHV (cf. *HHV*).

M

Mafic

See *Ultramafic*

Macroporous

A porous medium having pores with a diameter >50 nanometers

Martensite

Martensite is a metastable crystallization phase of iron formed by the rapid cooling, or quenching, of austenite (q.v.). Rapid cooling prevents carbon atoms from diffusing out of the iron crystal lattice, resulting in a body-centered tetragonal structure. Martensitic steels contain 12–18% chromium, up to 1% carbon, making the steel hard but brittle, and may also include small quantities (0.2–2%) of nickel, molybdenum, vanadium, or tungsten.

Mesoporous

A porous medium having pores with a diameter in the range of 2–50 nanometers.

Microporous

A porous medium having pores with a diameter <2 nanometers.

Miscibility

Two substances are miscible if they can be mixed in all proportions to form a homogeneous solution—that is, there is no limit to the solubility of either component in the other.

Mixed ionic electronic conductors (Mixed metal oxides)

See *Perovskite*

Moiety

A specific fragment of a molecule (e.g., an amine moiety), typically a charge-carrying species

P

Perovskite

A class of crystalline minerals with the general chemical formula ABO_3, in which A is a lanthanoid element and B is a transition metal. Metal cations A and B are typically arranged in two offset cubic structures linked together by oxide anions. The first perovskite discovered was $CaTiO_3$. Mixed ionic electronic conductors (MIECs) are perovskite-type materials in which one or both of the A and B lattices are doped with other cations to improve stability and performance; examples are so-called LSFO ($La_{0.5}Sr_{0.5}FeO_{3}$–δ) and LSFCO

($La_{0.2}Sr_{0.8}Fe_{0.8}Cr_{0.2}O_{3}$–δ), where δ represents the proportion of oxide vacancies in the structure. These materials are important for oxygen transport membranes and solid oxide fuel cells.

Pervaporation

Pervaporation is a liquid separation process in which one component of a liquid mixture permeates through a membrane and evaporates into a vapor phase on the permeate side of the membrane. The word, like the process, is a combination of permeation and evaporation.

Physical absorption

An absorption process in which the sorbate does not chemically combine with the sorbent but is held within the sorbate by physical forces such as Van der Waals force.

Pressure-swing adsorption

Pressure swing adsorption (PSA) refers to an adsorption–desorption cycle that is driven by the drop in sorbent carrying capacity with decreasing pressure. Sorbent regeneration is achieved using a low pressure or vacuum to release the sorbate.

R

Rankine cycle

A thermodynamic cycle that describes the operation of a steam engine. The cycle comprises pressurizing the working fluid in the liquid state, vaporization by heating in a boiler, expansion against a piston or turbine, typically without condensation, and cooling and condensing to return the working fluid to the initial liquid state.

Recalcitrant carbon

In the spectrum of susceptibility to decomposition, recalcitrant carbon occupies the middle ground between labile (easily changed) and inert material. Its resistance to decomposition is due to physical or chemical protection, for example as a result of adsorption onto mineral surfaces (physical recalcitrance) or the blocking of catabolic pathways due to the absence of essential enzymes or oxidants (chemical recalcitrance).

S

SAPO (Silicoaluminophosphate)

A synthetic ceramic membrane material that is an analog of the naturally occurring zeolite chabazite ($(Ca,Na_2,K_2,Mg)Al_2Si_4O_{12}\cdot 6H_2O$) and is used as a molecular sieve.

Soil taxonomy (USDA)

A hierarchical classification of soil types according to their properties and the conditions under which they were formed. Taxonomic differentiation is based on a wide range of characteristics including physical structure (particle size, aggregation, layering), organic versus mineral content, and chemical and electrical properties.

Solid oxide fuel cell

A solid oxide fuel cell (SOFC) is a device that generates electricity directly from the electrochemical oxidation of the fuel by oxygen ions (O^{2-}) that diffuse through a solid oxide electrolyte. In the case of hydrogen as fuel, electrochemical oxidation at the anode of the cell produces water and releases two electrons from the oxygen ion, which are conducted back to the cathode through the external circuit.

Specific area

The total surface area of a material per unit mass (m^2 per kg). The rate of chemical reactions that take place on the surface of solid particles can be increased by increasing the specific area of a material, for example by grinding into smaller particles.

Spinel

Spinels are a class of minerals of the general form AB_2O_4 where A and B may be any divalent, trivalent, or quadrivalent cation, including aluminum, chromium, iron, magnesium, manganese, titanium, silicon, or zinc. The formation of a surface manganese–chromium spinel layer ($MnCr_2O_4$) contributes to the corrosion resistance of alloy steels.

Steam cycle

See *Rankine cycle*

Steric hindrance

Steric hindrance occurs when the presence and position of chemical groups within a large molecule prevent or hinder chemical reactions that proceed more easily in smaller molecules. Steric hindrance is exploited in the amine absorption reaction to increase the absorption capacity of the solvent, but has the associated disadvantage that the sorption rate is very low.

Substrate

In the context of biochemical processes in soils, a substrate is an organic compound such as a root exudate or biomass decomposition product that is available for metabolism by a heterotroph.

Supercritical conditions

A thermodynamic process is termed supercritical if its operating temperature and pressure are above the critical point of the working fluid. For example, a supercritical Rankine steam cycle operates with a live steam pressure greater than P_c = 22.12 MPa and temperature above T_c = 374.15°C.

Syngas

Syngas is the product of a gasification or steam reforming process and consists of a mixture of hydrogen and carbon monoxide.

T

Temperature-swing adsorption

Temperature-swing adsorption (TSA) refers to an adsorption–desorption cycle that is driven by the drop in sorbent carrying capacity with increasing temperature. Sorbent regeneration is achieved by heating to drive off the sorbate.

Thermal efficiency

The thermal efficiency of a system is the ratio of the amount of useful work or heat produced divided by the amount of heat input or the heat content of the fuel consumed in the system. For a power plant, the efficiency is the amount of electrical energy produced divided by the heat value, typically the net calorific value or LLV, of the fuel consumed. See *HHV*, *LHV*.

Thermocline

The thermocline is the boundary layer that separates the warmer surface layers of the ocean (the epilimnion) from colder deeper water (the hypolimnion). In tropical oceans water temperature can drop from 25°C at surface to ca. 10°C below the thermocline, at 100 m to 200 m below sea level.

Thermodynamic cycle

A thermodynamic cycle is a sequence of changes in the state of a system such that at the end of the sequence all properties of the system (e.g., pressure, temperature, entropy) are returned to the values they had at the initial state of the cycle.

Transesterification

The process used in biodiesel production where triglycerols present in the lipids extracted from biomass feedstock, such as oilseeds crops or algae, are converted to methyl or ethyl esters by reaction with methanol or ethanol, typically in the presence of a strong basic catalyst, such as NaOH.

U

Ultramafic rocks

Mafic or ultramafic rocks are igneous rocks that are relatively poor in silica and are composed predominantly or, in the case of ultramafic rocks, almost entirely of iron and magnesium minerals. Typically, although not exclusively, ultramafic rocks contain $<45\%$ silica and are a prospective source material for CO_2 sequestration by mineral carbonation.

Ultrasupercritical conditions

Ultrasupercritical conditions are not defined by a specific temperature and pressure point (unlike *supercritical conditions*, q.v.) but they are generally taken to mean a supercritical pressure and a steam temperature $>650°C$. These conditions require a step-change in materials capabilities compared with typical supercritical boiler conditions (24 MPa/540°C/540°C) that have been common in the power industry since the 1980s.

V

VPDB carbonate standard

The Cretaceous limestone formation at Peedee in the Vienna Basin in South Carolina (derived from the marine fossil *Belemnitella americana*) is used as an international reference standard for the relative abundances of carbon and oxygen isotopes. For example, for ^{13}C the deviation of a sample from the standard is defined as:

$$\delta^{13}C_{sample} = 1000\{(^{13}C/^{12}C_{sample})/(^{13}C/^{12}C_{standard}) - 1\}$$

where the VPDB standard ratio of ^{13}C to ^{12}C is 0.0112372.

W

Water–gas shift

In the water-gas shift (WGS) reaction, steam reacts with the carbon monoxide to produce CO_2 and hydrogen. Separation of CO_2 from the WGS offgas then yields a pure hydrogen stream.

Z

Zeolites

A group of porous minerals composed of sodium, potassium, calcium, or magnesium aluminosilicate.

Index

Note: Page numbers followed by 'f' indicate figures 't' indicate tables.

D